ICE Core Concepts: Soil Mechanics

emerald
PUBLISHING   ice
Publishing

ICE Core Concepts

Soil Mechanics

Second Edition

Sanjay Kumar Shukla
School of Engineering, Edith Cowan University, Joondalup,
Perth, Australia

Published by Emerald Publishing Limited, Floor 5, Northspring, 21-23 Wellington Street, Leeds LS1 4DL.

ICE Publishing is an imprint of Emerald Publishing Limited

Other ICE Publishing titles:
ICE Core Concepts: Geotechnical Engineering
Sanjay Kumar Shukla ISBN: 978-0-72775-859-0
ICE Core Concepts: Low Carbon Cements and Concrete for Construction
Monower Sadique ISBN: 978-1-83608-477-8
ICE Core Concepts: Hydraulics for Civil Engineers
Peter Wynn ISBN: 978-0-72776-679-3

A catalogue record for this book is available from the British Library

ISBN 978-1-83608-519-5

Cover photo: Mehmet Cetin/Shutterstock.com
Commissioning Editor: Michael Fenton
Content Development Editor: Ryan Molyneux
Books Production Lead: Benn Linfield

Typeset by KnowledgeWorks Global Limited
Index created by David Gaskell

Contents

Preface

Soil mechanics is the study of how soil masses respond to various loads, such as those from structures, gravity and natural events like earthquakes. Composed of solid, liquid and gas phases, soil exhibits complex behaviour that requires integrating principles from solid and fluid mechanics. However, due to natural variability and uncertainties, simplifying assumptions and sound judgement are often necessary. Understanding soil behaviour is important for designing stable foundations and other ground structures in infrastructure, energy and environmental systems. The performance and durability of these systems depend heavily on the strength and stability of the supporting soil mass, making soil mechanics a critical field not only in civil engineering but also in disciplines such as mining, agriculture and aquaculture.

As a fundamental subject in civil engineering undergraduate programmes worldwide, soil mechanics introduces students to the physical properties of soils and the behaviour of soil masses under various load conditions. Its principles are not limited to natural soils but extend to soil-like materials such as coal ash and mine tailings, which are increasingly utilised in engineering applications. By applying these concepts, engineers can design sustainable and efficient solutions for a range of projects, ensuring the safety and reliability of structures built on or within the ground.

Over the years, many books on soil mechanics have been written, covering both fundamental and advanced concepts. Each is tailored to meet the needs of students and practising engineers. This textbook, as its title indicates, emphasises the core principles and concepts of soil mechanics. It comprises nine chapters, thoughtfully designed and developed for a one-semester course in undergraduate civil engineering programmes. The content of the chapters is presented clearly in plain English, with an optimal balance of text, illustrations, examples and practice questions. Each chapter includes valuable references cited within the text and listed at the end of the chapter, providing resources for further study and deeper understanding.

Chapter 1 presents the basic description of soil, including the fundamentals of soil phases, phase relationships, weathering and soil formation, and clay mineralogy. Chapter 2 describes soil particle size and shape, index properties of soil, and soil classification and description. Chapter 3 covers stresses within soil, including effective stress and stress induced by loading. Chapter 4 contains a description of flow through soil by explaining the basics of fluid flow, permeability, seepage forces, flow equation and flow nets. Chapter 5 deals with the consolidation and compressibility aspects of soil. Chapter 6 presents the details of shear strength of soil and explains the principles of different shear tests. The details of laboratory and field compaction of soil are given in Chapter 7. The basic concepts of lateral earth pressure, slope stability and bearing capacity of soil are described in Chapter 8. Chapter 9 covers the principles of specialised topics, including expansive soils, reinforced soils, thermal and electrical properties of soils and soil liquefaction.

This introductory textbook has been developed as a student-centred learning resource, drawing on my over 30 years of teaching, research and consultancy experience in geotechnical engineering. Civil engineering students will find it particularly useful for studying soil mechanics and understanding the fundamentals of the subject with minimal external assistance. For teachers, this textbook eliminates the need to prepare handwritten lecture materials, as the content has been refined and used effectively in my own classroom teaching for many years. Beyond students and teachers, this textbook also serves as a valuable reference for practising engineers and professionals in related fields, helping them refresh their understanding of the core principles and concepts of soil mechanics when addressing soil-related challenges in their projects.

Revised and updated, this second edition offers the following key features.

- A new chapter (Chapter 9) focused on principles of special topics, including expansive soil, reinforced soil, thermal and electrical properties of soil and soil liquefaction.
- Enhancements to existing chapters, with the addition of new sections in Chapters 2, 3, 4, 5 and 7, providing expanded coverage and greater depth.
- Comprehensive updates throughout the book, featuring additional illustrative examples, new and updated illustrations, a fresh set of multiple choice questions, numerical practice problems and conceptual questions. This edition also includes updated test standards, codes of practice and references to ensure relevance and to support enhanced learning and practical application.

I would like to express my heartfelt gratitude to the team at Emerald/ICE Publishing, London, especially Michael Fenton, Ryan Molyneux, Benn Linfield, S. Rajachitra, Alison Gilmour and Cathy Sellars, for their strong support and cooperation throughout the various stages of preparing and producing this textbook. Their dedication and professionalism have been invaluable.

I extend my sincere appreciation to my wife, Sharmila, for her constant encouragement and support during the preparation of the manuscript. I am equally grateful to my daughter, Sakshi, and my son, Sarthak, for their patience and understanding as I worked on this textbook at home.

Finally, I warmly welcome suggestions from readers and users of this textbook to further enhance its content in future editions. Your feedback is invaluable in making this resource even more useful and comprehensive.

Sanjay Kumar Shukla
Perth, Australia, 2025

About the author

Dr Sanjay Kumar Shukla is a globally renowned expert in Civil (Geotechnical) Engineering. He is the Founding Editor-in-Chief of the *International Journal of Geosynthetics and Ground Engineering*, published by Springer, Switzerland, and the Founding Leader of the Geotechnical and Geoenvironmental Engineering Research Group at Edith Cowan University, Joondalup, Perth, Australia. He holds distinguished professorships at several universities, including the prestigious Delhi Technological University, Delhi, India, and Southern Illinois University, Carbondale, USA.

Dr Shukla is a Chartered Professional Engineer in Civil and Geotechnical Engineering, registered with Engineers Australia. He also holds the designation of Asia Pacific Economic Cooperation (APEC) Engineer in Civil Engineering and is recognised as an International Professional Engineer in Civil Engineering by the International Engineering Association.

He is a distinguished Fellow of the American Society of Civil Engineers and Engineers Australia, as well as a Life Fellow of the Institution of Engineers (India) and the Indian Geotechnical Society. His prolific academic contributions include 28 books and over 320 research articles, earning him recognition among the world's top 2% of scientists by Elsevier, and among the top 0.5% globally by ScholarGPS.

Dr Shukla has received numerous accolades, including the 2021 ECU Aspire Award and the prestigious IGS Award 2018 from the International Geosynthetics Society, USA. In 2024, the Consulate General of India in Perth honoured him with the Distinguished Honour for his outstanding academic contributions to Geotechnical Engineering.

His pioneering works, such as his generalised expressions for seismic active thrust (2015) and passive resistance (2013), along with the innovative Shukla's wraparound reinforcement technique, are widely used in engineering practice and form integral components of engineering education worldwide. His Seven Research Mantras, introduced in 2022, have inspired sustainable research practices and influenced researchers globally.

A highly regarded speaker, Dr Shukla frequently delivers keynote talks and short courses internationally. He is also widely consulted by researchers and practising engineers for his expertise in advancing practical engineering solutions.

emerald
PUBLISHING

ice
Publishing

Sanjay Kumar Shukla
ISBN 978-1-83608-519-5
https://doi.org/10.1108/978-1-83608-516-420252001

Chapter 1
Basic description of soil

Learning aims

This chapter explores the core concepts of the following topics

- soil, soil mechanics and geotechnical structures
- soil phases, phase relationships and inter-relationships
- weathering and soil formation
- clay mineralogy
- types of soil and soil structures.

1.1.   Introduction

The Earth is composed of three well-defined layers: the crust, the mantle and the core, which is divided into the outer core and the inner core (Figure 1.1). The crust is the topmost layer, which has a thickness of about 30–35 km in continents and 5–6 km in oceans. The materials that constitute the Earth's crust are arbitrarily divided by civil engineers into the following two categories for practical purposes

- soil (Figure 1.2(a))
- rock (Figure 1.2(b)).

Soil comprises all the materials in the surface layer of the Earth's crust that are loose enough to be moved by a spade or shovel. These materials include natural systems such as sediments or other unconsolidated accumulations of solid particles, and they may also contain organic matter. According to Karl Terzaghi, widely regarded as the father of soil mechanics, soil is a natural aggregate of mineral grains or particles that can be separated by gentle means, such as agitation in water (Terzaghi et al., 1996). Rock, on the other hand, is a natural aggregate of mineral grains, connected by strong and permanent internal cohesive forces, occurring in large masses and fragments. The distinction between soil and rock is not very clear; the dividing line between them is a matter of convenience and varies depending on the purpose under consideration.

Soil mechanics is the branch of science that deals with the application of the laws of mechanics and hydraulics to problems related to soil. It includes the study of the physical properties of soil and the behaviour of soil masses in connection with their practical applications. The core principles of soil mechanics are also applied to other soil-like granular materials, such as coal ashes (fly ash, bottom ash), furnace slags and mine tailings.

Figure 1.1 Structure of the Earth

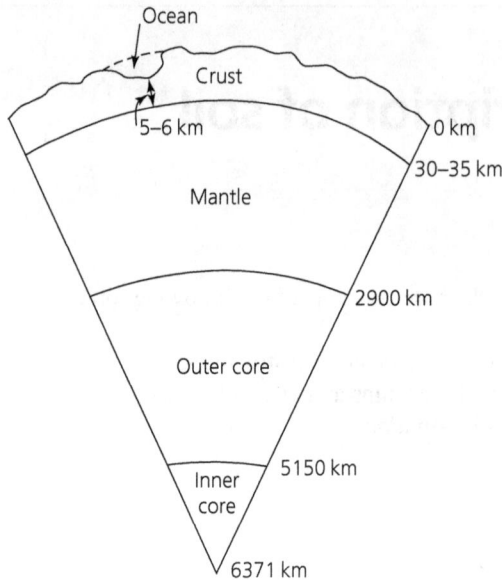

Note: not to scale

Soil supports structural foundations and serves as a construction material in civil engineering and other infrastructure projects. Soil mechanics provides scientific information for the analysis, design, construction, maintenance and renovation of geotechnical and ground structures, forming the basis of geotechnical engineering, as presented by Shukla (2025). Figure 1.3 illustrates seven basic types of geotechnical structures: foundation, slope, embankment, earth dam, retaining wall, tunnel and pavement, which are the most commonly recognised terms. Other geotechnical structures can be considered variations or combinations of these basic types.

The design and stability of geotechnical structures are largely governed by the properties of soil, primarily permeability (ability of a soil to conduct liquid or gas), compressibility (property of a soil pertaining to its susceptibility to change in volume when subjected to loading or unloading) and shear strength (maximum resistance of a soil to shearing stresses). These properties are described in detail in Chapters 4, 5 and 6, respectively.

This chapter introduces the basic description of soil, focusing on its phases, origin and types as considered in civil engineering practice.

1.2.   Phases in a soil mass

An element of soil (Figure 1.4(a)) consists of

- solid, usually mineral particles
- liquid, usually water
- gas, usually air and/or water vapour.

A phase in a soil mass is its one part (solid, liquid or gas) which is physically and chemically different from its other parts. Soil is thus a multiphase system consisting of, in general, three phases: solid, liquid and gas.

Figure 1.2 Field project sites with (a) soil formation and (b) rock formation

(a)

(b)

The space in a soil mass occupied by liquid and/or gas is known as the void. The soil is described as

- dry if the void volume is full of gas
- fully saturated or simply saturated if the void volume is full of liquid
- partially saturated or unsaturated if the void volume contains both gas and liquid.

For the solution of geotechnical engineering problems, it is often necessary to know the proportions by weight (or mass) and volume of the various soil phases within a soil element. Therefore, it is convenient to use a soil phase diagram (Figure 1.4(b)) which separately shows the phases within a soil element.

Figure 1.3 Basic types of geotechnical structures: (a) foundation: (i) shallow foundation, (ii) deep foundation; (b) slope; (c) embankment; (d) earth dam; (e) retaining wall; (f) tunnel and (g) pavement

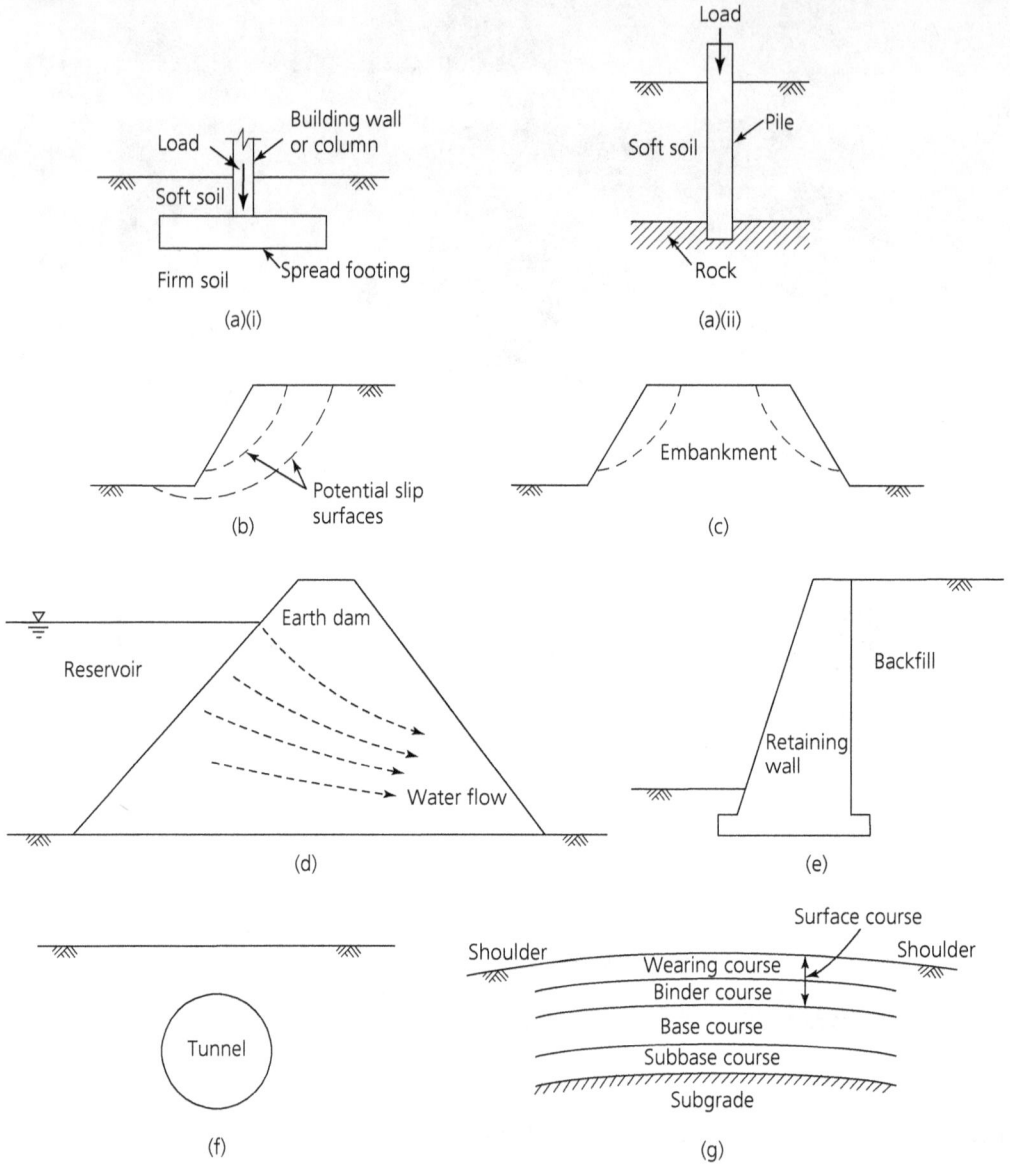

In Figure 1.4(b), the volumes of three phases are indicated on the left side and the corresponding weights on the right side. For geotechnical engineering applications, it is generally assumed that the weight of gas phase (air and water vapour) present in the voids of soil is zero. In some books and other literature, mass (M) is used in place of weight (W) in the phase diagram. In dealing with geotechnical engineering applications at various locations of the Earth, weight is preferred. The SI

Figure 1.4 (a) Element of soil with its phases in the natural state; (b) element of soil separated into its phases

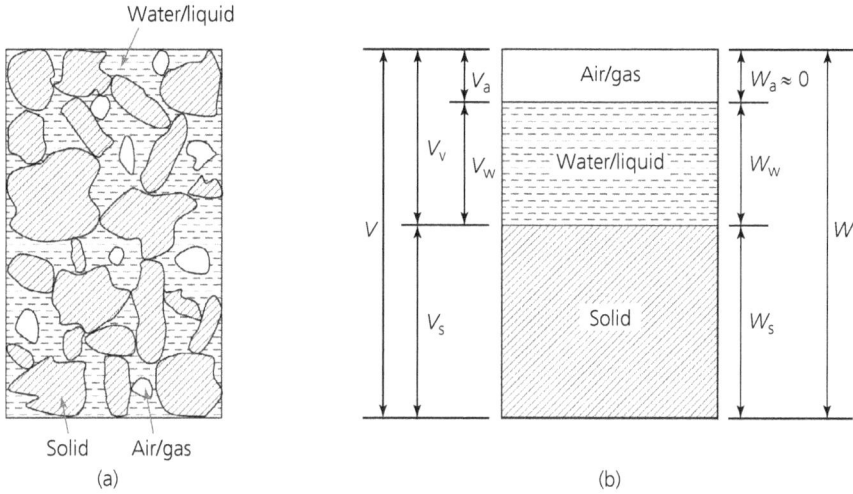

Note: V and W denote volume and weight, respectively; subscripts a, w, s and v refer to air, water, solid and void, respectively.

units of mass and weight are kilogram (kg) and newton (N), respectively. Newton's second law of motion establishes the following relationship between mass and weight

$$W = Mg \qquad (1.1)$$

where g denotes the acceleration due to gravity, which is usually taken as approximately $9.81\,\mathrm{m/s^2}$ or sometimes $10\,\mathrm{m/s^2}$, although it varies somewhat over the surface of the Earth depending on location and altitude relative to mean sea level. A 10 kg soil sample collected from a construction site will have a weight of 98.1 N. However, by rounding $9.81\,\mathrm{m/s^2}$ to $10\,\mathrm{m/s^2}$, the weight of the 10 kg soil sample becomes 100 N, introducing an error of approximately 2%. In professional practice, some engineers use a value of g equal to $10\,\mathrm{m/s^2}$ to simplify calculations, assuming the resulting error is negligible and does not compromise the safety of engineering designs.

1.3. Phase relationships

1.3.1 Volume relationships
There are three important volume relationships: porosity, void ratio and degree of saturation.

The porosity of a soil element is the ratio of void volume, V_v, to total volume, V. Therefore, if n denotes the porosity, then

$$n = \frac{V_v}{V} \qquad (1.2)$$

Porosity is normally expressed as a percentage.

The void ratio of a soil element is the ratio of void volume, V_v, to solid volume, V_s. Therefore, if e denotes the void ratio, then

$$e = \frac{V_v}{V_s} \tag{1.3}$$

Void ratio is normally expressed as a decimal.

The degree of saturation of a soil element is the ratio of water volume, V_w, to void volume, V_v. Therefore, if S denotes the degree of saturation, then

$$S = \frac{V_w}{V_v} \tag{1.4}$$

Degree of saturation is normally expressed as a percentage.

1.3.2   Weight relationships
There is only one weight relationship, namely water content (also known as moisture content), which is a crucial phase relationship in soil mechanics.

Water content of a soil element, w, is defined as the ratio of weight (or mass) of water, W_w, present in the voids to weight (or mass) of soil solids or particles (mineral matter), W_s. Thus,

$$w = \frac{W_w}{W_s} \tag{1.5}$$

The water content of soil is typically expressed as a percentage and is commonly determined in the laboratory using the oven-drying method. In this method, the soil specimen is dried generally at a temperature of about 110°C in a thermostatically controlled oven to a constant weight. The water loss due to drying is considered to represent the water in the voids of the soil specimen, while the dry weight of the soil specimen is regarded as the weight of the soil solids. This test requires the following three measurements

W_1 = weight of container
W_2 = weight of container and moist soil specimen
W_3 = weight of container and oven-dry soil specimen.

To calculate w, Equation 1.5 is used with $W_w = W_2 - W_3$, and $W_s = W_3 - W_1$, resulting in

$$w = \frac{W_2 - W_3}{W_3 - W_1} \tag{1.6}$$

Note that if the test soil contains organic materials or materials having a significant amount of hydrated water (e.g. gypsum, $CaSO_4 \cdot 2H_2O$), it may be desirable to dry such soils at a temperature much lower than 110°C, such as 65°C, or in a desiccator at room temperature. The procedure for determining the water content of soil is detailed in the standards of several countries, such as AS 1289.2.1.1 (Standards Australia, 2005), ASTM D2216-19 (ASTM, 2019), BS 1377-2:2022 (BSI, 2022) and IS 2720(Part 2):1973 (Bureau of Indian Standards, 2020).

1.3.3   Volume–weight relationships

The most commonly used volume–weight relationships, defined to serve the purpose of geotechnical projects, are as follows: total unit weight (also known as bulk, wet or moist unit weight), dry unit weight, unit weight of solids, saturated unit weight and submerged unit weight (also known as buoyant unit weight or effective unit weight). All these unit weights are expressed in SI units of N/m³ but are typically represented in practice as kilonewtons per cubic metre (kN/m³).

The total unit weight of a soil element, γ, is the weight of the entire soil element, W, divided by its total volume, V. Thus,

$$\gamma = \frac{W}{V} \tag{1.7}$$

Example 1.1

A cubical soil sample with an edge length of 20 cm was collected from the base level of a proposed shallow foundation. The mass of the sample was found to be 16.1 kg. Determine the total unit weight of the soil.

Solution

From Equation 1.1, the weight of soil

$$W = (16.1)(9.81) = 157.9\,\text{N}$$

Since the soil sample is cubical, its volume

$$V = (0.2)^3 = 0.008\,\text{m}^3$$

From Equation 1.7, the total unit weight

$$\gamma = \frac{157.9}{0.008} = 19737.5\,\text{N/m}^3 \approx 19.74\,\text{kN/m}^3$$

Using Equation 1.1 with M as the total mass of the entire soil element, Equation 1.7 can be expressed as

$$\gamma = \frac{Mg}{V} = \left(\frac{M}{V}\right)g = \rho g \tag{1.8}$$

where

$$\rho = \frac{M}{V} \tag{1.9}$$

is the total density (also known as bulk, wet or moist density) of soil, expressed in kg/m³.

It should be noted that some engineers report the total density of soil rather than its total unit weight. Therefore, you should be familiar with how to obtain the total unit weight using Equation 1.8, which is often required in most geotechnical design and stability calculations, including calculations for stresses and settlements.

Example 1.2

In a geotechnical site investigation report, the total density of the foundation soil is reported to be 1728 kg/m³. What will be the total unit weight of soil?

Solution

From Equation 1.8, the total unit weight of soil

$$\gamma = (1728)(9.81) = 16951.7\,\text{N/m}^3 \approx 16.95\,\text{kN/m}^3$$

The dry unit weight of a soil element, γ_d, is the weight of the soil solids in the entire soil element, W_s, divided by its total volume, V. Thus,

$$\gamma_d = \frac{W_s}{V} \tag{1.10}$$

The unit weight of solids of a soil element, γ_s, is the weight of the solids in the entire soil element, W_s, divided by the volume of the mineral matter, V_s. Thus,

$$\gamma_s = \frac{W_s}{V_s} \tag{1.11}$$

The saturated unit weight of a soil element, γ_{sat}, is the weight of the entire soil element in saturated condition, W_{sat}, divided by its total volume, V. Thus,

$$\gamma_{sat} = \frac{W_{sat}}{V} \tag{1.12}$$

When a soil element exists in nature in a submerged condition – that is, it exists below the ground-water table – it experiences an upward force. Consequently, the weight of the soil element in submerged condition is diminished by an amount equal to the weight of water equal in volume to the soil element. In view of this fact, the submerged (or buoyant) unit weight of a soil element is the weight of the entire soil element in submerged condition, W', divided by its total volume, V. Therefore, if γ' denotes the submerged unit weight of the soil element, then

$$\gamma' = \frac{W'}{V} = \frac{W_{sat} - V\gamma_w}{V} = \gamma_{sat} - \gamma_w \tag{1.13}$$

where $\gamma_w (= \rho_w g)$ is the unit weight of water with ρ_w as its density, which is 1000 kg/m³ or 1 Mg/m³ or 1 t/m³ at standard conditions. The unit weight of water at standard conditions is approximately 9810 N/m³ or 9.81 kN/m³. For most civil engineering purposes, the unit weight of water may be taken as 10 kN/m³.

For each unit weight, a corresponding density can be defined, as shown in Table 1.1.

Table 1.1 Unit weight and density terms with their relationships

Unit weight terms	Density terms	Relationship
Total/bulk/wet/moist unit weight, γ	Bulk/wet/moist density, ρ	$\gamma = \rho g$
Dry unit weight, γ_d	Dry density, ρ_d	$\gamma_d = \rho_d g$
Unit weight of solids, γ_s	Density of solids, ρ_s	$\gamma_s = \gamma_s g$
Saturated unit weight, γ_{sat}	Saturated density, ρ_{sat}	$\gamma_{sat} = \rho_{sat} g$
Submerged unit weight, γ'	Submerged density, ρ'	$\gamma' = \rho' g$
Unit weight of water, γ_w	Saturated density, ρ_w	$\gamma_w = \rho_w g$

As mentioned earlier, in geotechnical design and stability calculations, the weight of soil is preferred over its mass. Therefore, the core principles of soil mechanics and geotechnical engineering calculations will generally be explained in terms of soil unit weights rather than densities.

The total unit weight of soil, whether in its natural or compacted state, can be measured directly in the field. This measured value is referred to as the in situ total unit weight of the soil. The in situ total unit weights are required for the following: stability analysis of geotechnical structures, such as foundations, retaining walls, embankments and slopes; field compaction of soils; and estimation of the cut-and-fill quantities in earthworks. The three most common methods for determining in situ total and dry unit weights are

- core cutter method (also known as drive cylinder method) (ASTM D2937-17e2 (ASTM, 2024))
- sand replacement method (also known as sand cone method) (ASTM D1556 (ASTM, 2007))
- rubber balloon method (ASTM D2167 (ASTM, 2008)).

The core cutter method is suitable for natural and compacted cohesive soils that do not contain significant amounts of particles larger than 4.75 mm. This method involves driving a steel cylinder with a hardened cutting edge into the ground using a protective dolly and a specially designed steel rammer. The cutter is then dug out and the soil trimmed off flush at each end. Since the volume of the cutter is known and the contained weight of the soil can be found by weighing, the in situ total unit weight can be determined using Equation 1.7. At the same time, small specimens of soil are taken from either end from which the water content is determined to calculate in situ dry unit weight using the phase inter-relationship presented in Section 1.4. This method may not be applicable for soft, highly plastic, saturated or other soils which are easily deformed, compressed during sampling, or which may not be retained in the core cutter.

The sand replacement method and rubber balloon method are limited to unsaturated soils that do not deform under the pressure imposed during the test. In these tests, a test hole is made and the total weight of the soil removed from the hole and its water content are determined. The volume of the hole is determined by filling the hole with a standard sand of known density and measuring the mass of sand required (sand replacement method), or by expanding a rubber balloon into the hole with water and measuring the volume of water required to fill the hole (rubber balloon method). In the sand replacement method, the sand size should be large enough to avoid entering the voids of the soil. Knowing the soil weight, water content and volume of the hole, the in situ total and dry unit weights can be calculated.

A rapid, nondestructive, indirect method for making in situ unit weight determination involves the use of a nuclear meter. Both the total unit weight and the water content of the in situ soil can be measured with the nuclear meter using controlled radiation techniques.

1.3.4 Unit weight relationship

The most commonly used unit weight (or density) relationship in geotechnical engineering is specific gravity, which is defined as the ratio of unit weight, γ_s (or density, ρ_s) of solids or particles (mineral matter) in the soil element to unit weight, γ_w (or density, ρ_w) of pure water at 4°C. Therefore, if G denotes the specific gravity of the soil solids, then

$$G = \frac{\gamma_s}{\gamma_w} = \frac{\rho_s}{\rho_w} \tag{1.14}$$

The specific gravity of pure water at 4°C is 1.00. At a temperature other than 4°C, the specific gravity of water will be lower than 1.00 because the density of water is maximum at 4°C. For most geotechnical applications, the use of a value of unity at any temperature typically provides satisfactory results.

The specific gravity of soil solids is commonly used to determine void ratio, degree of saturation and unit weight of moist soils. A density bottle or any other suitable constant volume bottle (or pycnometer) of capacity varying from 50 mL to 1 L can be used to determine the specific gravity of soil solids. For fine-grained soils, a 50-mL or 100-mL density bottle with the stopper (Figure 1.5) works well. The test requires the following four measurements

W_1 = weight of density bottle with stopper
W_2 = weight of density bottle with stopper and dry soil
W_3 = weight of density bottle with stopper, dry soil and water
W_4 = weight of density bottle with stopper when full of water only.

Figure 1.5 Density bottle with stopper, vacuum pump and wash bottle for the specific gravity test

With W_1, W_2, W_3 and W_4, Equation 1.14 becomes

$$G = \frac{\gamma_s}{\gamma_w} = \frac{W_2 - W_1}{\left(W_4 - W_1\right) - \left(W_3 - W_2\right)} \tag{1.15}$$

The test procedure for determining the specific gravity of soil particles is described in detail in the following standards: AS 1289.3.5:1 (Standards Australia, 2006), ASTM D854-23 (ASTM International 2023), BS 1377-2:2022 (BSI, 2022) and IS 2720 (Part 3) (Bureau of Indian Standards, 2021). The values of specific gravity of some common minerals are given in Table 1.2.

The following points regarding phase relationships are noteworthy.

- The porosity (n) of soils ranges between 0 and 100%. However, practically, the range of porosity that soils exhibit is narrow, typically 12–55%.
- The void ratio (e) of soil may range from 0 to ∞. Therefore, the void ratio of soil can exceed unity, although it rarely does so for sandy soils. Micaceous sand may have a void ratio of 0.4 to 1.2. The typical values of void ratio of clayey soils vary from 0.2 to 1.5. The void ratio of some organic soils may be higher than that for common clayey soils.
- Both porosity and void ratio indicate the relative portion of void volume in a soil element and are used in soil mechanics. However, the void ratio is more useful than the porosity, particularly when describing the compression of a soil element. The reason behind this fact is that during compression of the soil element, both the numerator and the denominator of the porosity decrease, whereas only the numerator of the void ratio decreases.
- The degree of saturation (S) may vary from 0 to 100%. For a dry soil $S = 0\%$ and for a fully saturated soil $S = 100\%$. For a partially saturated soil S can be between 0 and 100%.

Table 1.2 Specific gravity of some common minerals

Minerals	Specific gravity
Kaolinite	2.62–2.66
Illite	2.66–2.72
Montmorillonite	2.75–2.78
Halloysite	2.55
Calcite	2.72
Quartz	2.65
Feldspars	2.54–2.76
Muscovite	2.7–3.1
Biotite	2.8–3.2
Gypsum	2.32
Hematite	4.72
Magnetite	5.2

Note: In the absence of measured values, it is a common practice to assume $G = 2.65$ for a sand and $G = 2.70$ for a clay.

▨ The natural water content (w) of soils can exceed 100%, although it is well under 100% for most soils. For dry soil, $w = 0\%$. The water content of granular or cohesionless soils such as sands varies from less than 0.1% for air-dry state to more than 40% for saturated loose state. Some marine clays and organic soils can have water content up to 500% or higher. Air-dry clay can absorb enough moisture from the atmosphere for its water content to rise to 2% or more. Montmorillonite, which is a clay mineral (see Section 1.6.3), can reach a water content as high as 20% by absorbing moisture from the atmosphere.

▨ The unit weight of solids (γ_s) is constant for a given soil element, whereas dry unit weight (γ_d) is not constant, as it depends on the total volume (V) of the soil element, which may vary in practice.

▨ The dry unit weight (γ_d) of a soil element is lower than its total unit weight (γ), except when the soil element is dry, for which both are equal.

▨ The typical values of saturated unit weight of some common soils are as follows: 19–24 kN/m³ for sands and gravels; 14–21 kN/m³ for silts and clays; 10–11 kN/m³ for peats.

▨ The ratio of saturated unit weight of a soil to its submerged unit weight (γ_{sat}/γ') can vary from approximately 1 to 2. This ratio can be unity for peats.

▨ The typical values of dry unit weight of some common soils are as follows: 15–23 kN/m³ for sands and gravels; 6–18 kN/m³ for silts and clays; 1–3 kN/m³ for peats.

▨ As discussed earlier, water content and specific gravity are easily determined in the laboratory, while the total unit weight is readily measured in the field. These parameters are used to calculate several other properties that are challenging to determine experimentally. For this purpose, phase inter-relationships, as developed in Section 1.4, are commonly applied.

1.4.   Phase inter-relationships

Phase relationships, as defined in the previous section, can be used to derive several useful expressions, particularly for estimating relationships that cannot be determined experimentally through simple procedures. These inter-relationships are also invaluable for understanding the mechanics of soil masses. Some commonly used phase inter-relationships are given below.

$$n = \frac{e}{1+e} \tag{1.16a}$$

or

$$e = \frac{n}{1-n} \tag{1.16b}$$

$$Se = wG \tag{1.17}$$

Special case

For fully saturated soils, $S = 100\%$; therefore, Equation 1.17 reduces to

$$e = wG \tag{1.18}$$

$$\gamma = \left(\frac{G + Se}{1+e} \right) \gamma_w \tag{1.19}$$

Equation 1.18 is widely used to determine the void ratio of saturated soils because both the water content, w, and the specific gravity, G, of soil solids can be measured accurately and easily in the laboratory.

Example 1.3

Saturated sodium montmorillonite under low confining pressure can exist with a void ratio exceeding 25. Determine the water content of this clay, given a void ratio of $e=25$ and a specific gravity of clay particles of $G=2.78$.

Solution

From Equation 1.18, the water content

$$w = \frac{e}{G} = \frac{25}{2.78} = 8.99 \text{ or } 899\%$$

Special cases

(a) Using Equation 1.17, Equation 1.19 reduces to

$$\gamma = \left(\frac{G+wG}{1+e}\right)\gamma_w = \left(\frac{1+w}{1+e}\right)G\gamma_w \qquad (1.20)$$

(b) For dry soil, $S=0\%$, $w=0\%$ and $\gamma=\gamma_d$. Then Equation 1.19 or Equation 1.20 reduces to

$$\gamma_d = \frac{G\gamma_w}{1+e} \qquad (1.21a)$$

or, using Equation 1.14

$$\gamma_d = \frac{\gamma_s}{1+e} \qquad (1.21b)$$

(c) Equations 1.20 and 1.21a result in

$$\gamma_d = \frac{\gamma}{1+w} \qquad (1.22)$$

(d) For saturated soils, $S=100\%$ and $\gamma=\gamma_{sat}$. Then Equation 1.19 gives

$$\gamma_{sat} = \left(\frac{G+e}{1+e}\right)\gamma_w \qquad (1.23)$$

(e) Using Equations 1.13 and 1.23

$$\gamma' = \gamma_{sat} - \gamma_w = \left(\frac{G+e}{1+e}\right)\gamma_w - \gamma_w = \left(\frac{G-1}{1+e}\right)\gamma_w \qquad (1.24)$$

The phase inter-relationships (Equations 1.16 to 1.24) can be derived, referring to Figure 1.4, with basic definitions of phase relationships as presented in Section 1.3. For instance, the derivations of Equations 1.16a, 1.16b, 1.17 and 1.19 are provided below.

Derivations of Equations 1.16a and 1.16b
From Equation 1.2

$$n = \frac{V_v}{V}$$

$$= \frac{V_v}{V_s + V_v} \left(\because V = V_v + V_s, \text{ see Figure 1.4} \right)$$

$$= \frac{\dfrac{V_v}{V_s}}{1 + \dfrac{V_v}{V_s}}$$

$$= \frac{e}{1+e} \left(\text{using Equation 1.3} \right)$$

Thus,

$$n = \frac{e}{1+e}$$

or

$$e = \frac{n}{1-n}$$

Derivations of Equation 1.17
From Equation 1.5

$$w = \frac{W_w}{W_s}$$

$$= \frac{V_w \gamma_w}{V_s \gamma_s} \left(\text{using Equation 1.11 and } \gamma_w = W_w / V_w \right)$$

$$= \frac{V_w \gamma_w}{V_s (G\gamma_w)} \left(\text{using Equation 1.14} \right)$$

$$= \frac{SV_v}{V_s G} \left(\text{using Equation 1.4} \right)$$

or

$$w = \frac{Se}{G} \left(\text{using Equation 1.3} \right)$$

or

$$Se = wG$$

14

Derivations of Equation 1.19

From Equation 1.7

$$\gamma = \frac{W}{V}$$

$$= \frac{W_s + W_w}{V_s + V_v} \left(\because W = W_s + W_w \text{ and } V = V_s + V_v, \text{ see Figure 1.4} \right)$$

$$= \frac{\left(W_s / V_s \right) + \left(W_w / V_s \right)}{1 + V_v / V_s}$$

$$= \frac{\gamma_s + \left(W_w / V_s \right)}{1 + e} \left(\text{using Equations 1.3 and 1.11} \right)$$

$$= \frac{G\gamma_w + \left(V_w \gamma_w \right) / V_s}{1 + e} \left(\text{using Equation 1.14} \right)$$

$$= \left[\frac{G + \left(V_w / V_v \right) \left(V_v / V_s \right)}{1 + e} \right] \gamma_w$$

or, using Equations 1.3 and 1.4

$$\gamma = \left(\frac{G + Se}{1 + e} \right) \gamma_w$$

Example 1.4

For a partially saturated soil deposit at a construction site, water content $w = 15\%$, void ratio $e = 0.6$ and specific gravity of soil particles $G = 2.67$. Determine the following

(a) degree of saturation
(b) dry unit weight
(c) weight of water required to fully saturate $5\,\text{m}^3$ of soil.

Solution

(a) From Equation 1.17, the degree of saturation

$$S = \frac{wG}{e} = \frac{(0.15)(2.67)}{0.6} = 0.668 = 66.8\%$$

(b) From Equation 1.19, the total unit weight

$$\gamma = \left(\frac{G + Se}{1 + e} \right) \gamma_w = \left[\frac{2.67 + (0.668)(0.6)}{1 + 0.6} \right] (9.81) = 18.83 \text{ kN/m}^3$$

From Equation 1.22, the dry unit weight

$$\gamma_d = \frac{\gamma}{1 + w} = \frac{18.83}{1 + 0.15} = 16.37 \text{ kN/m}^3$$

(c) For full saturation, $S = 100\%$. From Equation 1.23, the soil will have saturated unit weight

$$\gamma_{sat} = \left(\frac{G+e}{1+e}\right)\gamma_w = \left(\frac{2.67+0.6}{1+0.6}\right)(9.81) = 20.05 \text{ kN/m}^3$$

Weight of water required to saturate $5\,\text{m}^3$ of soil

$$W_w = (\gamma_{sat} - \gamma)(5) = (20.05 - 18.83)(5) = 6.1 \text{ kN}$$

1.5.   Weathering and soil formation

Weathering is the process of decay, disintegration and decomposition of rocks at or near the Earth's surface under the influence of certain physical, chemical and/or biological agencies. The climatic conditions play an important role in initiation and intensity of weathering. In arid regions, physical or mechanical disintegration of rocks is due to changes in temperature. In temperate regions, the combined effect of freezing and thawing is the main cause of disintegration. In tropical regions, a warm and humid climate supports chemical weathering of rocks. Chemical actions such as oxidation, hydration, carbonation and solution in the presence of water and/or organic acids cause the chemical decay and decomposition of rocks. Biological agencies such as plants, animals and organisms can cause physical disintegration and/or chemical decomposition. Generally, silt-sized particles (2–75 μm) and above are formed by the physical weathering of rocks; clay-sized (<2 μm) particles are formed by the chemical weathering of rocks. Uncemented or very lightly cemented aggregates of these particles in various proportions form different types of soil.

The rate of weathering and the nature of its products, such as the types of sediment or soil, are significantly influenced by the following five factors, commonly referred to as soil-forming factors

- parent materials or types of source rock
- topography
- climate
- organisms
- time.

The parent materials include rocks present at or near the Earth's surface. They establish the basic characteristics of soils. Topography that is described mainly in terms of ground slope, elevation, vegetation and drainage controls the rate of erosion. Steep grounds have little or no soil because of mainly high rates of erosion. Climate (temperature and rainfall) influences factors such as precipitation availability, freeze–thaw cycles, wetting and drying history, wind and more. Organisms, through their bacterial activity, loosen and mix the soil and, in some cases, create an entirely new soil profile. Time is the most subjective of the soil-forming factors because the influence of time is difficult to quantify. The formation of soil to just a few centimetres in thickness requires several thousand years.

The properties of soil are significantly influenced by the type of minerals derived from the parent rock, as well as the type and amount of liquid and gas present in the voids between solid particles. They also depend on the size, shape and arrangement of the mineral particles.

1.6. Clay mineralogy

A significant part of many soils consists of mineral particles smaller than $2\,\mu m$ (0.002 mm), which are principally clay minerals. There are three principal clay minerals: kaolinite, illite and montmorillonite (Grim, 1968; Lambe and Whitman, 1979). They can be seen only with an electron microscope. Clay minerals are a group of complex alumino-silicates, mainly formed during the chemical weathering of primary minerals. Most clay mineral particles are of 'plate-like' form having a high specific area (surface area per unit mass), with the result that their properties are influenced significantly by surface forces rather than gravitational body forces. Long 'needle-shaped' particles can also occur (e.g. halloysite) but are rare. Soil that mainly consists of clay minerals is called clayey soil or simply clay.

X-ray diffraction and microscopic studies indicate that most clay mineral particles consist of stacking of combinations of basic structural or crystal sheets, namely silica sheets and alumina sheets, with different forms of bonding between them. Basic structural sheets are formed from a silica tetrahedron and an alumina octahedron as the basic structural or crystal units (Figure 1.6). A silica tetrahedron consists of four oxygen (O) atoms nestled around a silicon (Si) atom (Figure 1.6(a)). An alumina octahedron consists of six hydroxyls (OH) enclosing an aluminium (Al) atom (Figure 1.6(b)). In these basic units, silicon (Si) and aluminium (Al) may be partially replaced by elements with lower electrovalence such as magnesium (Mg) and iron (Fe), a process known as isomorphous substitution. This substitution may result in the following two effects.

▨ A net unit charge deficiency results per substitution.
▨ A slight distortion of the crystal lattice occurs since the ions are not of identical size.

A combination of silica tetrahedrons gives a silica sheet while the combination of alumina octahedrons gives an alumina or gibbsite sheet. Figure 1.7 illustrates the symbolic or schematic representation of silica and alumina sheets.

The study of clay minerals is the foundation of the field of clay mineralogy. Grim (1968) provided comprehensive scientific insights into this subject.

Figure 1.6 Basic structural units: (a) silica tetrahedron; (b) alumina octahedron

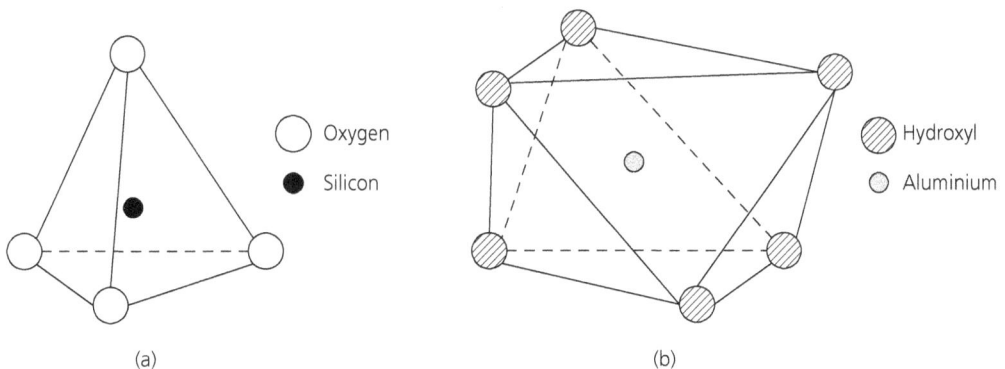

(a)   (b)

Figure 1.7 Symbolic representation of basic structural sheets: (a) silica (Si) sheet; (b) alumina (Al) sheet

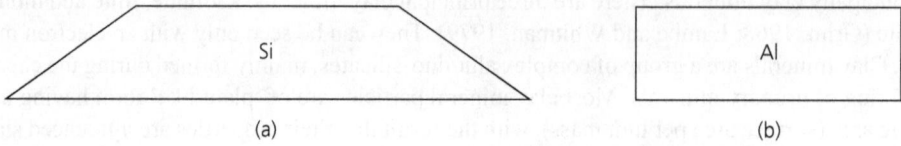

(a)   (b)

1.6.1   Kaolinite

The basic structure or unit of kaolinite consists of a combination of one silica sheet and one alu-mina sheet (Figure 1.8), and therefore this mineral is called a 'two-layer' mineral. The thickness of the basic structure is about 7.2 Å (0.72 nm). A kaolinite particle or crystal may consist of over one hundred stacks of the basic structure. There is a very limited isomorphous substitution in kaolinite. The combined basic structures are held together fairly tightly by hydrogen bonding. As the hydrogen bond is very strong, it prevents hydration; hence kaolinite does not swell and shrink significantly with changes in water content. A typical kaolinite particle is about $10\,000\,\text{Å} \cdot 1000\,\text{Å}$ thick. Kaolinite is significantly present in China clay. Another two-layer clay mineral, called hal-loysite, has essentially the same composition and structure as kaolinite. However, the presence of water between the basic sheets results in the formation of tubular particles.

1.6.2   Illite

The basic structure or unit of illite consists of an alumina sheet sandwiched between two silica sheets (Figure 1.9). The thickness of the basic structure is about 10 Å (1.0 nm). In the alumina

Figure 1.8 Symbolic representation of the structure of kaolinite

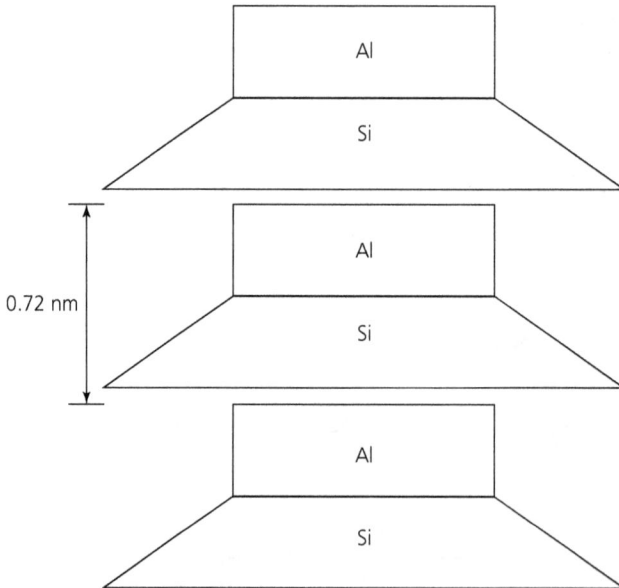

Note: Al is the alumina sheet; Si is the silica sheet

Figure 1.9 Symbolic representation of the structure of illite

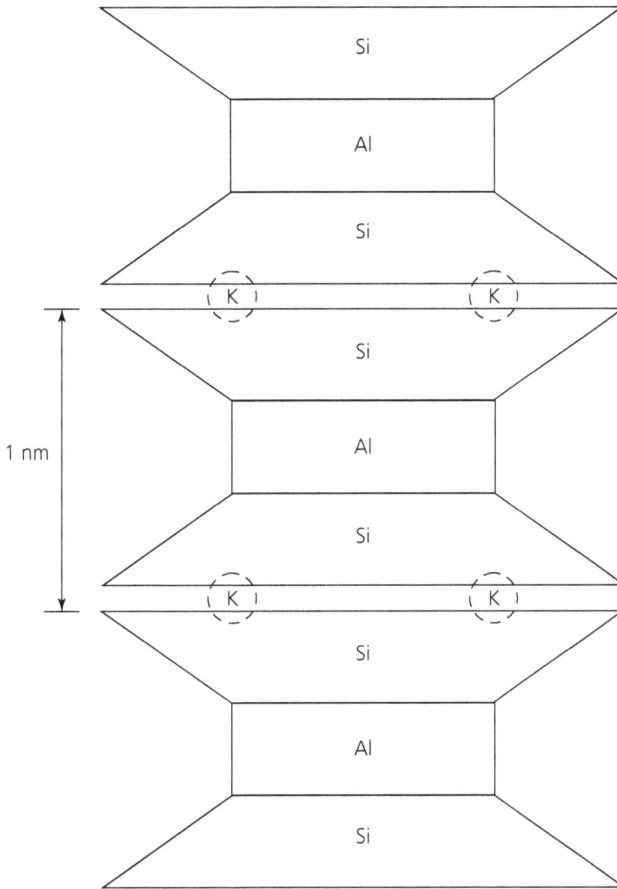

Note: Al is the alumina sheet; Si is the silica sheet; K is the potassium atom

sheet, there is a partial substitution of aluminium by magnesium and iron, and in the silica sheet there is a partial substitution of silicon by aluminium. The net charge created by substitutions is balanced by potassium (K) ions, which serve to link the three-layer sandwiches together by fairly weak bonding. The crystal structure of illite is similar to that of mica minerals but with less potassium and less isomorphous substitutions. Illite is chemically much more active than any mica mineral.

1.6.3   Montmorillonite

Similar to illite, montmorillonite has a 'three-layer' basic structure, consisting of an alumina sheet sandwiched between two silica sheets (Figure 1.10). The thickness of the basic structure is about 9.6 Å (0.96 nm). In the alumina sheet, there is a partial substitution of aluminium by magnesium, resulting in a net negative charge deficiency in the alumina sheet. The space between the basic structures or units is easily occupied by water molecules (nH_2O) and exchangeable cations other than potassium, thus resulting in a very weak bond formed by van der Waals' forces between the basic structures. A typical montmorillonite particle is about 1000 Å · 10 Å thick. Therefore,

19

Figure 1.10 Symbolic representation of the structure of montmorillonite

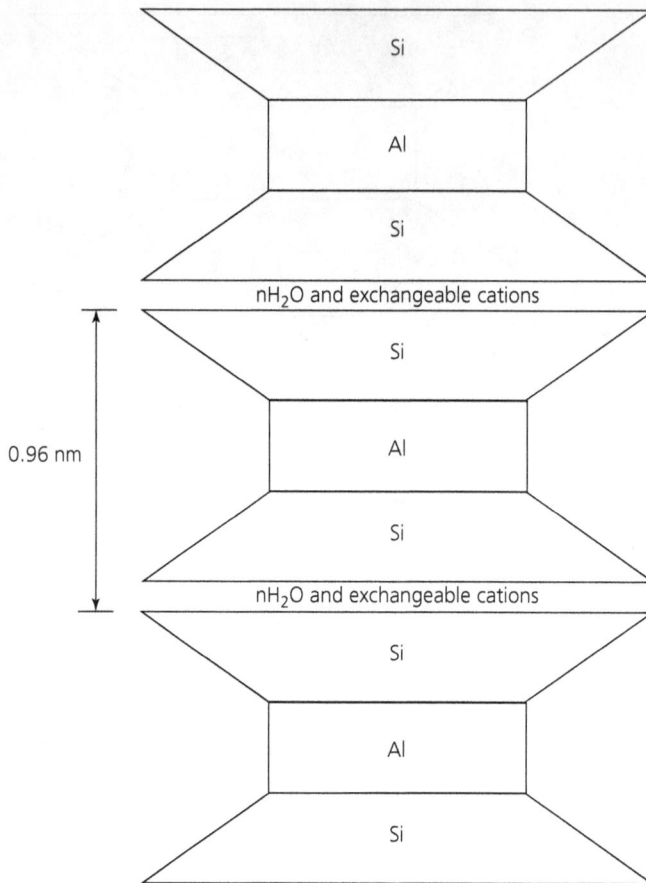

Note: Al is the alumina sheet; Si is the silica sheet; nH_2O is the water molecules

montmorillonite is very susceptible to swelling and shrinkage with changes in water content. Montmorillonite is a major component of bentonite clay, which is commercially used in various applications. For instance, it is used as a drilling fluid or mud to stabilise boreholes and slurry trenches, as part of a liner with common clayey soils in landfills, and as a barrier to reduce seepage through porous or cohesionless soils.

1.6.4   Cation exchange and electric diffuse double layer

The surfaces of clay mineral particles carry net negative charges that arise mainly from isomorphous substitution of aluminium or silicon atoms by atoms of lower electrovalence in their basic structures. Some other factors, which may also bring negative charges on the clay mineral particle surfaces, are surface disassociation of hydroxyl ions, absence of cations in the crystal lattice, adsorption of anions and presence of organic matter.

A clay mineral particle in nature attracts ions to neutralise its net negative charge. As these attracted ions are usually weakly held on the particle surface and can be readily replaced by other ions, they are termed exchangeable ions, and the phenomenon is referred to as cation exchange. Calcium (Ca) is a very common exchangeable ion in soils.

A water molecule, due to its bipolar nature, can form strong bonds with both the surface of clay mineral particles and the exchangeable cations surrounding them. One to four molecular layers of water are typically attached to the particles by way of hydrogen bonding as adsorbed water. Since the soil particle is surrounded by water, the exchangeable cations are not attached to it. However, the cations are attracted to a clay mineral particle because of the negative electrical surface charges, but at the same time they tend to thermally diffuse toward the lower cation concentration away from the particle. The net effect is that cations form a dispersed layer adjacent to the particle. The cation concentration decreases with increasing distance from the surface until the concentration becomes equal to that in the normal water in the void space. The soil particle is thus surrounded by a domain known as an electric diffuse double layer (Mitchell and Soga, 2005; van Olphen 1977). The inner layer of the double layer is the negative charge on the surface of the soil particle, and the outer layer is the excess of cations and deficiency of anions with respect to the concentration in free water not influenced by the force field of the particle (Terzaghi et al., 1996). The distribution of ions adjacent to the surface of the clay particle as the electric diffuse double layer is shown schematically in Figure 1.11.

1.7.   Types of soil and soil structure

The majority of rocks and soils present today, at or near the Earth's surface, formed during the last one-eighth of geological time, approximately 4500 million years, which is the age of the Earth and its moon. The processes of soil formation are complex and they directly affect the engineering properties of the resulting soil mass. Based on their method of formation, soils are primarily classified as

- sedimentary soils
- residual soils
- fills
- organic soils.

Figure 1.11 Electric diffuse double layer according to the Gouy–Chapman theory (Chapman, 1913; Gouy, 1910; Mitchell and Soga, 2005; Verwey and Overbeek, 1948)

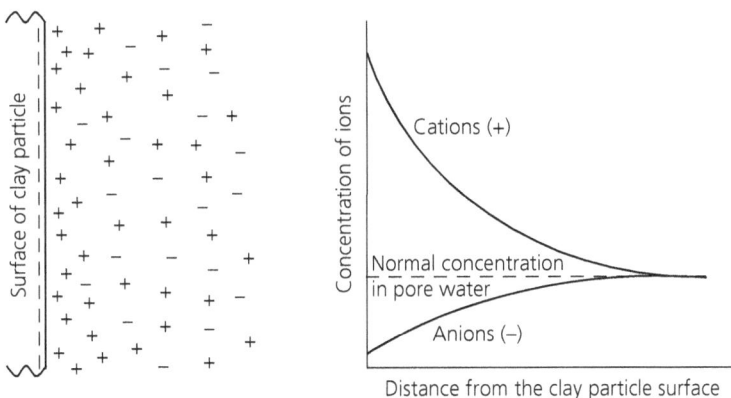

The formation of sedimentary soils involves the following three processes

▪ sediment formation resulting from the weathering of rocks
▪ sediment transport by water, wind, ice, gravity and organisms, collectively referred to as transporting agents
▪ sediment deposition in different environments.

Transportation of sediments impacts them in the following two significant ways

▪ alteration in particle shape, size and texture by abrasion, grinding, impact and solution
▪ sorting of the particles.

The transportation by water and wind results in rounding of particles and their considerable sorting. Ice and gravity do not cause any rounding of particles; thus, the particles remain in angular shape. Very little sorting can be noticed in sediments transported by ice and organisms. As a transporting agent, gravity does not sort the particles. The final process of sediment deposition makes a sedimentary soil. The main causes of sediment deposition in water are decrease in velocity of particles being carried by transporting agents, decrease in solubility and increase in electrolyte concentration. The orientation and distribution of particles in a soil mass, commonly referred to as the soil structure (or soil fabric), are governed by the deposition environment. There are two types of soil structure, namely

▪ flocculated structure (Figure 1.12(a))
▪ dispersed structure (Figure 1.12(b)).

The flocculated structure (Figure 1.12(a)) results in a saltwater environment, where the soil particles have edge-to-face or edge-to-edge contacts and there is a net attraction.

The dispersed structure (Figure 1.12(b)) forms in a freshwater environment, where the soil particles tend to assume a face-to-face orientation and there is a net repulsion.

The engineering behaviour of soil is significantly controlled by the type of structure. In general, compared to an element of soil with dispersed structure, an element of the flocculated soil at the same void ratio has higher permeability, lower compressibility and higher strength. If the flocculated soil is subjected to horizontal shear displacement, the particles will tend to line up in the dispersed structure. Table 1.3 provides a list of sedimentary soils along with their corresponding transporting agents and depositional environments.

Figure 1.12 Soil structure: (a) flocculated; (b) dispersed

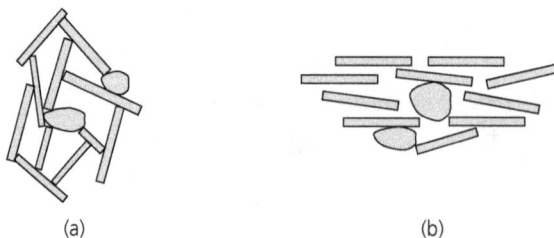

(a)   (b)

Table 1.3 Sedimentary soils with their corresponding transporting agents and depositional environments

Sedimentary soil	Transporting agent or mode of transport	Depositional environment
Alluvium/alluvial soil	Water	Flowing water
Marine soil		Quiet saline water
Lacustrine soil		Fresh water
Aeolian soils (loess[a]/dune sand)	Wind	Arid and coastal lands
Glacial till/boulder clay	Ice	Glacial
Colluvium/colluvial soil/talus	Gravity	Below slide areas, base of hills/ mountains
Muck/peat	In situ	Swamp and marshy

[a]A loess deposit consists of silt-sized particles, weakly cemented with calcareous and/or argillaceous/clayey materials. This deposit is formed in arid and semi-arid regions and can stand in a nearly vertical bank or slope.

Residual soils are products of the in situ weathering of rocks. They are formed mainly in humid, warm regions that are favourable to chemical weathering of rocks and have sufficient vegetation to keep the weathering products from being easily transported as sediments. Their grading and mineralogy depend on the parent material, depth and type of weathering, and drainage conditions. They are commonly situated above the groundwater table and, therefore, they are often unsaturated. Since these soils are typically found in areas with less developed economies, their engineering behaviour has not been studied as extensively as that of sedimentary soils, which are commonly located in urban and industrial regions. Typical depths of residual soils over the Earth vary from a few centimetres to 25 m.

Fill is a human-made soil and the process of its formation is referred to as filling. Essentially, a fill is a sedimentary soil deposit created entirely through human activity. Soil, referred to as borrow, is obtained from a source or made by blasting, transported by vehicle and deposited by dumping. The fill can be left as dumped or can be processed and densified or compacted as required in the field by rolling, vibration or impact in a specific number of layers.

Organic soils, such as peats, are derived from the composition of organic materials, such as decayed vegetation, including leaves and tree roots.

Note that the soil is not inert but rather very much alive and sensitive to its environment. The properties of soil as observed at the start of the project may keep changing during the design life of the project. The factors that contribute to changes in soil behaviour are (Lambe and Whitman, 1979)

- water
- stress
- disturbance
- environment
- time.

The presence of water causes the attractive forces between clay particles to decrease. An increase in water content of a soil generally reduces its strength. In general, an increase in stress on a soil element results in decreased permeability, reduced compressibility and increased shear strength. Conversely, a reduction in stress produces the opposite effects, although the changes are typically very limited. A sedimentary soil at any given elevation can experience an increase in stress during its formation due to additional deposition or loading, and a decrease in stress due to removal of soil overburden, such as through erosion or unloading. Thus, it is important to consider the stress history in the design and construction of geotechnical structures.

The nature of the pore fluid and the temperature are two environmental parameters that influence soil behaviour. The gradual leaching of the salt present in the pore fluid of clayey soils such as marine clays lying above the sea level over many thousands of years can cause a reduction in shear strength. Marine clays have flocculated structure during initial formation, but with salt leaching, they acquire dispersed structure, which can easily be noticed when they are subjected to enough disturbance, resulting in a soil–water slurry with almost zero shear strength. Marine clays without presence of salt in the pore water are called 'quick clays'. An increase in temperature of a clayey soil normally causes an expansion of the soil and can remove some air dissolved in the pore water.

Stress, water and environmental conditions are time-dependent, making time a crucial factor in controlling soil behaviour. Additionally, time plays a significant role in chemical weathering by allowing chemical reactions to occur.

The following points regarding soil types and soil structure are noteworthy.

- The clay content, even a small amount, say 20–30%, in a soil greatly controls its engineering behaviour.
- The element of a soil with flocculated structure has a much larger void ratio and water content.
- A tendency toward flocculation is typically caused by an increase in one or more of the following within the pore water: exchangeable cation concentration, cation valance, temperature.
- A sedimentary soil is generally bedded and layered, reflecting changes in the depositional environment.
- Engineers must consider how soil properties might change over the design life of a structure.

Chapter summary

1. Soil comprises all materials in the surface layer of the Earth's crust that are loose enough to be moved by a spade or shovel. It is a multiphase system generally consisting of three phases: solid, liquid and gas.
2. Porosity, void ratio, degree of saturation, water content, total unit weight, dry unit weight, submerged unit weight and specific gravity are the commonly used phase relationships.
3. There are several phase relationships and inter-relationships. The expression $Se = wG$ is useful to check computations of the various relationships.
4. Physical, chemical and biological weathering of rocks result in the formation of soil. Generally, silt-sized particles (2–75 μm) and larger are formed through physical

weathering of rocks, while clay-sized particles ($<2\,\mu m$) are typically formed through chemical weathering.

5. Kaolinite, illite and montmorillonite are the principal clay minerals present in clayey soils. Montmorillonite is very susceptible to swelling and shrinkage with changes in water content.

6. Based on their method of formation, soils are classified as sedimentary soils, residual soils, fills and organic soils. Sedimentary soils can have flocculated or dispersed structures, depending on their depositional environment.

Questions for practice

Select the most appropriate answer for each of the multiple choice questions from Q1.1 to Q1.10. Questions Q1.11 to Q1.23 require more detailed answers.

1.1 The calculation of unit weight γ of soil from its density ρ is done using the relationship $\gamma = \rho g$. If $g = 9.81\,m/s^2$ is rounded to $g = 10\,m/s^2$, the error in the calculated value of γ is approximately
   (a) 1%
   (b) 2%
   (c) 3%
   (d) 4%.

1.2 The ratio of water volume to void volume in a soil element is called
   (a) void ratio
   (b) porosity
   (c) degree of saturation
   (d) water content.

1.3 The ratio of the submerged unit weight of clayey soil to the unit weight of water can be approximated as
   (a) 1
   (b) 2
   (c) 3
   (d) 4.

1.4 If the void ratio and specific gravity of the soil particles are 0.60 and 2.70, respectively, what will be the submerged unit weight of the soil?
   (a) $9.81\,kN/m^3$
   (b) $10.42\,kN/m^3$
   (c) $22.68\,kN/m^3$
   (d) none of the above.

1.5 The dry unit weight (γ_d), unit weight of solids (γ_s) and porosity (n) of a soil element have the following relationship:
   (a) $\gamma_d = n\gamma_s$
   (b) $\gamma_s = (1 - n)\gamma_d$
   (c) $\gamma_d = (1 + n)\gamma_s$
   (d) $\gamma_d = (1 - n)\gamma_s$

1.6 For clayey and silty soils, the specific gravity of particles typically varies from
  (a) 1.6 to 1.9
  (b) 2.6 to 2.9
  (c) 3.6 to 3.9
  (d) none of the above.

1.7 Which type of rock weathering can result in clay-sized particles ($<2\,\mu m$)?
  (a) physical weathering
  (b) chemical weathering
  (c) biological weathering
  (d) both (b) and (c).

1.8 Which of the following clay minerals has a much greater attraction for exchangeable cations?
  (a) montmorillonite
  (b) illite
  (c) kaolinite
  (d) both (b) and (c).

1.9 The transporting agent responsible for the formation of talus is
  (a) ice
  (b) gravity
  (c) water
  (d) wind.

1.10 Select the incorrect statement.
  (a) Water content of a soil can be greater than 100%.
  (b) A loess deposit consists of silt-sized particles.
  (c) In the outer layer of the electric diffuse double layer of a clay mineral particle, there is an excess of anions.
  (d) An increase in temperature of a cohesive soil normally causes an expansion of the soil.

1.11 A sample of saturated soil weighs 1435 g in its natural state and 947 g after drying. If the specific gravity of soil solids is 2.70, what is the void ratio of the soil?

1.12 Establish the following relationship:
  (a) $\gamma = \gamma_s(1 - n)(1 + w)$
  (b) $\gamma = (1 - n)\gamma_s + nS\gamma_w$

1.13 A saturated clay compressed under high stress (e.g. 70 MPa), which can occur at great depths in the ground, may have a void ratio of less than 0.2. Determine the water content of this soil, considering the void ratio, $e = 0.2$, and the specific gravity of soil particles, $G = 2.78$.

1.14 The dry unit weight of a sand sample consisting of mainly quartz mineral is 15.46 kN/m³. Determine the following for the sand sample:
  (a) void ratio
  (b) total unit weight when saturated.

1.15 An in situ soil has a total unit weight of $19.05 \, kN/m^3$, and a water content of 29%. Determine the following for the soil:
(a) void ratio
(b) degree of saturation.

Assume that the specific gravity of soil solids is 2.69.

1.16 For a soil deposit at a construction site, void ratio, $e = 0.7$, water content, $w = 17\%$, and specific gravity of soil solids, $G = 2.69$. Determine the following for the soil:
(a) degree of saturation
(b) total unit weight
(c) dry unit weight.

1.17 Identify some geotechnical structures in your area and list your engineering observations.

1.18 Differentiate between the void ratio and porosity of a soil. How can the void ratio of a soil be determined?

1.19 List the factors that control the rate of weathering of rocks. In your opinion, which factor has the most significant influence on soil behaviour?

1.20 What is isomorphous substitution? Explain briefly.

1.21 Discuss the engineering behaviour of soils containing the primary clay minerals kaolinite, illite and montmorillonite.

1.22 How does transported soil differ from residual soil?

1.23 Draw a world map and mark the locations of major soil deposits.

REFERENCES

ASTM (2007) ASTM D1556: Standard test method for density and unit weight of soil in place by sand-cone method. ASTM International, West Conshohocken, PA, USA.

ASTM (2008) ASTM D2167: Standard test method for density and unit weight of soil in place by the rubber balloon method. ASTM International, West Conshohocken, PA, USA.

ASTM (2019) ASTM D2216-19: Standard test method for laboratory determination of water (moisture) content of soil and rock by mass. ASTM International, West Conshohocken, PA, USA.

ASTM (2023) ASTM D854-23: Standard test methods for specific gravity of soil solids by water pycnometer. ASTM International, West Conshohocken, PA, USA.

ASTM (2024) ASTM D2937-17e2: Standard test method for density of soil in place by the drive-cylinder method. ASTM International, West Conshohocken, PA, USA.

BSI (2022) BS 1377-2:2022: Method of test for soils for civil engineering purposes – classification tests and determination of geotechnical properties. BSI, London, UK.

Bureau of Indian Standards (2020) IS 2720 (Part 2):1973: Determination of water content. Bureau of Indian Standards, New Delhi, India.

Bureau of Indian Standards (2021) IS 2720 (Part 3): Determination of specific gravity: Section 1 – fine grained soils, Section 2 – fine, medium and coarse grained soils. Bureau of Indian Standards, New Delhi, India.

Chapman DL(1913) A contribution to the theory of electrocapillarity. *Philosophical Magazine* **25(6)**: 475–481.

Gouy G (1910) Sur la constitution de la charge electrique à la surface d'un electrolyte. *Anniue Physique (Paris)* **9(4)**: 457–468. (In French.)

Grim RE (1968) *Clay Mineralogy*, 2nd edn. McGraw-Hill, New York, NY, USA.

Lambe TW and Whitman RV (1979) *Soil Mechanics, SI Version*. Wiley, New York, NY, USA.

Mitchell JK and Soga K (2005) *Fundamentals of Soil Behaviour*, 3rd edn. Wiley, Hoboken, NJ, USA.

Shukla SK (2025) *ICE Core Concepts: Geotechnical Engineering*, 2nd edn. Emerald/ICE Publishing, London, UK.

Standards Australia (2005) AS 1289.2.1.1: Determination of the moisture content of a soil – oven drying method (standard method). Standards Australia, Strathfield, Australia.

Standards Australia (2006) AS 1289.3.5.1: Determination of the soil particle density of a soil – standard method. Standards Australia, Strathfield, NSW, Australia.

Terzaghi K, Peck RB and Mesri G (1996) *Soil Mechanics in Engineering Practice*, 3rd edn. Wiley, New York, NY, USA.

van Olphen H (1977) *An introduction to Clay Colloid Chemistry*, 2nd edn. Wiley, New York, NY, USA.

Verwey EJW and Overbeek JThG (1948) *Theory of the Stability of Lyophobic Colloids*. Elsevier, Amsterdam, the Netherlands.

FURTHER READING

Atkinson J (2007) *The Mechanics of Soils and Foundations*, 2nd edn. CRC Press, London, UK.

Das BM (2022) *Principles of Geotechnical Engineering*, 10th edn. Cengage, Boston, MA, USA.

Holtz RD, Kovacs WD and Sheahan TC (2023) *An Introduction to Geotechnical Engineering*, 3rd edn. Pearson Education, Hoboken, NJ, USA.

Sanjay Kumar Shukla
ISBN 978-1-83608-519-5
https://doi.org/10.1108/978-1-83608-516-420252002
Emerald Publishing Limited: All rights reserved

Chapter 2
Particle size and shape, consistency and classification of soil

Learning aims

This chapter explores the core concepts of the following topics

- soil particle size analysis
- soil particle shape
- consistency limits and indices of fine-grained soils
- relative density of coarse-grained soils
- soil classification and description.

2.1.   Introduction

Chapter 1 introduced a basic description of soil, namely that soil is a multiphase system generally consisting of three phases: solid, liquid and gas. The solid phase of a soil mass comprises
mineral particles and/or organic matter. The size of particles within a soil mass ranges from very
small ($<2\,\mu m$ or 0.002 mm) to very large (>300 mm) in diameter. Depending on the particle size
distribution and particle shape, a soil mass exhibits specific engineering properties, such as permeability (see Chapter 4), compressibility (see Chapter 5) and shear strength (see Chapter 6). Prior to
directly determining the specific properties of the soil at a construction site, it is common practice
to perform simple, low-cost tests, known as index tests, such as particle size analysis and tests for
soil consistency (the degree of resistance a soil offers to its deformation). The results of index tests
help categorise soil into groups with similar engineering behaviour, a process known as soil classification. Understanding soil classification and description offers several advantages, which are
explored in this chapter.

2.2.   Particle size analysis of soil
2.2.1   Soil particle size

Soil particles are rarely perfect spheres; thus, the diameter of a soil particle typically refers to an
equivalent diameter determined through particle size analysis. Clay, silt, sand, gravel, cobble and
boulder are names of particle sizes, also used to describe soils, particularly when they contain
smaller quantities of other particle sizes. Table 2.1 shows the size divisions based on common
engineering classification systems (AASHTO M 146-91 (AASHTO, 1991a); ASTM D2487-17
(ASTM, 2020); BS 5930:2015+A1:2020 (BSI, 2020); IS 1498:1970 (Bureau of Indian Standards, 2021); Canadian Foundation Engineering Manual (CGS, 2023); AS 1726:2017 (Standards
Australia, 2017)). The particle size classification of the unified soil classification system (USCS)

Table 2.1 Particle size classification

Name of particle size		Range of particle size: mm				
		AASHTO M 146-91 (AASHTO, 1991a)	AS 1726:2017 (Standards Australia, 2017)	ASTM D2487-17, USCS (ASTM, 2020)	BS 5930:2015 (BSI, 2020)	IS 1498:1970 (reaffirmed 2021) (Bureau of Indian Standards, 2021)
Fines (silt and clay)		0.001–0.075	<0.075	<0.075	<0.063	<0.075
Colloids		<0.001	–	–	–	–
Clay		0.002–0.001	–	–	<0.002	<0.002
Silt	Fine	0.002–0.075	–	–	0.002–0.0063	0.002–0.075
	Medium		–	–	0.0063–0.02	
	Coarse		–	–	0.02–0.063	
Sand	Fine	0.075–0.425	0.075–0.2	0.075–0.425	0.063–0.2	0.075–0.425
	Medium	–	0.2–0.6	0.425–2	0.2–0.63	0.425–2
	Coarse	0.425–2	0.6–2.36	2–4.75	0.63–2	2–4.75
Gravel	Fine	2–9.5	2.36–6	4.75–19	2–6.3	4.75–20
	Medium	9.5–25	6–20	–	6.3–20	–
	Coarse	25–75	20–63	19–75	20–63	20–80
Cobble		75–305	63–200	75–300	63–200	80–300
Boulder		>305	>200	>300	>200	>300

Note: According to BS 5930:2015 (BSI, 2020), a particle size larger than 630 mm is classified as a large boulder.

ASTM D2487-17 (ASTM, 2020), which is explained in more detail in Section 2.5, is widely used worldwide. This system will be followed throughout this book whenever required. Based on particle size, soil is generally divided into the following two fractions

- fine-grained fraction of soil: soil with particles equal to or smaller than 75 μm (0.075 mm) in size
- coarse-grained fraction of soil: soil with particles larger than 75 μm (0.075 mm) in size.

Note that 75 μm is approximately the smallest particle size visible to the naked eye under normal vision. Silts and clays are fine-grained soils, while sands, gravels, cobbles and boulders are coarse-grained soils. According to the USCS, a soil sample with 50% or more particles smaller than 75 μm is classified as fine-grained soil. Similarly, a sample with more than 50% particles larger than 75 μm is classified as coarse-grained soil.

2.2.2   Particle size distribution of soil

The distribution by weight (or mass) of different particle sizes in a soil sample is determined by its particle size analysis, which consists of

- sieve analysis for the coarse fraction of soil
- sedimentation analysis for the fine fraction of soil.

In sieve analysis, the dry coarse-grained fraction of the soil sample is separated into various size fractions by passing it through a series of standard sieves with openings arranged from coarsest to finest, such as 80 mm, 20 mm, 4.75 mm, 2 mm, 425 μm and 75 μm. The sieves consist of woven-wire square meshes (Figure 2.1(a)). The quantity of soil specimen for the sieve analysis depends on the maximum particle size present in the specimen, and may vary from 100 g to 60 kg, depending on the maximum particle size in the sample – for example, if the maximum size of the particle is 4.75 mm, 100 g of the soil sample may be sufficient for the test. A 60 kg soil sample may be required if the maximum particle size is 75 mm. The soil sample is shaken manually or mechanically at the coarsest sieve size for about 10–15 minutes. Manual sieving is quite convenient for soil fractions retained on the 4.75 mm sieve. Figure 2.1(b) shows an arrangement for the sieving operation, including a series of sieves and a sieve shaker. At the end of the sieving operation, the mass of the soil retained on each sieve size is weighed, and the cumulative percentage by weight (or mass) passing is calculated as the results of the sieve analysis.

Figure 2.1 Sieve analysis: (a) typical sieves; (b) series of sieves with a sieve shaker

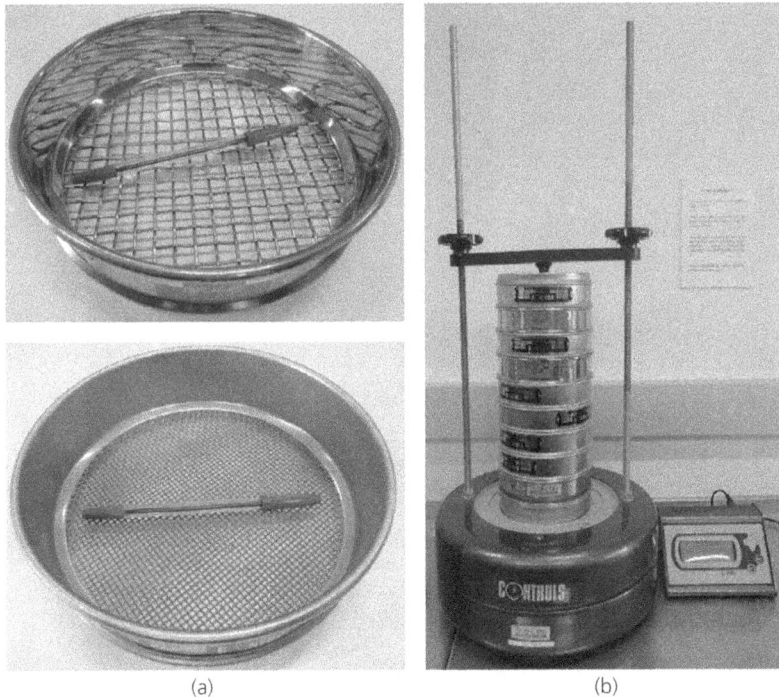

(a)   (b)

Sedimentation analysis for the fine-grained fraction of the soil sample uses a hydrometer (Figure 2.2(a)) to determine the density of the soil–water suspension at various time intervals. The soil–water suspension is prepared with a small amount of a chemical dispersing agent, usually sodium hexametaphosphate, in a glass sedimentation cylinder. Figure 2.2(b) shows an arrangement for the sedimentation analysis. This analysis is based on Stokes's law, which relates to the terminal velocity of spheres (or particles) in a viscous fluid (or suspension), their density (or unit weight) and the density (or unit weight) of the fluid. A hydrometer is graduated to read either specific gravity of the suspension or grams per litre of suspension. The analysis helps determine the percentage of particles with a specific equivalent particle diameter at a given time. Note that Stokes's law applies only to spherical particles although the particles of silts and clays are rarely spherical; thus, this analysis does not give a true picture of particle sizes of the fine-grained fraction (silts and clays) of the soil sample. Silt particles are typically angular and clay particles flat. Further details on particle size analysis can be found in ASTM D422-63(2007) (ASTM, 2014) and BS 1377-2:2022 (BSI, 2022).

Sedimentation analysis can be time consuming, and the results often vary for the same soil when conducted by different individuals. Fortunately, the particle size information based on the sedimentation analysis is not necessarily required for several applications, including soil classification.

Figure 2.2 Sedimentation analysis: (a) hydrometer; (b) test arrangement

(a) (b)

The distribution (or proportion) of different particle sizes in a soil mass affects its engineering behaviour, such as permeability, compressibility and shear strength. The particle size distribution significantly affects the engineering behaviour when the soil mass consists predominantly of coarse-grained soils. In contrast, the engineering behaviour of a soil mass composed mainly of fine-grained soils is strongly influenced by the presence of water. For coarse-grained soils, particularly fine sand and coarse silt deposits, water can significantly influence their engineering behaviour under loose, saturated conditions when subjected to dynamic or seismic loading, primarily through a phenomenon known as soil liquefaction (see Chapters 4 and 9).

Example 2.1

The tests for particle size analysis of a soil sample are conducted in the soil mechanics laboratory, and the results are presented in Table E2.1.

Table E2.1

Particle size analysis	Particle diameter (or sieve opening size): mm	Percentage finer by weight (or mass)
Sieve analysis	4.75	100
	2	87
	1.18	75
	0.425	45
	0.150	29
	0.075	21
Sedimentation analysis	0.03	14
	0.01	8
	0.005	6
	0.002	5

Draw a semi-logarithmic plot of percentage finer by weight (or mass) against particle diameter for the soil.

Solution

The semi-logarithmic plot, with particle diameter on the x-axis and percentage finer by weight on the y-axis, is shown in Figure E2.1. A spreadsheet was used to complete this task.

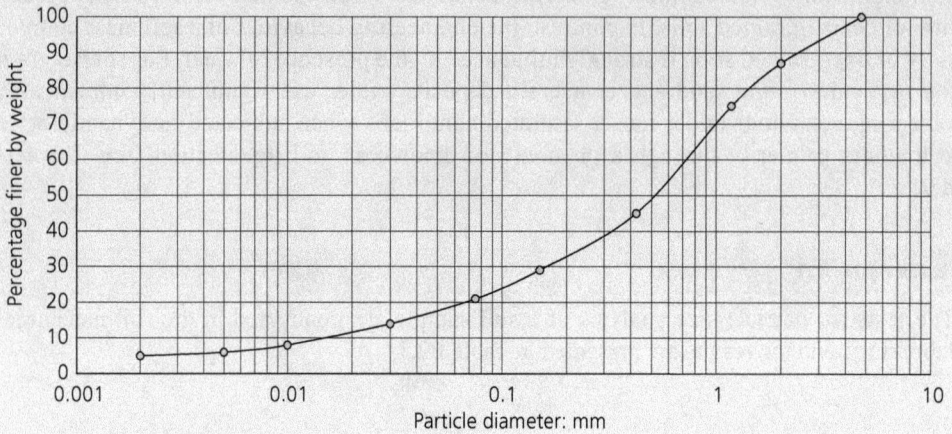

Figure E2.1

In Figure E2.1, the curve drawn as a semi-logarithmic plot of percentage finer by weight (or mass) against particle diameter is called the particle size distribution curve or grading curve of soil.

From the particle size distribution curve, three commonly used particle diameters, D_{10}, D_{30} and D_{60}, are defined as

D_{10} = particle diameter (in mm) corresponding to 10% finer by weight

D_{30} = particle diameter (in mm) corresponding to 30% finer by weight

D_{60} = particle diameter (in mm) corresponding to 60% finer by weight.

In Figure E2.1, D_{10} = 0.016 mm, which means that 10% of the particles in soil are smaller than 0.016 mm in diameter. The diameter, D_{10}, is called the effective size of the soil.

The diameters D_{10}, D_{30} and D_{60} are used to define two shape parameters: the coefficient of uniformity (C_U) and the coefficient of curvature (C_C)

$$C_U = \frac{D_{60}}{D_{10}} \tag{2.1}$$

$$C_C = \frac{(D_{30})^2}{D_{10} \times D_{60}} \tag{2.2}$$

C_U provides an indication of the spread or range of the particle sizes present in a soil and typically ranges from 1 to 1000. $C_U \approx 1$ refers to the soil consisting of particles of nearly the same size, which is called uniformly graded soil (curve A in Figure 2.3). If a soil has an excess of certain particle sizes and a deficiency of other sizes, such as $C_U < 6$, it is referred to as a poorly graded soil (curve B in Figure 2.3). A soil is called a well-graded soil if the distribution of the particle sizes extends over a large range, such as $C_U \geq 6$ (curve C in Figure 2.3). C_C provides an indication of the shape of particle size distribution curve between diameters D_{10} and D_{60}. For a well-graded soil, C_C lies between 1 and 3. A soil may have a deficiency in a specific intermediate range of particle sizes, indicated by a relatively flat section of the particle size distribution curve; such a soil is called gap-graded soil (curve D in Figure 2.3).

Figure 2.3 Types of particle size distribution curves

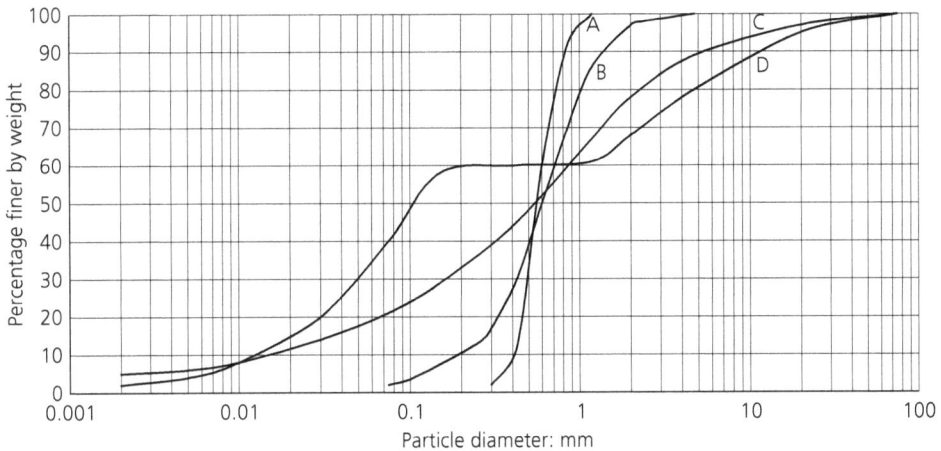

Note that the most important characteristic of a semi-logarithmic plot is that the particle size distribution curves of soils having equal uniformity are identical in shape, regardless of the average particle size. Additionally, the horizontal distance between two curves of the same shape is equal to the logarithm of the ratio of the average particle sizes of the respective curves. The shape of the particle size curve of a soil often reflects its geological origin. This information is used to provide a more correct interpretation of the soil data obtained by laboratory and field tests, resulting in safer and more economical geotechnical designs.

Example 2.2

For the particle size distribution curve shown in Figure E2.1, determine

(a) the coefficient of uniformity
(b) the coefficient of curvature.

Based on the above values, classify the soil as uniformly graded, poorly graded or well-graded.

Solution

In Figure E2.1, particle diameters D_{10}, D_{30} and D_{60}, correspond to 10%, 30% and 60% finer by weight, respectively. These are

$D_{10} = 0.016\,\text{mm}$, $D_{30} = 0.17\,\text{mm}$ and $D_{60} = 0.72\,\text{mm}$

(a) From Equation 2.1, the coefficient of uniformity

$$C_U = \frac{D_{60}}{D_{10}} = \frac{0.72}{0.016} = 45$$

(b) From Equation 2.2, the coefficient for curvature

$$C_C = \frac{(D_{30})^2}{D_{10} \times D_{60}} = \frac{(0.17)^2}{(0.016)(0.72)} = 2.509$$

Since $C_U > 6$ and C_C lies between 1 and 3, the soil is described as well-graded.

Note that the particle size distribution curve is only approximate due to several practical and theoretical factors. However, despite these limitations, the curves, particularly for sands and silts, hold significant practical value (Lambe and Whitman, 1979). Both theory and laboratory experiments show that soil permeability and capillarity of sands and silts are related to some effective particle diameter (see Chapter 4). Particle size distribution curves are also used in the design of granular filters and drains, as well as for assessing the susceptibility of sand and silt to frost damage.

The particle size curves of clays have limited value in geotechnical applications because their engineering behaviour usually depends much more on mineralogical composition, geological history and structure rather than on their particle size. The presence of water significantly affects the engineering behaviour of clays, such as their ability to be moulded, called plasticity. In this sense, clays are plastic materials, whereas silts, sands and gravels are non-plastic materials. Water also helps clay particles stick together, a property known as cohesiveness. In this context, clays are considered cohesive soils, whereas silts, sands and gravels are classified as cohesionless soils because their particles do not stick together in the presence of water.

Cohesionless soils are often referred to as granular soils. They are frictional because, according to Coulomb's law of friction, the resistance to sliding at the points of contact between soil particles is proportional to the normal force at those contacts. The interlocking of soil particles also contributes to the resistance to sliding. Thus, granular soil is a frictional material, but with a deviation from purely frictional behaviour (Lambe and Whitman, 1979).

2.3. Soil particle shape

Soil particles are rarely equidimensional or bulky as a cube or sphere. However, in general, most silt-sized and coarser particles are considered approximately equidimensional. Most clay-sized particles are plate-like or platy in shape. Rods and laths are also found in soils, generally in their clay-sized fractions.

Figure 2.4 Degree of particle roundness: (a) angular; (b) subangular; (c) subrounded; (d) rounded; and (e) well-rounded (after Lambe and Whitman, 1979; Pettijohn, 1949)

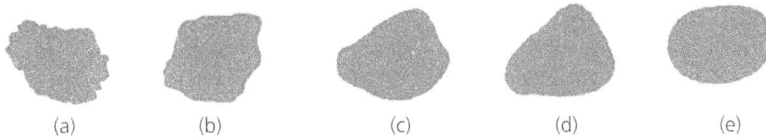

| (a) | (b) | (c) | (d) | (e) |

Soils with equidimensional particles and those with non-equidimensional particles behave very differently under structural loads. Compression is relatively much lower in the case of soils with equidimensional particles.

The sharpness of edges and corners of a coarse-grained particle is presented in terms of the degree of roundness, which is described in five levels as shown in Figure 2.4. Particles having sharp edges and/or relatively plane sides with unpolished surfaces are described as angular. Well-rounded particles have smooth, curved sides with no edges.

The shape of coarse-grained particles significantly controls the shear strength of granular soils through friction and/or interlocking.

2.4.   Consistency of soil

The degree of resistance a soil offers to deformation, or the relative ease with which it can be deformed, is known as its consistency.

2.4.1   Consistency limits and indices

The consistency of the fine-grained fraction of soil or cohesive soil is generally described using consistency limits. These limits are based on the concept that the fine-grained fraction can exist in any of the following four consistency states: liquid state, plastic state, semi-solid state and solid state, depending on the water content, as shown in Figure 2.5.

The presence of water that can move freely within the voids of a soil reduces the interaction between the adjacent particles and causes the soil to behave like a liquid. Liquid limit (w_L), plastic limit (w_P) and shrinkage limit (w_S), which are the water contents distinguishing the boundaries between the adjacent consistency states, are the consistency limits, often called the Atterberg limits (Lambe and Whitman, 1979).

The liquid limit of a soil is its water content at the boundary between the liquid and plastic states, where it begins to flow like a liquid as the water content increases.

The plastic limit of a soil is its water content at the boundary between the plastic and semi-solid states, where it ceases to be plastic as the water content decreases and becomes brittle.

The shrinkage limit of a soil is its water content at the boundary between the semi-solid and solid states, where any further reduction in water content no longer results in a decrease in volume.

Soil in its solid state behaves as a brittle material, similar to rock. However, when its water content is between the liquid limit and plastic limit, it can be moulded into a desired shape, exhibiting plasticity with putty-like properties.

The range of water content over which a soil behaves plastically is called the plasticity index, I_P (or PI), which is numerically the difference between the liquid limit and the plastic limit. Therefore

$$I_P = w_L - w_P \tag{2.3}$$

For plastic soils, I_P is significantly large. A plastic soil, such as a clay, generally retains its shape on drying and exhibits considerable strength when air dry. A silt, which exhibits little or no strength when air dry, is non-plastic or very slightly plastic with I_P less than 4 or plotting below the A-line on the plasticity chart (see Section 2.5). Cohesionless or granular soils, such as sands and gravels, are non-plastic.

The nearness of a natural soil to its liquid limit is expressed by the liquidity index or relative water content, I_L (or LI), which is defined as

$$I_L = \frac{w - w_P}{I_P} \tag{2.4}$$

where w is the natural water content of soil. A soil behaves plastically so that it can be moulded into any desired shape if its I_L lies between 0 and 1 (Figure 2.5). For $I_L < 0$, the soil is in its semi-solid or solid state. The soil behaves like a viscous liquid or liquid if its $I_L > 1$.

The nearness of a natural soil to its plastic limit is expressed by the consistency index or relative consistency, I_C (or CI), which is defined as

$$I_C = \frac{w_L - w}{I_P} \tag{2.5}$$

A soil behaves plastically so that it can be moulded into any desired shape if its I_C lies between 0 and 1 (Figure 2.5). For $I_C < 0$, the soil behaves like a viscous liquid or liquid. The soil is in its semi-solid or solid state if its $I_C > 1$.

The degree of plasticity of the clay-sized fraction of a soil can be expressed by the ratio of the plasticity index to the percentage by weight of the clay-sized fraction, called the activity of the clay (Skempton, 1953). If A represents the activity of a clay, then

Figure 2.5 Consistency limits or Atterberg limits

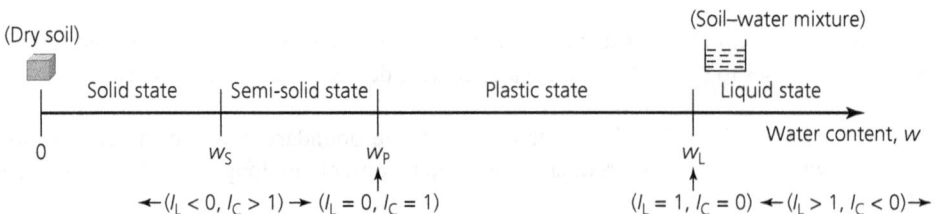

$$A = \frac{I_\mathrm{P}}{\% \text{ by weight finer than } 2\,\mu\mathrm{m}} \qquad (2.6)$$

For the principal clay minerals, the typical values of activity are (Mitchell and Soga, 2005; Skempton, 1953)

$A = 0.38$ for kaolinite

$= 0.9$ for illite

$= 7.2$ for Na-montmorillonite

$= 1.5$ for Ca-montmorillonite

For London clay, the reported activity is 0.95.

The activity is closely related to the specific surface, which is defined as the total surface area of all particles per unit mass. The specific surface values for kaolinite, illite and montmorillonite are $10\text{--}20\,\mathrm{m}^2/\mathrm{g}$, $65\text{--}200\,\mathrm{m}^2/\mathrm{g}$ and up to $800\,\mathrm{m}^2/\mathrm{g}$, respectively. For comparison, note that clean sand has a specific surface of approximately $0.0002\,\mathrm{m}^2/\mathrm{g}$.

2.4.2   Determination of consistency limits

The liquid limit of a soil is determined in the laboratory using the Casagrande liquid limit device, as shown in Figure 2.6. This device consists of a brass cup suspended from a carriage designed to control its free drop by 10 mm on a hard rubber base. A portion of the remoulded soil specimen is spread in the brass cup and is divided into two parts by a grooving tool. The cup is lifted and dropped by turning the crank at a rate of about 2 drops per second until the two halves of the soil pat come in contact at the bottom of the groove along a specified length (13 mm as per ASTM D4318-17e1 (ASTM, 2018a) and BS 1377-2:2022 (BSI, 2022)), and the number of blows/drops, N, required to close the groove is recorded. A total of three to five trials are performed with

Figure 2.6 The Casagrande liquid limit device

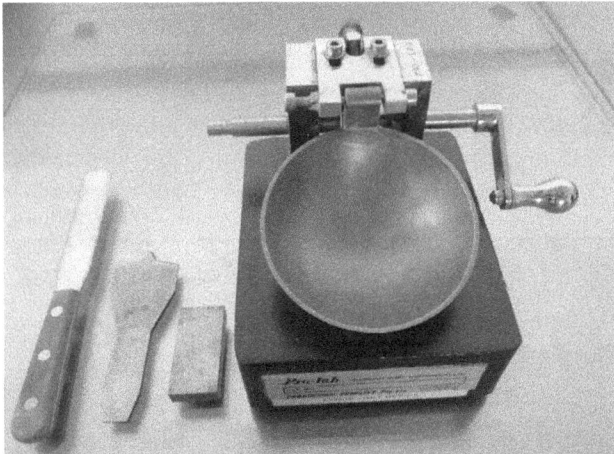

increasing water content of soil with the number of blows lying between 15 and 40. The water content is plotted as ordinates on the arithmetic scale, with the corresponding number of blows as abscissas on a logarithmic scale. The best-fit line through the plotted points is called the flow curve, which is a straight line, and its slope is called flow index, I_F, defined as

$$I_F = \frac{w_1 - w_2}{\log_{10}\left(\dfrac{N_2}{N_1}\right)} \qquad (2.7)$$

where w_1 and w_2 are the water contents in percent corresponding to N_1 and N_2 blows, respectively.

The flow index indicates the relationship between the change in water content and the corresponding change in shear strength of soil. The water content, rounded to the nearest whole number, corresponding to the intersection of the curve with the 25-blow abscissa is reported as the liquid limit of the soil. Details of the test procedure are provided in ASTM D4318-17e1 (ASTM, 2018a) and BS 1377-2:2022 (BSI, 2022).

If the groove in the Casagrande liquid limit device gets closed for the number of blows, N, between 20 and 30, the liquid limit of soil can be calculated using the following expression (ASTM, 2018a)

$$w_L = w_N \left(\frac{N}{25}\right)^{0.121} \qquad (2.8)$$

where w_N is the water content corresponding to the number of blows, N.

This test, which involves the measurement of one water content only, is called the one-point liquid limit test.

The ratio of the plasticity index (I_P) to the flow index (I_F) is called the toughness index (I_T). Thus,

$$I_T = \frac{I_P}{I_F} \qquad (2.9)$$

The toughness index indicates the shear strength of soil at its plastic limit. For most soils, it typically ranges from 0 to 3. If $I_F < 1$, the soil at its plastic limit is friable (Means and Parcher, 1963).

Example 2.3

The test for the liquid limit of soil is conducted using the Casagrande device and the following results are obtained.

Table E2.3

Trial number	Number of blows, N	Water content, w: %
1	15	49.8
2	21	47.6
3	27	45.7
4	34	44.4
5	39	43.2

Determine the following

(a) liquid limit
(b) flow index.

Solution

(a) Plot w against $\log_{10} N$, as shown in Figure E2.3. From the flow curve, the liquid limit is obtained as the water content corresponding to $N = 25$, and is $w_L = 46.3\%$.
(b) From Equation 2.7 and Figure E2.3, the flow index

$$I_F = \frac{w_1 - w_2}{\log_{10}\left(\dfrac{N_2}{N_1}\right)} = \frac{49.8 - 43.1}{\log_{10}\left(\dfrac{40}{15}\right)} = 15.73$$

Figure E2.3 Water content against log of the number of blows

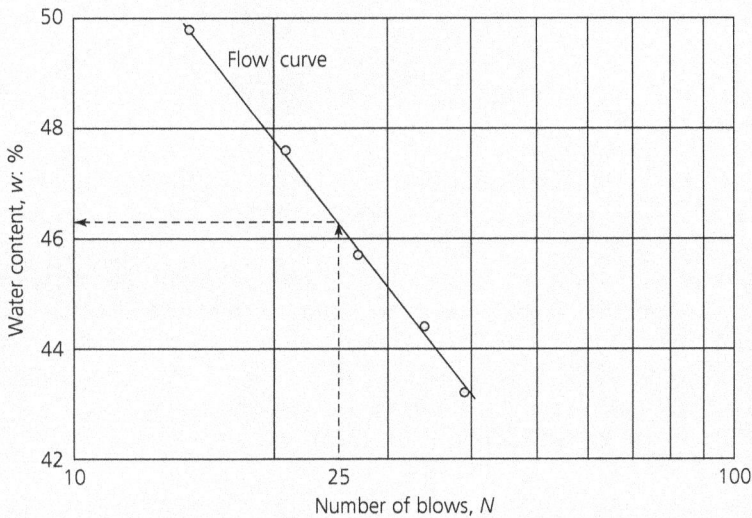

An alternative test to determine the liquid limit of a soil is the cone penetration method (BS 1377-2:2022 (BSI, 2022). The apparatus consists of a metal cone of a certain weight (80 g) and apex angle (30°) suspended from a support (Figure 2.7). The cone tip is lowered down to just barely touch the surface of the remoulded soil specimen placed in a cylindrical cup, and is then allowed to drop freely into the soil. The depths of penetrations are recorded for different water contents by repeating the test. The plot of the relationship between the water content as ordinates and the cone penetration as abscissas on arithmetic scales is used to determine the liquid limit as the water content, rounded to the nearest whole number, corresponding to a specified depth of 20 mm. It is observed that the liquid limit obtained from the cone penetration method agrees closely with the liquid limit determined using the Casagrande device for values up to about 100. However, for

Figure 2.7 Cone penetration liquid limit device

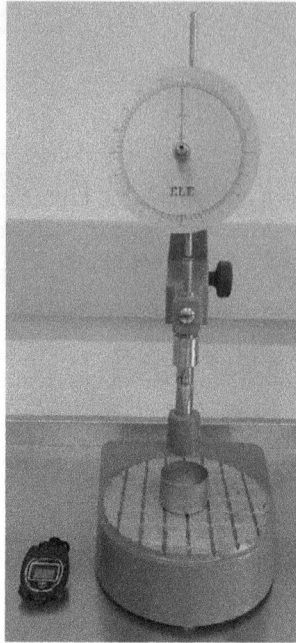

Figure 2.7 Cone penetration liquid limit device

values above 100, the cone penetration method tends to yield lower liquid limit values compared to the Casagrande method (Holtz *et al.*, 2023; Karlsson, 1977).

The plastic limit of a soil is determined by rolling a small portion of soil into a 3 mm-diameter thread until its water content is reduced as a result of evaporation to a point at which the thread splits and crumbles. The water content of this thread is measured, and another two test trials are conducted. The average of the three water contents, rounded to the nearest whole number, is reported as the plastic limit of soil. Details of the test procedure are provided in ASTM D4318-17e1 (ASTM, 2018a) and BS 1377-2:2022 (BSI, 2022).

The descriptive terms for the consistency of cohesive soils are provided in Table 2.2, along with the respective ranges of SPT values (also known as N-values) (ASTM D1586/D1586M-18e1 (ASTM, 2022)), unconfined compressive strength and undrained shear strength.

As stated earlier, the shrinkage limit of a soil is its water content at the boundary between its semi-solid and solid states, where a reduction in the water content does not cause any further decrease in its volume. Thus, the shrinkage limit represents the water content needed to just fill the voids of a cohesive soil at its minimum void ratio, achieved through oven drying. ASTM D4943-18 (ASTM, 2024) provides the test procedure for determining the shrinkage limit by the water submersion method. The shrinkage limit can be used to evaluate the shrinkage potential, crack development potential and swell potential of cohesive soils in earthwork applications. The shrinkage limit is also used to calculate shrinkage ratio, R (ratio of change in volume of soil expressed as a percentage of dry volume to corresponding change in water content above the shrinkage limit), volumetric

Table 2.2 Description of consistency of cohesive soils

Descriptive term	N-value[a] (blows/300 mm)	Unconfined compressive strength[b], q_u: kPa	Undrained shear strength[b], c_u (=q_u/2), kPa	Visual/field identification
Very soft	<2	<25	<12.5	Extrudes between fingers when squeezed; the fist penetrates several centimetres
Soft	2–4	25–50	12.5–25	Easily moulded by light finger pressure; the thumb penetrates several centimetres
Firm	4–8	50–100	25–50	Moulded by strong finger pressure; the thumb penetrates several centimetres with moderate effort
Stiff	8–15	100–200	50–100	Cannot be moulded by fingers; the thumb penetrates only with great effort
Very stiff	15–30	200–400	100–200	Can be readily indented by thumbnail
Hard	>30	>400	>200	Can be indented by thumbnail with difficulty

[a]For more details, see Shukla (2025).
[b]For more details, see Chapter 6.

shrinkage, V_S (decrease in volume of soil expressed as a percentage of its dry volume when water content, w, from a given value is reduced to its shrinkage limit, w_S), and linear shrinkage, L_S (decrease in one dimension of the soil element expressed as a percentage of the original dimension when water content, w, from a given value is reduced to its shrinkage limit, w_S). Following the definitions, the following expressions can be derived.

$$w_S = \left(\frac{1}{R} - \frac{1}{G} \right) \times 100 \tag{2.10}$$

where G is specific gravity of soil solids.

$$V_S = R(w - w_S) \tag{2.11}$$

and

$$L_S = 100 \left[1 - \left(\frac{100}{V_S + 100} \right)^{1/3} \right] \tag{2.12}$$

2.4.3 Relative density

The consistency of the coarse-grained fraction of soil, or cohesionless/granular soil, is generally described in terms of relative density (also known as index density or index unit weight). Relative density is defined using the minimum and maximum possible void ratios or dry unit weights/dry densities as

$$D_r = \left(\frac{e_{max} - e}{e_{max} - e_{min}} \right) \times 100\%$$

(2.13)

or

$$D_r = \left(\frac{\dfrac{1}{\gamma_{d\,min}} - \dfrac{1}{\gamma_d}}{\dfrac{1}{\gamma_{d\,min}} - \dfrac{1}{\gamma_{d\,max}}} \right) \times 100\,\% = \left(\frac{\gamma_d - \gamma_{d\,min}}{\gamma_{d\,max} - \gamma_{d\,min}} \right) \left(\frac{\gamma_{d\,max}}{\gamma_d} \right) \times 100\,\%$$

(2.14)

where e is the in situ (or in place) void ratio of soil; e_{min} is the minimum void ratio (i.e. the void ratio in the densest possible state of soil); e_{max} is the maximum void ratio (i.e. the void ratio in the loosest possible state of soil); γ_d is the in situ (or in place) dry unit weight of soil; $\gamma_{d\,min}$ is the minimum dry unit weight (i.e. the dry unit weight in the loosest possible state of soil); $\gamma_{d\,max}$ is the maximum dry unit weight (i.e. the dry unit weight in the densest possible state of soil).

The void ratio (or dry unit weight) of a vibrated and/or heavily loaded soil sample is taken as the minimum void ratio (or maximum dry unit weight), while the void ratio (or dry unit weight) of a very loosely poured oven-dry soil sample in a container is taken as the maximum void ratio (or minimum dry unit weight). The laboratory test details for the determination of minimum and maximum dry unit weights/densities are provided in the ASTM standards: ASTM D4253-16e1 (ASTM, 2019) and ASTM D4254-16 (ASTM, 2016). The tests are applicable to clean, free-draining cohesionless (granular) soils containing a maximum of 15% passing the 75 mm sieve.

It should be noted that laboratory tests do not provide the absolute values of minimum and maximum dry unit weights and void ratios. Therefore, the test results are referred to as minimum and maximum index dry unit weights and index void ratios, which are sufficient for geotechnical engineering purposes. In fact, unit weights greater than the so-called maximum unit weight can be achieved. Unit weights significantly lower than the so-called minimum unit weight can also be obtained, particularly with very fine sands and silts, by slowly sedimenting the soil in water or fluffing it with a small amount of moisture (Lambe and Whitman, 1979).

Note that minimum and maximum dry unit weights of a cohesionless soil can be considered as its consistency limits, equivalent to the liquid limit and plastic limit, respectively, of a cohesive soil (Atkinson, 2007). Descriptive terms for the consistency of cohesionless soils are provided in Table 2.3 with the respective ranges of relative density.

The engineering behaviour of cohesionless (granular) soil deposits is strongly affected by the relative density. The penetration resistance values obtained from penetration tests (ASTM, 2022) are correlated with relative density. Note that penetration tests are commonly conducted at construction sites at desired depths due to the difficulty of obtaining undisturbed soil samples, especially for

Table 2.3 Description of consistency of cohesionless soils

Descriptive term	Relative density, D_r : %
Very loose	0–15
Loose	15–35
Medium dense	35–65
Dense	65–85
Very dense	85–100

cohesionless soils. These tests help determine the engineering properties required for the analysis and design of geotechnical structures (Shukla, 2025).

The following points regarding soil consistency are noteworthy.

- Most standards recommend that tests for consistency limits should be performed only on the portion of soil passing the 425 μm sieve. About 150 to 200 g of the soil passing the 425 μm sieve is required for the test. Wherever possible the soil used for the consistency limits tests should not be dried prior to testing. Comparison of the liquid limit of a soil before and after oven drying can be used as a qualitative measure of its organic matter content. The liquid limit of a soil containing substantial amounts of organic matter decreases dramatically when the soil is oven dried before testing (ASTM, 2018a).
- The undrained shear strength of a soil at its liquid and plastic limits is considered to be approximately 1.7 kPa and 170 kPa, respectively.
- The flow index as determined from the liquid limit test based on the Casagrande liquid limit device generally lies between 2 and 20.
- The consistency limits of a soil are generally reported to the nearest whole number. The consistency limits and plasticity index are often reported without the percentage sign (%) – that is, a soil has a liquid limit of 30, rather than 30%.
- For most soils, $w_P < 40$ and $w_L < 100$.
- If the liquid limit or plastic limit of a soil cannot be determined, or if the plastic limit of a soil is equal to or greater than its liquid limit, the soil is described as non-plastic (NP). Clean sand is a non-plastic soil.
- The consistency limits and related indices, such as the plasticity index and liquidity index, are very useful for identifying and classifying the fine-grained fractions of soils. These limits are often incorporated directly into construction specifications as quality control measures for soils used as engineered fills. They are also used to correlate or estimate the engineering behaviour of soils in their natural moist state and are applied in some semi-empirical geotechnical designs.
- The plasticity index of a cohesive soil is related to its maximum volume change – that is, its compressibility.
- The difference between the maximum and minimum void ratios or dry unit weights of a cohesionless soil is related to its relative compressibility.

Example 2.4

The in situ dry unit weight of the sand deposit at a construction site was determined to be 16.92 kN/m^3. The minimum and maximum dry unit weights of sand, determined from laboratory tests, were 14.52 kN/m^3 and 18.34 kN/m^3, respectively. Determine the relative density of the sand deposit at the construction site.

Solution

Given $\gamma_d = 16.92 \, \text{kN}/\text{m}^3$, $\gamma_{dmin} = 14.52 \, \text{kN}/\text{m}^3$ and $\gamma_{dmax} = 18.34 \, \text{kN}/\text{m}^3$

From Equation 2.14, the relative density

$$D_r = \left(\frac{\gamma_d - \gamma_{dmin}}{\gamma_{dmax} - \gamma_{dmin}}\right)\left(\frac{\gamma_{dmax}}{\gamma_d}\right) = \left(\frac{16.92 - 14.52}{18.34 - 14.52}\right)\left(\frac{18.34}{16.92}\right) \times 100\% = 68.1\%$$

2.5. Soil classification and description

2.5.1 Soil classification

Soil classification is the process of placing a soil into a group of soils that exhibit similar engineering behaviour, and the set of such groups created for a specific engineering purpose or application is called a soil classification system. The main purposes of the soil classification are

- to obtain a consistent and nationally or internationally recognised basic description of the soil
- to facilitate the interchange of general information about similar soils from any geographic location, forming a basis for decisions on further tests required for the solution to a particular engineering problem
- to solve soil problems in a cost-effective manner. This has been possible with developments of empirical correlations through considerable experience of the engineering properties of soils with their index properties for different soil groups in a soil classification system.

Most countries have developed their own soil classification systems. Some of these systems are listed below and should be followed for geotechnical projects conducted in the respective countries.

- American Association of State Highway and Transport Officials (AASHTO) system ((AASHTO M 145-91 (AASHTO (1991b))
- Australian soil classification system (ASCS) (AS 1726:2017 (Standards Australia, 2017))
- British soil classification system (BSCS) (BS 5930:2015+A1:2020 (BSI, 2020))
- Indian soil classification system (ISCS) (IS 1498:1970 (reaffirmed 2021) (Bureau of Indian Standards, 2021))
- Unified soil classification system (USCS) (ASTM D2487-17 (ASTM, 2020)).

Presented here are the concepts and salient features of the USCS, which is the best-known and most widely used soil classification system worldwide. Particle size distribution, primarily based on sieve analysis, and the Atterberg limits, particularly the plastic and liquid limits, serve as useful index properties for classifying soils. The particle size distribution results from the sedimentation analysis, which can be relatively time consuming, are not necessarily required for soil classification

(Holtz *et al.*, 2023). A soil classification system loses its value if the index tests become more complicated than the test to measure directly the fundamental property needed in the project.

For descriptive purposes, a group of soils of similar engineering behaviour are designated by two letters of the English alphabet. The first letter indicates the main soil type and the second letter denotes a qualifying subdivision to give a more accurate description. The commonly used letters in USCS, and in most other classification systems, are given in Table 2.4.

In most soil classification systems (e.g. ASCS, ISCS, USCS), the given soil is classified as coarse-grained soil or fine-grained soil, depending on whether more than 50% of particles are larger or smaller than 0.075 mm in size. In the BSCS, the given soil is classified as coarse-grained soil or fine-grained soil, depending on whether more than 35% of particles are larger or smaller than 0.063 mm in size. If the soil consists of primarily organic matter and is dark in colour, having an organic odour, it is classified as highly organic soil.

As per USCS, the soil group symbols and their names, along with the criteria for identifying the soils of similar behaviour for the purpose of making soil groups, are provided in Table 2.5. The following points are worth mentioning.

▨ Only the particle size distribution based on sieve analysis, liquid limit (w_L) and plasticity index (I_p) are required to precisely classify a soil.
▨ The classification system identifies three major soil divisions as coarse-grained soils, fine-grained soils and highly organic soils; these are subdivided into a total of 15 basic soil groups.
▨ For the classification of fine-grained soils and the fine-grained fractions of coarse-grained soils, the plasticity chart shown in Figure 2.8 is used. This chart, originally proposed by

Table 2.4 Main and qualifying terms with symbols used in USCS (ASTM, 2020)

Soil type/fraction	Term used to describe the soil		Symbol
Coarse-grained soils (more than 50% retained on 75 μm sieve)	Main terms	Gravel	G
		Sand	S
	Qualifying terms	Well-graded	W
		Poorly graded	P
Fine-grained soils (50% or more passes 75 μm sieve)	Main terms	Silt	M
		Clay	C
		Organic	O
	Qualifying terms	Low plasticity ($w_L < 50\%$)	L
		High plasticity ($w_L > 50\%$)	H
Highly organic soils (primarily organic matter, dark in colour, having organic odour)	Main term	Peat	PT

Casagrande (1948), presents the variation of plasticity index with liquid limit of naturally occurring soils. In this chart, the A-line separates clayey soils from silty soils, and the U-line is an upper limit for natural soils. Any test results that plot above or to the left of the U-line should be verified. Note that the U-line is vertical at $w_L = 16\%$ because the liquid limit values for natural soils below 16 are generally not determined correctly. In the USCS, the '$w_L = 50$' line separates soils of low (L) plasticity from soils of high (H) plasticity.

Table 2.5 Unified soil classification system (USCS) (after Casagrande, 1948; ASTM D2487-17 (ASTM, 2020)) (continued on next page)

Criteria for assigning group symbols and group names using laboratory tests[a]				Soil classification	
				Group symbol	Group name[b]
Coarse-grained soils (more than 50% retained on 75 µm sieve)	**Gravels** (more than 50% of coarse fraction retained on 4.75 mm sieve)	Clean gravels ($<5\%$ fines[c,d])	$C_u \geq 4$ and $1 \leq C_c \leq 3$[f]	GW	Well-graded gravel[g]
			$C_u < 4$ and/or $1 > C_c > 3$[f]	GP	Poorly graded gravel[g]
		Gravels with fines ($>12\%$ fines[d])	Fines classify as ML or MH	GM	Silty gravel[g,h,i]
			Fines classify as CL or CH	GC	Clayey gravel[g,h,i]
	Sands (50% or more of coarse fraction passes 4.75 mm sieve)	Clean sands ($<5\%$ fines[e])	$C_u \geq 6$ and $1 \leq C_c \leq 3$[f]	SW	Well-graded sand[j]
			$C_u < 6$ and/or $1 > C_c > 3$[f]	SP	Poorly graded sand[j]
		Sands with fines ($>12\%$ fines[e])	Fines classify as ML or MH	SM	Silty sand[h,i,j]
			Fines classify as CL or CH	SC	Clayey sand[h,i,j]
Fine-grained soils (50% or more passes 75 µm sieve)	**Silts and clays** $w_L < 50\%$	Inorganic	$I_p > 7$ and plots on or above A-line[k]	CL	Lean clay[l,m,n]
			$I_p < 4$ or plots below A-line[k]	ML	Silt[l,m,n]
		Organic	$\dfrac{w_{L(\text{oven dry})}}{w_{L(\text{natural})}} < 0.75$	OL	Organic clay[l,m,n,o] Organic silt[l,m,n,p]
	Silts and clays $w_L \geq 50\%$	Inorganic	I_p plots on or above A-line	CH	Fat clay[l,m,n]
			I_p plots below A-line	MH	Elastic silt[l,m,n]
		Organic	$\dfrac{w_{L(\text{oven dry})}}{w_{L(\text{natural})}} < 0.75$	OH	Organic clay[l,m,n,q] Organic silt[l,m,n,r]

Table 2.5 Continued

Criteria for assigning group symbols and group names using laboratory tests[a]		Soil classification	
		Group symbol	Group name[b]
Highly organic soils	Primarily organic matter, dark in colour and having organic odour	PT	Peat

[a]Based on the material passing the 75 mm sieve.
[b]If the field sample contains cobbles and/or boulders, add 'with cobbles and boulders' to the group name.
[c]Fines refer to silts and clays.
[d]Gravels with 5−12% fines require dual symbols:
  GW-GM well-graded gravel with silt
  GW-GC well-graded gravel with clay
  GP-GM poorly graded gravel with silt
  GP-GC poorly graded gravel with clay.
[e]Sands with 5−12% fines require dual symbols:
  SW-SM well-graded sand with silt
  SW-SC well-graded sand with clay
  SP-SM poorly graded sand with silt
  SP-SC poorly graded sand with clay.
[f]For the definitions of C_U and C_C, see Equations 2.1 and 2.2, respectively.
[g]If the soil contains $\geq$ 15% sand, add 'with sand' to the group name.
[h]If the fines classify as CL-ML, use dual symbol GC-GM or SC-SM.
[i]If the fines are organic, add 'with organic fines' to the group name.
[j]If the soil contains $\geq$ 15% gravel, add 'with gravel' to the group name.
[k]If I_p plots in the hatched area on the plasticity chart (Figure 2.8), soil is a CL-ML, silty clay.
[l]If the soil contains 15−29% particles larger than 75 μm, add 'with sand' or 'with gravel', whichever is predominant, to the group name.
[m]If the soil contains $\geq$ 30% particles larger than 75 μm, predominantly sand, add 'sandy' to the group name.
[n]If the soil contains $\geq$ 30% particles larger than 75 μm, predominantly gravel, add 'gravelly' to the group name.
[o]$I_p \geq 4$ and plots on or above the A-line on the plasticity chart (Figure 2.8)
[p]$I_p < 4$ or plots below the A-line on the plasticity chart (Figure 2.8).
[q]I_p plots on or above the A-line on the plasticity chart (Figure 2.8).
[r]I_p plots below the A-line on the plasticity chart (Figure 2.8).

- In any soil group symbol, such as SM, the major soil fraction 'sand (S)' appears first. However, in its group name, it is mentioned at the end as 'silty sand'.
- The soil group symbols and their names are also used in a similar form in most other soil classification systems. You need to see the details of the classification system being adopted at your workplace.

2.5.2 Soil description

Soil classification that places any soil in a soil group with its symbol and name does not provide a complete description of the soil. Soil classification is based on the soil material characteristics, whereas soil description uses both the soil material characteristics and the in situ characteristics of soil mass. Thus, soil classification and soil description are not identical. The complete description of soil includes

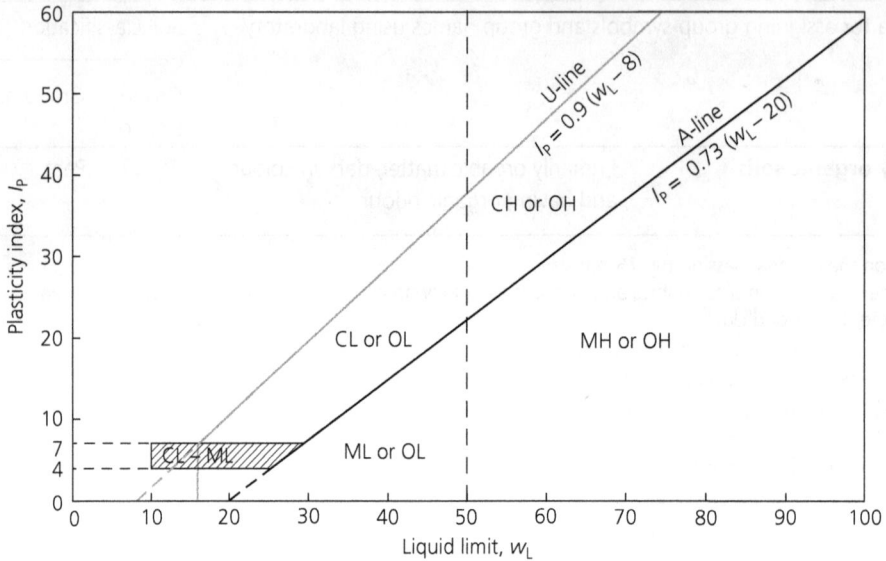

Figure 2.8 Soil plasticity chart (after Casagrande, 1948; ASTM D2487-17 (ASTM, 2020)

- soil material characteristics, such as colour, particle shape (equidimensional/platy/laths/ elongated), particle angularity/degree of roundness (angular/subangular/subrounded/rounded/ well-rounded), hardness, reaction with hydrochloric acid (HCl) (none/weak/strong), plasticity (non-plastic/low/medium/high), dry strength (none/low/medium/high/very high), dilatancy (none/slow/quick), toughness (low/medium/high), soil group symbol and name, and odour
- soil mass/in situ characteristics, such as moisture condition (dry/moist/wet), consistency (very soft/soft/firm/hard/very hard), cementation (weak/moderate/strong), structure (stratified/ laminated/fissured/slickensided/blocky/lensed/homogeneous) and geologic origin.

A soil is generally described and identified using its classification details and the information based on visual examination and some simple manual tests such as dry strength (resistance to crushing in dry condition), dilatancy (reaction to shaking in saturated/moist condition) and toughness (consistency near the plastic limit) tests for the fine-grained fraction of the soil.

For the dry strength test, after removing particles larger than 0.2 mm, mould a pat of soil into a ball of about 12 mm in diameter with the consistency of a putty by adding water, if necessary. Allow the ball to dry completely in air or sun, or by putting it in an oven. Test the strength by breaking and crumbling between fingers. The descriptive terms for dry strength are none, low, medium, high and very high. Strength is a measure of the character and quantity of the clay fraction of the soil. Highly plastic clays give very high dry strength, and inorganic silts and silty fine sands possess low dry strength.

For the dilatancy test, after removing particles larger than 0.2 mm, prepare a pat of the moist soil with a volume of about 10 cm³. Add enough water as necessary to make the soil soft but not sticky.

Place the soil pat in the open palm of one hand and shake horizontally, striking vigorously against the other hand several times. Squeeze the pat by closing the hand or pinching the soil between fingers. The appearance and disappearance of the water with shaking and squeezing is referred to as a reaction. This reaction is called quick or rapid if water appears and disappears rapidly; slow, if water appears and disappears slowly; and none, if the water condition does not appear to change. Very fine clean sands give the quickest reaction, inorganic silts show a moderate reaction and plastic clay has no reaction.

For the toughness test, after removing particles larger than 0.2 mm, mould a pat of soil about 10 cm^3 in size with the consistency of a putty by adding water, if necessary. The soil pat is elongated in shape and rolled by hand on a smooth surface or between the palms into a thread about 3 mm diameter. Fold the threads and reroll repeatedly until the thread crumbles at a diameter of about 3 mm, which happens when the soil is near its plastic limit. The resistance or stiffness felt in terms of pressure required to roll the soil at its plastic limit is called the toughness, which is described as low, medium or high. Relatively more resistance to moulding indicates higher clay content in the soil. Highly organic clays have a very weak and spongy feel at the plastic limit.

More details of the description and identification of soils can be found in ASTM D2488-17e1 (ASTM, 2018b), BS 5930:2015+A1:2020 (BSI, 2020), IS 1498:1970 (Bureau of Indian Standards, 2021) and AS 1726:2017 (Standards Australia, 2017). A typical example of the soil description is given below.

> Clayey gravel with sand and cobbles (GC): 50% fine to coarse, hard, subrounded reddish gravel; 25% clayey fines with high plasticity, high dry strength, non-dilatancy and medium toughness; 20% fine to coarse, hard, rounded sand; 5% hard, subrounded cobbles, maximum dimension 200 mm; weak reaction with hydrochloric acid (HCl); firm, moist, homogeneous deposit.

The following points regarding the classification and description of soils are worth mentioning.

- In cases where the liquid limit exceeds 100, or the plasticity index exceeds 60, the plasticity chart may be expanded by maintaining the same scale on both axes and extending the A-line at the indicated slope (ASTM, 2020).
- Soils grouped as well-graded gravel (GW) can have a dry unit weight of 20–22.6 kN/m^3, while others have usually lower values. The dry unit weight of soils can be as low as 10.4 kN/m^3 for soils grouped as OH (Bureau of Indian Standards, 2021).
- Soils grouped as GW, GP, GM, GC and SW are good foundation materials; hence heavier structures can be supported on such soils in stable condition. Soils grouped as ML are very poor foundation materials because of their susceptibility to liquefaction. Soils grouped as PT should be replaced by good foundation materials.
- Soils grouped as GW, GP, GM, GC and SW have good to excellent workability as construction materials; others have fair to poor workability.
- Soils grouped as GW, GP, SW and SP are highly permeable materials; hence water can flow through them easily.
- Soils grouped as MH, CH and OH have very low permeability and high compressibility in saturated condition; hence large and long-term settlements of structures founded on these soils are well expected. However, these materials can be used as liners and barriers for seepage control in canals, ponds, reservoirs, dams and landfills.

Example 2.5

A soil sample consists of sand with 8% fines. For the sand, $C_u = 5$ and $C_c = 0.95$. For the fines, plastic limit, $w_P = 15$ and liquid limit, $w_L = 25$. Classify the soil according to the USCS.

Solution

As the fines content (8%) is between 5 and 12%, and $C_u < 6$ and $1 > C_c > 3$, the sand will require dual symbols (see note e in Table 2.5) of SP-SM (poorly graded sand with silt) or SP-SC (poorly graded sand with clay).

From Equation 2.3, the plasticity index of soil

$$I_P = w_L - w_P = 25 - 15 = 10$$

In the plasticity chart (Figure 2.8), $I_P \, (= 10)$ against $w_L \, (= 25)$ plots above the A-line; therefore, the fines are classified as lean clay (CL). This indicates that the soil should be classified as poorly graded sand with clay (SP-SC).

Chapter summary

1. Clay, silt, sand, gravel, cobble and boulder are the names of particle sizes, and they are also used to describe soils, especially when they contain limited quantities of particles of other sizes. The curve drawn as a semi-logarithmic plot of the percentage finer by weight (or mass) against particle diameter is called the particle size distribution curve of the soil.

2. Most of the silt-sized and coarser particles are generally considered approximately equidimensional. The most clay-sized particles are platy/plate-like.

3. The engineering behaviour of clayey soils is greatly controlled by the water content. The particle size distribution of these soils has relatively negligible influence. Silty soils are usually non-cohesive and non-plastic, and their behaviour may be affected by the presence of water. Water does not significantly affect the engineering behaviour of sandy soils unless they are in loose condition, and dynamic loadings act.

4. The consistency of the cohesive soil is generally described in terms of consistency limits: liquid limit, plastic limit and shrinkage limits.

5. The consistency of the coarse-grained fraction of a soil or the cohesionless (granular) soil is generally described in terms of relative density, which is defined in terms of the minimum and maximum possible void ratios or dry unit weights/ dry densities

6. The index tests (particle size analysis by sieving, and liquid and plastic limit tests), and some simple manual tests (dry strength test, dilatancy test, toughness test) are used for classification and description of soils. Soils grouped as GW, GP, GM, GC and SW are good foundation materials.

Questions for practice

Select the most appropriate answer for each of the multiple choice questions from Q2.1 to Q2.10. Questions Q2.11 to Q2.30 require more detailed answers.

2.1 The particles of which of the following cannot be seen with the normal naked eye?
(a) sand
(b) silt
(c) clay
(d) both (b) and (c).

2.2 A vertical line as a particle size distribution curve represents a
(a) uniformly graded soil
(b) well-graded soil
(c) poorly graded soil
(d) gap-graded soil.

2.3 For a well-graded sand
(a) $C_U \geq 6$ and $1 \geq C_C \geq 3$
(b) $C_U \geq 6$ and $1 \leq C_C \leq 3$
(c) $C_U \leq 6$ and $1 \leq C_C \leq 3$
(d) $C_U \leq 6$ and $1 \geq C_C \geq 3$.

2.4 If Figure Q2.4 illustrates the particle size distribution curve of a soil, the sand content of the soil is
(a) 4%
(b) 50%
(c) 96%
(d) 100%.

Figure Q2.4

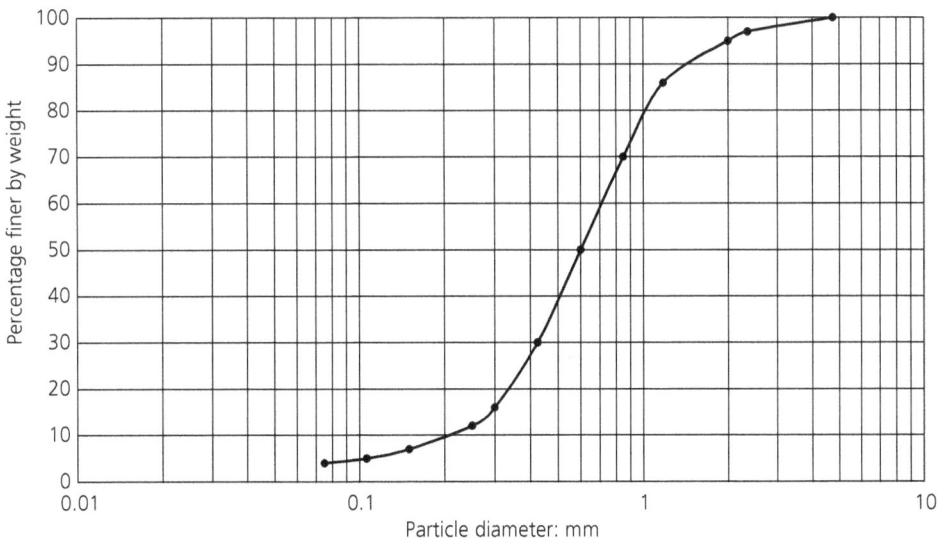

Percentage finer by weight vs Particle diameter: mm

2.5 The water content of soil, which represents the boundary between semi-solid state and plastic state is called
   (a) plastic limit
   (b) plasticity index
   (c) liquid limit
   (d) shrinkage limit.

2.6 If the liquidity index of a soil is unity, its undrained shear strength is approximately
   (a) 0 kPa
   (b) 1.7 kPa
   (c) 17 kPa
   (d) 170 kPa.

2.7 The group symbols assigned to silty sand and clayey sand are, respectively
   (a) SS and CS
   (b) MS and CS
   (c) SM and SC
   (d) SM and CS.

2.8 The soils which plot below the A-line in the plasticity chart are
   (a) silts
   (b) clays
   (c) organic soils
   (d) sands.

2.9 The relative density of compacted dense sand at a construction site can be approximately equal to
   (a) 20%
   (b) 50%
   (c) 90%
   (d) 110%.

2.10 If the void ratios in the densest, loosest and natural states of a sand deposit are 0.2, 0.95 and 0.57, respectively, its relative density is approximately
   (a) 31%
   (b) 51%
   (c) 61%
   (d) 81%.

2.11 What is the significance of liquid limit of a soil? The flow curves of two soils (soil I and soil II) are illustrated in Figure Q2.11. How does soil I differ from soil II? Explain briefly.

Figure Q2.11

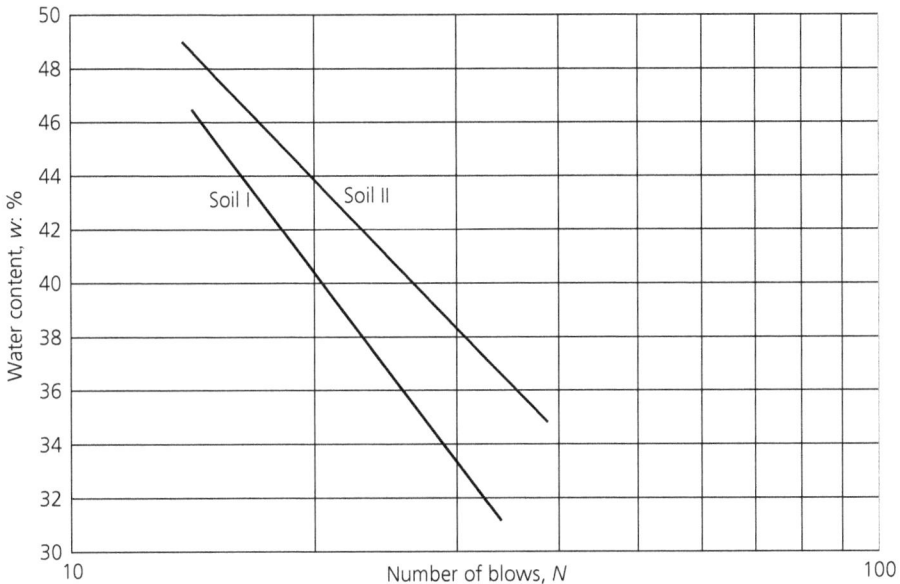

2.12 Assume that a soil consists of equal-sized spherical particles. These particles, like uniform spheres, can be packed in a container in several ways. Show the simple cubic packing, known as open packing, and the rhombic packing, known as closed packing, of uniform spheres. Calculate porosity and void ratio of the soil for these two packings, which result in the loosest and densest states, respectively.

2.13 For a sand, minimum dry unit weight, $\gamma_{dmin} = 15.12$ kN/m^3, and specific gravity of soil particles, $G = 2.68$. What maximum void ratio of sand can be obtained using the laboratory test? Can you determine the minimum void ratio using these given data?

2.14 The test for the liquid limit of a soil was conducted using the Casagrande liquid limit device, and the results are given in Table Q2.14. Determine the following
(a) liquid limit
(b) flow index.

Table Q2.14

Trial number	Number of blows, N	Water content, w: %
1	19	61.8
2	26	58.3
3	33	55.1
4	38	53.6

2.15 Two soil slopes are shown in Figure Q2.15. The liquid limits of soils of slope I and slope II are 35% and 47%, respectively. During the rainfall, which slope would fail first, slope I or slope II? Assume that all other conditions are the same.

Figure Q2.15

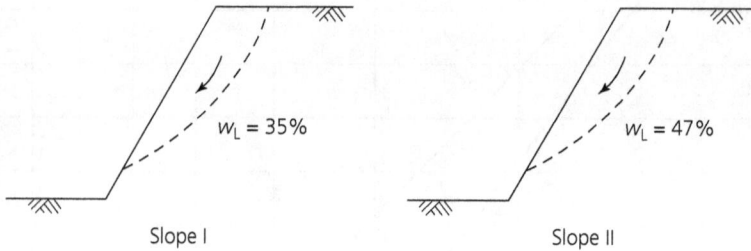

Slope I   Slope II

2.16 Table Q2.16 shows the results of a sieve analysis. The plasticity data obtained from the plastic and liquid limit tests conducted on the soil fraction passing 425 µm sieve are

liquid limit, $w_L = 28\%$ and plastic limit, $w_P = 19\%$

(a) Plot the particle size distribution curve using a spreadsheet.
(b) Classify the soil according to the Unified Soil Classification System.

Table Q2.16

Sieve size: mm	Weight/mass of soil retained: g
4.75	0
2	15.2
1.18	30.8
0.425	92.7
0.150	49.1
0.075	17.3
Pan	44.9

2.17 For a natural deposit of soil at a construction site, the following data are available

sand content = 32%

silt content = 41%

clay content = 27%

liquid limit $= 40\%$

plastic limit $= 21\%$

natural water content $= 26\%$

(a) Classify the soil according to the Unified Soil Classification System.
(b) Determine the liquidity and consistency indices.
(c) Determine the activity of the clay-sized fraction of soil.

2.18 If an individual soil particle is too heavy to lift with one hand and requires both hands, it may have a boulder size ($> 300\,$mm). Carry out similar experiments with particles of other sizes such as cobble, gravel, sand, silt and clay, and describe your observations.

2.19 Collect some gravel-size particles and describe their degree of roundness.

2.20 What are the index properties of soils? How are they determined and used?

2.21 Consider two different soils, A and B. Compared to soil B, soil A has a greater tendency to attract water to its particle surfaces. Which soil is likely to have higher liquid and plastic limits? Justify your answer.

2.22 How can you determine whether a given soil sample behaves as a brittle/friable solid, plastic material or liquid?

2.23 How can you distinguish between silt and clay?

2.24 What do you mean by consistency of soils? How is it determined?

2.25 What are the uses of consistency limits? Can you list their limitations, if any?

2.26 Explain the significance of liquidity index.

2.27 What is relative density? Is it applicable to all types of soil? Can it be greater than unity? Explain briefly.

2.28 What are the common classification tests? Are they expensive? Determine their cost in your local area.

2.29 How can you use field identification tests to determine soil types?

2.30 How is the plasticity chart used in the classification of soils?

REFERENCES

AASHTO (American Association of State Highway and Transport Officials) (1991a) AASHTO M 146-91: Standard Specification for terms relating to subgrade, soil-aggregate, and fill materials. AASHTO, Washington, DC, USA.

AASHTO (1991b) AASHTO M 145-91: Standard specification for classification of soils and soil-aggregate mixtures for highway construction purposes. AASHTO, Washington, DC, USA.

ASTM (2014) ASTM D422-63(2007): Standard test method for particle-size analysis of soils. ASTM International, West Conshohocken, PA, USA.

ASTM (2016) ASTM D4254-16: Standard test methods for minimum index density and unit weight of soils and calculation of relative density. ASTM International, West Conshohocken, PA, USA.

ASTM (2018a) ASTM D4318-17e1: Standard test methods for liquid limit, plastic limit and plasticity index of soils. ASTM International, West Conshohocken, PA, USA.

ASTM (2018b) ASTM D2488-17e1: Standard practice for description and identification of soils (visual-manual procedure). ASTM International, West Conshohocken, PA, USA.

ASTM (2019) ASTM D4253-16e1: Standard test method for maximum index density and unit weight of soils using a vibratory table. ASTM International, West Conshohocken, PA, USA.

ASTM (2020) ASTM D2487-17: Standard practice for classification of soils for engineering purposes (Unified Soil Classification System). ASTM International, West Conshohocken, PA, USA.

ASTM (2022) ASTM D1586/D1586M-18e1: Standard test method for standard penetration test (SPT) and split-barrel sampling of soils. ASTM International, West Conshohocken, PA, USA.

ASTM (2024) ASTM D4943-18: Standard test method for shrinkage factors of cohesive soils by the water submersion method. ASTM International, West Conshohocken, PA, USA.

Atkinson J (2007) *The Mechanics of Soils and Foundations*, 2nd edn. Taylor and Francis, London, UK.

BSI (2020) BS 5930:2015+A1:2020: Code of practice for ground investigations. BSI, London, UK.

BSI (2022) BS 1377-2:2022: Method of test for soils for civil engineering purposes – classification tests and determination of geotechnical properties. BSI, London, UK.

Bureau of Indian Standards (2021) IS 1498:1970: Classification and identification of soils for general engineering purposes. Bureau of Indian Standards, New Delhi, India.

Casagrande A (1948) Classification and identification of soils. *Transactions*, ASCE, **113**: 901–930.

CGS (Canadian Geotechnical Society) (2023) *Canadian Foundation Engineering Manual*, 5th edn. CGS, Surrey, Canada.

Holtz RD, Kovacs WD and Sheahan TC (2023). *An Introduction to Geotechnical Engineering*, 3rd edn. Pearson Education, Hoboken, NJ, USA.

Karlsson R (1977) *Consistency Limits*. Swedish Council for Building Research, Stockholm, Sweden, Document D6.

Lambe TW and Whitman RV (1979) *Soil Mechanics, SI Version.* Wiley, New York, , NY, USA.

Means RE and Parcher JN (1963) *Physical Properties of Soils.* Charles E Merrill, Columbus, OH, USA.

Mitchell JK and Soga K (2005) *Fundamentals of Soil Behaviour*, 3rd edn. Wiley, Hoboken, NJ, USA.

Pettijohn FJ (1949) *Sedimentary Rocks*, Harper, New York, NY, USA.

Shukla SK (2025) *ICE Core Concepts: Geotechnical Engineering*, 2nd edn. Emerald/ICE Publishing, London, UK.

Skempton AW (1953) The colloidal activity of clays. *Proceedings of the 3rd International Conference on Soil Mechanics and Foundation Engineering*, Switzerland, vol. I, pp. 57–61.

Standards Australia (2017) AS 1726:2017: Geotechnical site investigations. Standards Australia, Strathfield, NSW, Australia.

FURTHER READING

Das BM (2022) *Principles of Geotechnical Engineering*, 10th edn. Cengage, Boston, MA, USA.

emerald
PUBLISHING

ice
Publishing

Sanjay Kumar Shukla
ISBN 978-1-83608-519-5
https://doi.org/10.1108/978-1-83608-516-420252006
Emerald Publishing Limited: All rights reserved

Chapter 3
Stresses within soil

Learning aims

This chapter explores the core concepts of the following topics

- stress and geostatic state of stress
- the effective stress principle and its importance
- soil capillarity and effective stress
- ratio of horizontal to vertical stress
- stress induced by applied loads.

3.1. Introduction

Stresses within a soil are caused by its own weight and also by the external forces or loads applied to it. They give rise to displacements or strains, which must be within the permissible limits of the foundations and other geotechnical structures for their structural stability and performance. Thus, understanding stresses within soil is very important for civil engineers and will be discussed in this chapter.

3.2. Concept of stress

A stress at a point within a material is defined as the force or load acting on a unit area of the desired orientation centred at the point of interest. It is a tensor and requires three values to describe it fully, namely the magnitude of the force, the direction of the force and the area it acts on. In general, any force acting on a plane can be resolved into normal and shear/tangential force components, resulting in normal and shear stresses, respectively.

Figure 3.1 shows the following three components of a force, F, applied at an arbitrary angle to a horizontal plane: a vertical normal force N_z, and two shear forces, T_x and T_y, along the Cartesian coordinate directions z, x and y, respectively. When these forces are divided by the area of the plane, they result in normal stress, σ_z, and shear stresses, τ_{zx} and τ_{zy}, respectively.

In Figure 3.1, when F is normal to the plane, there are no shear stresses on the plane; in this case, the normal stress is called the principal stress, and the plane is called the principal plane. In general, for an infinitesimally small cubical element of a soil centred at the point of interest, there are three normal stresses, σ_x, σ_y and σ_z, and six shear stresses, τ_{xy}, τ_{yx}, τ_{yz}, τ_{zy}, τ_{zx} and τ_{xz}, acting against the sides of the cubical soil element. The moment equilibrium of the element provides $\tau_{xy} = \tau_{yx}$, $\tau_{yz} = \tau_{zy}$ and $\tau_{zx} = \tau_{xz}$. Thus, the state of stress at a point within a soil is completely described by six independent stresses as: σ_x, σ_y, σ_z, τ_{xy}, τ_{yz}, τ_{zx}, which are the components of the stress tensor at the point of interest.

Normal and shear force components of a force applied to a plane within a soil

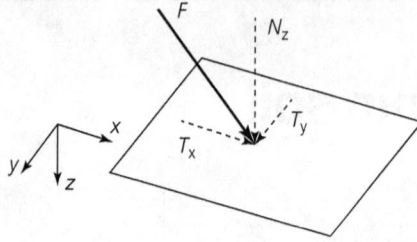

Stresses within a soil may be complicated. However, in sedimentary soil deposits, when the ground surface is horizontal and when the nature of the soil varies but little in the horizontal directions, the shear stresses caused by the weight of the soil on the horizontal and vertical planes within the soil are zero. In such a situation, the stresses are called the geostatic stresses (Lambe and Whitman, 1979). Thus, the horizontal and vertical planes within such soils are the principal planes, and the normal stresses are the principal stresses. Figure 3.2(a) shows a soil element at a depth, z, below the ground surface within a soil with geostatic state of stresses. Figure 3.2(b) shows the magnified view of the soil element with vertical normal stress, σ_v, acting on the horizontal plane, and horizontal normal stress, σ_h acting on the vertical plane; both are the principal stresses.

If $\sigma_v > \sigma_h$, σ_v is called the major principal stress, often denoted by σ_1, while σ_h is called the minor principal stress, often denoted by σ_3. Note that in three-dimensional (3D) space – that is, in the Cartesian coordinate system – the intermediate principal stress, σ_2, may be equal to σ_3. For most geotechnical problems, the state of stress in the plane that contains the major and minor principal stresses, σ_1 and σ_3, is considered, even though σ_2 may be between σ_1 and σ_3 in some specific problems.

The total vertical geostatic normal stress, or simply the total vertical stress, at a depth, z, can be computed by considering the weight, W, of the soil cylinder with an area of cross-section, A, above that depth. Thus,

$$\sigma_v = \frac{\text{Weight of soil cylinder of height } z}{\text{Area of cross-section of soil cylinder}} = \frac{W}{A} = \frac{(zA)\gamma}{A} = \gamma z \qquad (3.1)$$

(a) Soil profile under geostatic state of stresses and (b) magnified view of the soil element subjected to normal stresses

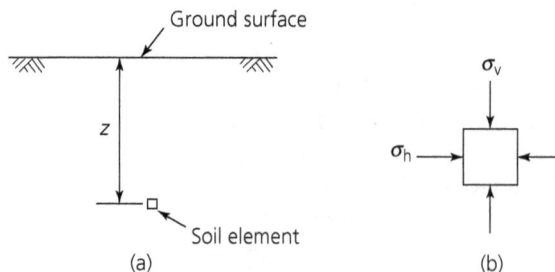

where γ is the total unit weight of soil.

For dry soil, $\gamma = \gamma_d$, and for saturated soil, $\gamma = \gamma_{sat}$. In the field, this total vertical stress can also be measured with an instrument known as an earth pressure cell.

Note that Equation 3.1 can be used to determine the total stress if γ is constant with depth.

If the soil deposit consists of several strata having different total unit weights, then the total vertical stress can be computed by means of the summation as

$$\sigma_v = \sum \gamma \Delta z \tag{3.2}$$

If γ varies continuously with depth, then the total vertical stress can be computed by means of the integral as

$$\sigma_v = \int_0^z \gamma dz \tag{3.3}$$

To determine the horizontal stress at any depth, an understanding of the coefficient of lateral stress or lateral stress ratio, K, is necessary; this is explained in Section 3.6.

Example 3.1

Figure E3.1 shows a layered soil profile. Determine the total vertical stress at point A.

Figure E3.1

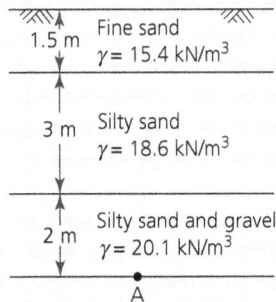

Solution

From Equation 3.2, the total vertical stress at point A

$$\sigma_v = \sum \gamma \Delta z = (15.4)(1.5) + (18.6)(3) + (20.1)(2) = 23.1 + 55.8 + 40.2 = 119.1 \text{kN/m}^2$$

Since $1 \text{N/m}^2 = 1$ pascal (Pa), this can also be written as

$$\sigma_v = 119.1 \text{ kPa}$$

3.3.   The effective stress principle for saturated soil

When engineers talk about the stress within a soil, the area used for its calculation must be large enough to include a representative number of soil particles and pass through soil solids only at points of contact between the particles (Figure 3.3). This is considered because the stress conditions within the soil particles are not of primary concern for the determination of deformation/strain and strength of the soil. Thus, the area, A, centred at the point of interest within the soil, which may be a wavy plane, such as XX in Figure 3.3, is the sum of the area of the voids, A_v, and the contact area between the particles, A_c, that is

$$A = A_v + A_c$$

or

$$A_v = A - A_c \tag{3.4}$$

Note that $A_c << A$, and A_c may be less than 1% of A.

In the definition of the total vertical stress, σ_v, in Equation 3.1, the area of cross-section of the soil cylinder consists of both the area of the voids, A_v, and the contact area between the particles, A_c. Thus, σ_v is not the stress at the points of contact between the particles. In fact, σ_v is the macroscopic stress, which is different from the contact stress. The contact stresses between soil particles are very large (of the order of 700 MPa), while the macroscopic stress will typically range from 10 to 10 000 kPa for most actual problems (Lambe and Whitman, 1979). In soil mechanics, when engineers use the word 'stress', they mean the macroscopic stress – that is, force (or load) divided by total area A, given by Equation 3.4.

If soil is saturated in Figure 3.2(a), then for equilibrium in the vertical direction, the total load or weight of the soil cylinder, W, will be balanced by the forces at the particle contacts through which the wavy plane passes – that is, intergranular contact forces, $\sum N$, and the hydrostatic force, $A_v u = (A - A_c)u$, where u is the pressure of water present in the voids of soil, called the pore water pressure, or simply the pore pressure. Thus,

Figure 3.3 Concept of the effective stress principle for fully saturated soil

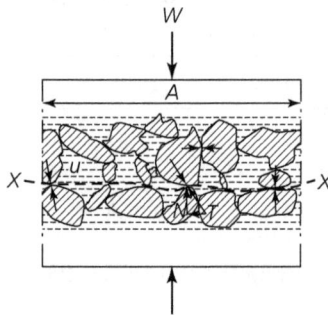

Note: u is the pore water pressure; N is the intergranular normal force between particles; T is the intergranular shear force between particles.

Figure 3.4 Subsoil profiles with: (a) water table below the ground surface and (b) water table above the ground surface

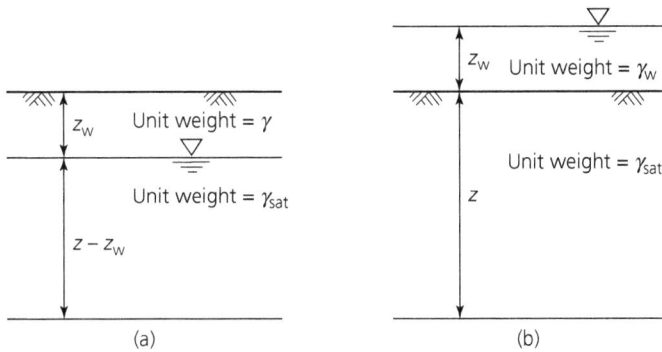

(a) z_w — Unit weight = γ; Unit weight = γ_{sat}; $z - z_w$

(b) z_w — Unit weight = γ_w; Unit weight = γ_{sat}; z

$$W = \sum N + (A - A_c)u \tag{3.5a}$$

Dividing both sides of Equation 3.5a by A gives

$$\frac{W}{A} = \frac{\sum N}{A} + \left(1 - \frac{A_c}{A}\right)u$$

or

$$\sigma_v = \sigma'_v + \left(1 - \frac{A_c}{A}\right)u \tag{3.5b}$$

where $\sigma'_v \left(= \dfrac{\sum N}{A}\right)$ is the effective vertical normal stress, or simply effective vertical stress.

If it is assumed that the contact area ratio, $A_c / A \to 0$, which is practically okay, Equation 3.5b reduces to

$$\sigma_v = \sigma'_v + u$$

or

$$\sigma'_v = \sigma_v - u \tag{3.6a}$$

In a similar way, the effective horizontal normal stress, or simply effective horizontal stress, may be expressed as

$$\sigma'_h = \sigma_h - u \tag{3.6b}$$

In general, within saturated soils, the effective stress, σ', may be defined as being equal to the total stress, σ, minus the pore water pressure, u. Thus,

$$\sigma' = \sigma - u \qquad (3.7)$$

Equation 3.7, which is the effective stress equation for saturated soils, was first proposed in the 1920s by Karl Terzaghi (Terzaghi, 1943).

The definition for effective stress and the fact that it correlates with soil behaviour combine to give the effective stress principle, which can be stated as follows (Lambe and Whitman, 1979; Terzaghi *et al.*, 1996).

- The effective stress is equal to the total stress minus the pore water pressure.
- The effective stress controls certain aspects of soil behaviour, notably compression, distortion and strength of soil.

In a saturated cohesionless or granular soil, the effective stress is approximately equal to the contact stress multiplied by the contact area ratio. In a highly plastic, saturated, dispersed clay, the effective stress is the net electrical stress transmitted between particles. Thus, the effective stress is closely related to the stress transmitted through the mineral skeleton. For this reason, effective stress is often called intergranular stress.

The pore water pressure at any point in a soil is determined by measuring the vertical distance to the water table (also known as the phreatic surface) above that point. This can be readily done by inserting a standpipe at the desired point in the soil and observing the level to which water rises in it. A standpipe is the simplest type of piezometer. There are some other types of piezometer for measuring the pore water pressure at the point of interest within a soil. Details of piezometers can be found in some advanced soil mechanics books (Lambe and Whitman, 1979).

If the water table coincides with the ground surface in Figure 3.2(a), then the pore water pressure at depth, z, is

$$u = \gamma_w z \qquad (3.8)$$

where γ_w is the unit weight of water.

Substituting σ_v and u from Equation 3.1 with $\gamma = \gamma_{sat}$ and Equation 3.8, respectively, into Equation 3.6a gives

$$\sigma'_v = \sigma_v - u = \gamma_{sat} z - \gamma_w z = (\gamma_{sat} - \gamma_w) z = \gamma' z \qquad (3.9)$$

where γ' is the submerged unit weight of soil.

Equation 3.9 can be used to determine the effective stress directly without using Equation 3.6 or Equation 3.7.

In terms of pore water pressure, the water table can be defined as the surface at which the pore water pressure, u, in the soil is equal to zero gauge pressure – that is, the atmospheric pressure, u_a.

Figure 3.4(a) shows a subsoil profile with the water table at a depth, z_w, below the ground surface. The expressions for the total vertical stress and the pore water pressure within the soil at a depth, z, below the ground surface are, respectively

$$\sigma_v = \gamma z_w + \gamma_{sat}(z - z_w) \tag{3.10a}$$

and

$$u = \gamma_w(z - z_w) \tag{3.10b}$$

Substituting σ_v and u from Equations 3.10a and 3.10b, respectively, into Equation 3.6a, gives

$$\sigma_v' = \sigma_v - u = \left[\gamma z_w + \gamma_{sat}(z - z_w)\right] - \gamma_w(z - z_w) = \gamma z_w + (\gamma_{sat} - \gamma_w)(z - z_w) = \gamma z_w + \gamma'(z - z_w)$$

or

$$\sigma_v' = (\gamma - \gamma')z_w + \gamma' z \tag{3.10c}$$

With a rise in the water table, z_w decreases, and, therefore, σ_v' from Equation 3.10c also decreases, while σ_v' increases with a fall in the water table – that is, with an increase in z_w.

Figure 3.4(b) shows a subsoil profile with the water table at a depth, z_w, above the ground level. The expressions for the total vertical stress and the pore water pressure within the soil at a depth, z, below the ground surface are, respectively

$$\sigma_v = \gamma_w z_w + \gamma_{sat} z \tag{3.11a}$$

and

$$u = \gamma_w(z + z_w) \tag{3.11b}$$

Substituting σ_v and u from Equations 3.11a and 3.11b, respectively, into Equation 3.6a gives

$$\sigma_v' = \sigma_v - u = \left(\gamma_w z_w + \gamma_{sat} z\right) - \gamma_w(z + z_w) = (\gamma_{sat} - \gamma_w)z = \gamma' z \tag{3.11c}$$

In Equation 3.11c, σ_v' is independent of z_w. This means that any change in the water table elevation above the ground surface does not affect the effective stress at any point within the soil.

Note that the placement of a fill over the large area may be seen as the placement of a surcharge load or elevating the ground surface – that is, shifting the ground surface up. Any such attempt increases both the total and the effective vertical stresses equally by an amount equal to the stress applied to the ground surface by the fill or surcharge load, provided the soil is drained immediately after placement of the fill or surcharge load, which generally happens in sand deposits. In the case of saturated clay deposits, an increase in effective vertical stress is a time-dependent process because of consolidation phenomenon, which is explained in detail in Chapter 5.

Example 3.2

Below the ground surface, a layer of sand, 4.5 m thick, rests over the clay deposit. The water table is 2.5 m below the ground surface. The saturated unit weights of the sand and clay are 19.3 kN/m³ and 18.5 kN/m³, respectively. The unit weight of sand above the water table is 16.2 kN/m³. Determine the total vertical stress, pore water pressure and effective vertical stress at elevations of 2.5 m, 4.5 m and 9 m below the ground surface, and plot their values against the elevation. If the water table rises up to an elevation of 1 m above the ground surface, what will be the total vertical stress, pore water pressure and effective vertical stress at an elevation of 9 m below the ground surface?

Solution

At 2.5 m depth

total vertical stress (Equation 3.1), $\sigma_v = \gamma z = (16.2)(2.5) = 40.5$ kPa

pore water pressure (Equation 3.8), $u = \gamma_w z = (9.81)(0) = 0$ kPa

effective vertical stress (Equation 3.6a), $\sigma'_v = \sigma_v - u = 40.5 - 0 = 40.5$ kPa

At 4.5 m depth

total vertical stress, $\sigma_v = \sum \gamma \Delta z = (16.2)(2.5) + (19.3)(2) = 40.5 + 38.6 = 79.1$ kPa

pore water pressure, $u = \gamma_w z = (9.81)(2) = 19.6$ kPa

effective vertical stress, $\sigma'_v = \sigma_v - u = 79.1 - 19.6 = 59.5$ kPa

At 9 m depth

total vertical stress,

$$\sigma_v = \sum \gamma \Delta z = (16.2)(2.5) + (19.3)(2) + (18.5)(4.5) = 40.5 + 38.6 + 83.2 = 162.3 \text{ kPa}$$

pore water pressure, $u = \gamma_w z = (9.81)(6.5) = 63.8$ kPa

effective vertical stress, $\sigma'_v = \sigma_v - u = 162.3 - 63.8 = 98.5$ kPa

The plots of σ_v, u and σ'_v are shown in Figure E3.2.

If the water table rises up to an elevation of 1 m above the ground surface, σ_v, u and σ'_v at an elevation of 9 m below the ground surface are determined as follows.

$$\sigma_v = \sum \gamma \Delta z = (9.81)(1) + (19.3)(4.5) + (18.5)(4.5) = 9.81 + 86.8 + 83.2 = 179.8 \text{ kPa}$$

$$u = \gamma_w z = (9.81)(10) = 98.1 \text{ kPa}$$

$$\sigma'_v = \sigma_v - u = 179.8 - 98.1 = 81.7 \text{ kPa}$$

Figure E3.2

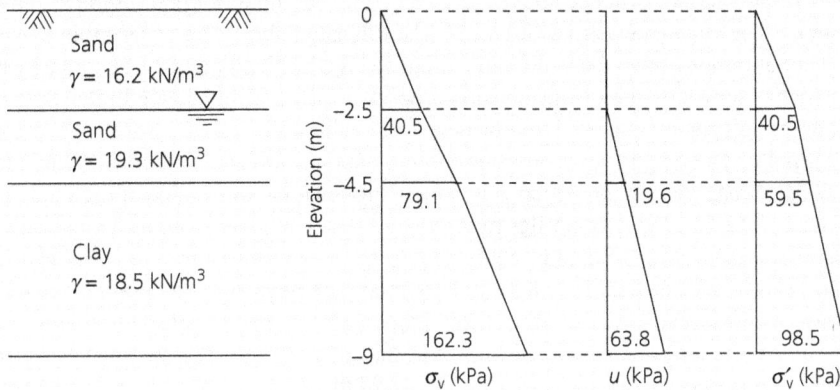

The following points regarding the effective stress principle are noteworthy.

- The effective stress, σ', does not represent the true contact stress $\left(= \left(\sum N \right) / A_c \right)$ between the particles.
- A change in pore water pressure produces practically no volume change and has practically no influence on the stress conditions for failure. For this reason, pore water pressure is known as neutral stress.
- Both the total stress and the pore water pressure can be readily calculated and measured in the field. The effective stress cannot be measured in the field, but it can be calculated using the effective stress principle.
- The pore water pressure acts in the water and in the solid in every direction equally. Within a dry soil, the total stress is equal to the effective stress – that is, $\sigma' = \sigma$.
- An increase in effective stress within a soil causes the soil particles to shift into a denser packing, resulting in an increase in density and strength.
- Equal increases in the total stress and the pore water pressure within a soil do not change the effective stress, thus causing no or little effect on the soil particle packing.

3.4.   The effective stress principle for partially saturated soil

If the air in a partially saturated soil exists within bubbles, Equations 3.6 and 3.7 can be used. If, however, there is an air channel, these equations should not be used. The effective stress equation for partially saturated soil is given below (Bishop, 1960; Bishop *et al.*, 1960; Mitchell and Soga, 2005).

$$\sigma' = \sigma - u_a + \chi(u_a - u) \tag{3.12}$$

where u_a is the pore air pressure and χ is a parameter related to the degree of saturation but also dependent on factors such as soil structure.

For a dry soil, $\chi = 0$, and for a saturated soil, $\chi = 1.0$. The usefulness of Equation 3.12 is limited in practice because of uncertainties about χ for intermediate degree of saturation.

3.5.   Soil capillarity and effective stress

The soil below the water table is fully saturated and the water pressure is positive – that is, more than the atmospheric pressure, u_a, which is generally taken as zero gauge pressure. The soil may be fully saturated for a certain distance above the water table, but the water pressure is negative – that is, less than u_a. The water present in the voids of soil located above the water table is called soil moisture. The saturation above the water table is caused by the phenomenon of capillarity or capillary rise, which is shown by the surface tension property of water in contact with the air/gas in soil. The voids in the soil can be assumed to represent space within tiny tubes whose walls are made of soil solid particles. The height, h_c, of a water column which a soil can support is called the height of capillary rise or capillary head, which is inversely proportional to the size of the void in the soil at the air–water interface. This is similar to the rise of water in a very small-diameter glass tube, called a capillary tube, with a diameter equivalent to the size of the pore spaces/voids in the soil. Figure 3.5(a) shows the capillary rise in a capillary tube with its lower end immersed in water. The upper surface of the water assumes almost the hemispherical shape, called the meniscus, which meets the walls of the tube at an angle, α, known as the contact angle. If T_s denotes the surface tension of water in force per unit length, for equilibrium of the water column in the tube, then

$$\left(\frac{\pi}{4} d^2 h_c\right) \gamma_w = (\pi d) T_s \cos \alpha$$

or

$$h_c = \frac{4 T_s \cos \alpha}{d \gamma_w} \tag{3.13a}$$

where d is the diameter of the tube and γ_w is the unit weight of water.

As T_s decreases slightly with increasing temperature, h_c also decreases with an increase in temperature. For pure water and a clean glass capillary tube, $\alpha \to 0$ and $\cos \alpha \to 1$; therefore, Equation 3.13a becomes

Figure 3.5 (a) Rise of water in a capillary tube and (b) capillary and hydrostatic pressures

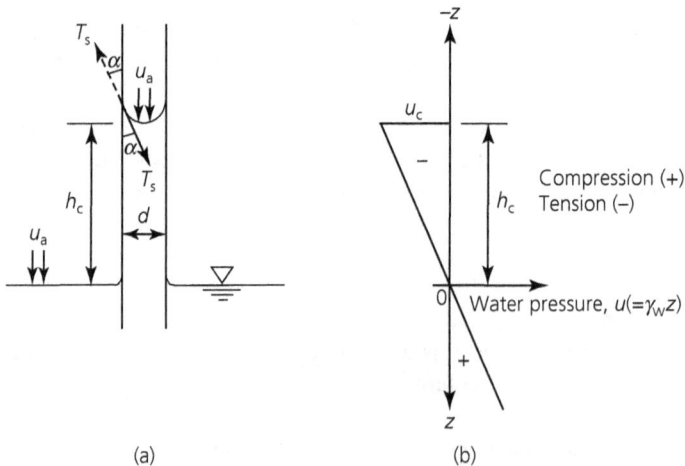

(a)   (b)

$$h_c = \frac{4T_s}{d\gamma_w} \qquad (3.13b)$$

Figure 3.5(b) shows the variation of hydrostatic pressure, u, in water. At any depth, z, below the free water surface, $u = z\gamma_w$; thus, u is positive. Above the free water surface, the hydrostatic water pressure in the tube, $u = -z\gamma_w$; thus, u is negative; this pressure is called the capillary pressure. Therefore, the water within the capillary rise is in tension or suction, while water below the free surface of the water body is in compression.

At the bottom of the meniscus, $z = h_c$, and the capillary pressure is

$$u_c = -\gamma_w h_c = -u_s \qquad 3.14)$$

where

$$u_s = \gamma_w h_c \qquad (3.15)$$

is the capillary suction, called the matric suction.

Substituting h_c from Equation 3.13a into Equation 3.14 results in

$$u_c = -\frac{4T_s \cos\alpha}{d} \qquad (3.16a)$$

Substituting h_c from Equation 3.13b into Equation 3.14 results in

$$u_c = -\frac{4T_s}{d} \qquad (3.16b)$$

In Figure 3.5(a), the atmospheric pressure, u_a, acts on the meniscus at the top of the water column in the tube as well as on the free water surface of the water body in which the capillary tube is inserted. Thus, u_a has no effect on h_c, u_c and u_s. Strictly speaking, the matric suction is defined as (Terzaghi et al., 1996)

$$u_s = u_a - u_c \qquad (3.17)$$

In very fine-grained soils such as clays, the difference in concentration of the cations in the electric double layer surrounding the particles and in the free water farther from the particles also generates suction, which is called osmotic suction. The total suction in clays is made up of matric suction and osmotic suction, although the two are not strictly additive (Nelson and Miller, 1992). In coarse- to fine-grained soils, the total suction is likely to be primarily matric suction. In most geotechnical problems, the effects of practical importance result primarily from changes in matric suction (Terzaghi et al., 1996).

In soils, for use of Equations 3.13 and 3.16, it is necessary to assume that the diameter, d, is some fraction of the effective particle size, D_{10}, say 1/5 of D_{10}. Terzaghi et al. (1996) suggest the following equation to calculate the value of h_c approximately in m

$$h_c = \frac{C}{eD_{10}} \qquad (3.18)$$

where e is the void ratio of soil; C is an empirical coefficient that ranges between 0.01 and 0.05 and is a function of the particle shape and surface impurities; and D_{10} is in mm.

When the capillary rise is considered to be above the water table within a soil, the effective vertical stress at the top level of the capillary zone, where the total vertical stress is σ_v, can be calculated from Equation 3.6a by substituting $u = u_c = -\gamma_w h_c$ as

$$\sigma_v' = \sigma_v - (-\gamma_w h_c) = \sigma_v + \gamma_w h_c \tag{3.19a}$$

If the soil gets saturated by capillarity up to the ground surface, which may happen near a sandy beach beside the sea or a lake, the effective stress within the sand close to the ground surface will be

$$\sigma_v' = 0 - (-\gamma_w h_c) = \gamma_w h_c \tag{3.19b}$$

At the water table level with the capillary rise, $\sigma_v' = \sigma_v$ because the pore water pressure is zero.

Thus, the effective vertical stress in soils with low void ratio can be very large because of the capillary pressure that basically applies a compressive force on the wall of the capillary tube made of soil particles by pulling them.

The following points regarding capillarity are noteworthy.

- In contrast to capillary tubes, the continuous voids in soils have a variable width. Therefore, it is possible that when the soil meets water from below, say from the water table, the lower part of the soil becomes completely saturated; however, in the upper part, the water may occupy only the narrowest voids, and the wider one may remain filled with air (Terzaghi *et al.*, 1996). Thus, the degree of saturation within the soil above the water table may not be equal to 100% throughout the capillary rise zone (Atkinson, 2007).
- As the soil particle size increases, the size of the voids – that is, the void ratio – also increases, and the height of the capillary rise decreases.
- In the capillary zone, the soil is assumed to be saturated but not submerged above the water table. This implies that the buoyant force does not act on the soil particles. In contrast, the soil below the water table is both saturated and submerged, with buoyant forces acting on the soil particles.
- The capillary water does not contribute to hydrostatic pressure below the water table. The only effect of capillary rise above the water table is to increase the total unit weight of the soil; therefore, both the total and the effective vertical stresses at any point at or below the water table are increased by the same amount.

Example 3.3

The ground at a construction site consists of a coarse sand deposit. The water table is 4 m below the ground surface. Determine the effective vertical stress at a depth of 10 m below the ground surface without considering capillary rise, and also when there is a capillary rise above the water table. For the sand, dry unit weight $= 15.5\,\text{kN/m}^3$, saturated unit weight $= 21.3\,\text{kN/m}^3$, effective particle size, $D_{10} = 0.11\text{mm}$, and void ratio, $e = 0.27$. Assume missing data, if any, suitably.

Solution

Without capillary rise

total vertical stress, $\sigma_v = \sum \gamma \Delta z = (15.5)(4) + (21.3)(6) = 62 + 127.8 = 189.8 \text{ kPa}$

pore water pressure, $u = \gamma_w z = (9.81)(6) = 58.9 \text{ kPa}$

effective vertical stress, $\sigma'_v = \sigma_v - u = 189.8 - 58.9 = 130.9 \text{ kPa}$

σ'_v can be calculated in one step as

$$\sigma'_v = \sum \gamma' \Delta z = (15.5)(4) + (21.3 - 9.81)(6) = 130.9 \text{ kPa}$$

Please note that (21.3–9.81) is the submerged unit weight of sand below the water table.

The height of capillary rise can be calculated using Equation 3.18 with C ranging between 0.01 and 0.05. Assume $C = 0.03$ for this problem.

$$h_c = \frac{C}{eD_{10}} = \frac{0.03}{(0.27)(0.11)} = 1.01 \text{ m}$$

With capillary rise

total vertical stress,

$$\sigma_v = \sum \gamma \Delta z = (15.5)(4 - 1.01) + (21.3)(1.01) + (21.3)(6) = 46.3 + 21.5 + 127.8 = 195.6 \text{ kPa}$$

pore water pressure, $u = \gamma_w z = (9.81)(6) = 58.9 \text{ kPa}$

effective vertical stress, $\sigma'_v = \sigma_v - u = 195.6 - 58.9 = 136.7 \text{ kPa}$

σ'_v can be calculated in one step as

$$\sigma'_v = \sum \gamma' \Delta z = (15.5)(4 - 1.01) + (21.3)(1.01) + (21.3 - 9.81)(6) =$$
$$46.3 + 21.5 + 68.9 = 136.7 \text{ kPa}$$

3.6. Coefficient of lateral stress

To determine the horizontal stress at any depth, an understanding of the coefficient of lateral stress, K, is necessary (also known as the lateral stress ratio or coefficient of lateral earth pressure), which is defined as the ratio of effective horizontal stress to effective vertical stress. Thus,

$$K = \frac{\sigma'_h}{\sigma'_v} = \frac{\sigma_h - u}{\sigma_v - u} \tag{3.20}$$

When there has been no lateral strain within the ground during the formation of a soil deposit, which usually happens during the formation of sedimentary soils over a large lateral area, the ratio of effective horizontal stress to effective vertical stress is called the coefficient of lateral stress at rest, K_0 (also known as the lateral stress ratio at rest, or coefficient of lateral earth pressure at rest). Thus, K_0 is a special case of K. The lateral stress ratio, K, for different ground movement cases will be discussed in some detail in Chapter 8.

The following points regarding coefficient of lateral stress are noteworthy.

- Note that Equation 3.20 is applicable regardless of whether the stresses are geostatic.
- K and K_0 are independent of the location of the groundwater table and are very important parameters for use in the design of geotechnical structures, such as earth-retaining structures, earth dams and slopes, and foundations. K_0 remains constant within a soil layer that has no variation in density; however, it significantly depends on the overburden and the geological and stress history.
- K_0 typically varies between 0.4 and 0.5 for sand deposits which have not been preloaded earlier. If a soil deposit has been heavily preloaded in the past, the effective horizontal stress can be greater than the effective vertical stress; therefore, K_0 may be greater than 1, and it may even reach a value of 3.
- Note that $K = 1$ within water – that is, at any point within water, the horizontal stress is exactly equal to vertical stress; this stress condition is called hydrostatic stress condition.

Example 3.4

The water table coincides with the ground surface. If the saturated unit weight of soil is $20.7\,\mathrm{kN/m^3}$, determine total and effective horizontal stresses at a depth of $5\,\mathrm{m}$ below the ground surface. Assume coefficient of lateral stress at rest, $K_0 = 0.5$. Also determine the ratio of total horizontal stress to total vertical stress and compare the value obtained with K_0.

Solution

total vertical stress, $\sigma_\mathrm{v} = \gamma z = (20.7)(5) = 103.5$ kPa

pore water pressure, $u = \gamma_\mathrm{w} z = (9.81)(5) = 49.0$ kPa

effective vertical stress, $\sigma'_\mathrm{v} = \sigma_\mathrm{v} - u = 103.5 - 49.0 = 54.5$ kPa

Using Equation 3.20 from the definition of K_0

$$K_0 = \frac{\sigma'_\mathrm{h}}{\sigma'_\mathrm{v}}$$

or

$$\sigma'_\mathrm{h} = K_0 \sigma'_\mathrm{v} = (0.5)(54.5) = 27.2 \text{ kPa}$$

From Equation 3.6b

$$\sigma_\mathrm{h} = \sigma'_\mathrm{h} + u = 27.2 + 49.0 = 76.2 \text{ kPa}$$

The ratio of total horizontal stress to total vertical stress is $\dfrac{\sigma_\mathrm{h}}{\sigma_\mathrm{v}} = \dfrac{76.2}{103.5} = 0.74$, which is not equal to K_0.

3.7.   Stresses within soil induced by applied loads

The total stresses within soil subjected to applied loads are determined by adding the increase in stresses induced by the applied loads to the initial geostatic stresses.

There are some field situations in which the increase in vertical stress caused by the applied load at the desired depth is equal to 100% of the applied stress at the ground surface. The two common examples of such loadings, called the one-dimensional (1D) loadings, are placement of a fill over a very large area, and loading over an area or a footing having width or diameter significantly greater than the thickness of the compressible soil layer.

There are several field situations in which the applied loads dissipate rapidly with depth in three dimensions, and the increase in vertical stress caused by the applied loads at the desired depth would be less than 100% of the applied stress at the ground surface. Examples of such loadings are point loads, line loads and loaded areas whose widths or diameters are equal to or less than the thickness of the compressible soil layer. Since these loadings induce stresses in three dimensions, they are called three-dimensional (3D) loadings. The stress distribution theories based on the theory of elasticity are commonly used to determine the increase in stresses induced by the 3D loadings within the soil. The soil is assumed to be elastic (stress is proportional to strain), homogeneous (properties are constant throughout) and isotropic (properties are identical in all directions at any point). Poulos and Davis (1974) have presented closed-form solutions to stress determination for a great variety of loadings, geometries and elastic properties. Here, the fundamental aspects of certain specific problems are presented using elastic theory.

3.7.1   Vertical stress induced by a point load

In 1885, Boussinesq developed expressions for calculating the normal (vertical, radial, tangential) and shear stresses within a homogeneous, isotropic, linearly elastic half-space induced by a point load acting perpendicular to the surface, as shown in Figure 3.6 (Boussinesq, 1885). A linearly elastic half-space means that the material below the surface extends to infinity in all directions and follows a linear stress–strain relationship. The expression for the vertical normal stress increase, or simply the vertical stress increase, $\Delta\sigma_v$, at a depth, z, and a horizontal distance, r, from the direction of vertical point load, Q, is

$$\Delta\sigma_v = \frac{3Q}{2\pi z^2 \left[1+\left(\dfrac{r}{z}\right)^2\right]^{5/2}} \tag{3.21}$$

Equation 3.21 is popularly known as Boussinesq's equation, and is commonly presented in terms of an influence factor, I, as

Figure 3.6 Vertical stress induced by a point load

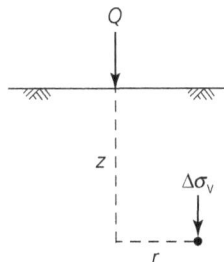

$$\Delta\sigma_v = \frac{Q}{z^2} I \tag{3.22}$$

where

$$I = \frac{3}{2\pi\left[1+\left(\dfrac{r}{z}\right)^2\right]^{5/2}} \tag{3.23}$$

Note that I is a function of the r/z ratio and is independent of the material properties. Some values of I are given in Table 3.1. Figure 3.7 shows the variation of $\Delta\sigma_v$ with r for two different values of z as $z = z_1$ and $z = z_2$, and also the variation of $\Delta\sigma_v$ with z for $r = 0$.

Table 3.1 Values of I for several values of r/z

r/z	I	r/z	I
0.00	0.4775	1.20	0.0513
0.10	0.4657	1.40	0.0317
0.20	0.4329	1.60	0.0200
0.30	0.3849	1.80	0.0129
0.40	0.3294	2.00	0.0085
0.50	0.2733	2.20	0.0058
0.60	0.2214	2.40	0.0040
0.70	0.1762	2.60	0.0029
0.80	0.1386	2.80	0.0021
0.90	0.1083	3.00	0.0015
1.00	0.0844	4.00	0.0004

Figure 3.7 Variation of vertical stress induced by a vertical point load

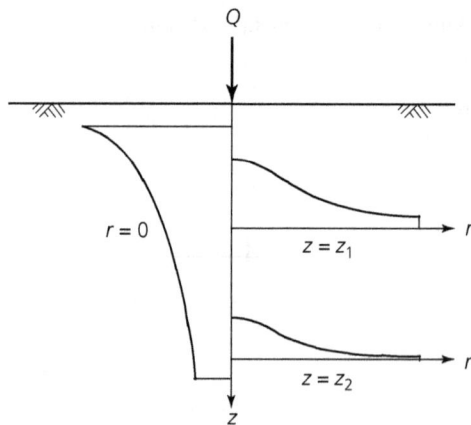

Boussinesq's equations for other stress components are also available, but they are not appropriate, as Equation 3.21 provides a more reasonable value for the vertical stress increase, although soils are not even approximately elastic, homogeneous and isotropic. This is mainly because other stress components largely depend on the stress–strain characteristics of the loaded soil (Terzaghi et al., 1996).

Example 3.5

A point load of 1000 kN is applied vertically to a horizontal ground surface. Determine the increase in vertical stress at

(a) a depth of 5 m below the point of load application
(b) a depth of 5 m below the point of load application and at a radial distance of 2.5 m from the line of action of the load.

Solution

(a) Refer to Figure 3.6. Here, $z = 5$ m, $r = 0$ and $Q = 1000$ kN. From Equation 3.23, the influence factor

$$I = \frac{3}{2\pi \left[1 + \left(\frac{r}{z}\right)^2\right]^{5/2}} = \frac{3}{2\pi} = 0.4775$$

From Equation 3.22, the increase in vertical stress

$$\Delta\sigma_v = \frac{Q}{z^2} I = \frac{1000}{(5)^2} \times 0.4775 = 19.10 \text{ kPa}$$

(b) Refer to Figure 3.6. Here, $z = 5$ m, $r = 2.5$ m and $Q = 1000$ kN. From Equation 3.22, the influence factor

$$I = \frac{3}{2\pi \left[1 + \left(\frac{r}{z}\right)^2\right]^{5/2}} = \frac{3}{2\pi \left[1 + \left(\frac{2.5}{5}\right)^2\right]^{5/2}} = 0.2733$$

From Equation 3.22, the increase in vertical stress

$$\Delta\sigma_v = \frac{Q}{z^2} I = \frac{1000}{(5)^2} \times 0.2733 = 10.93 \text{ kPa}$$

3.7.2 Vertical stress induced by vertical line and surface loads

The stresses induced within the soil by applied vertical line and surface loads can be obtained by discretising the loads into several point loads and then using Equation 3.21 with integration. Expressions for the increase in vertical stress, $\Delta\sigma_v$, due to uniform line, strip, circular and rectangular loads acting on the surface of a linearly elastic half-space are presented here without their derivations. Further details can be found in advanced soil mechanics books.

3.7.2.1   Uniform line load

The increase in vertical stress induced by the uniform line load (Figure 3.8) is

$$\Delta\sigma_v = \frac{2q}{\pi z\left[1+\left(\dfrac{x}{z}\right)^2\right]^2} \tag{3.24}$$

where q is the load per unit length; x is the horizontal distance of the point of interest from the direction of vertical line load; z is the depth of the point of interest below the surface.

3.7.2.2   Uniform strip load

The increase in vertical stress induced by the uniform strip load at a depth, z, below the ground surface (Figure 3.9) is

$$\Delta\sigma_v = \frac{q}{\pi}(\alpha+\sin\alpha\cos 2\beta) \tag{3.25}$$

where q is the uniform pressure on a strip area of width, B, and infinite length; α and β are angles defined in Figure 3.9.

Figure 3.8 Vertical stress induced by a line load

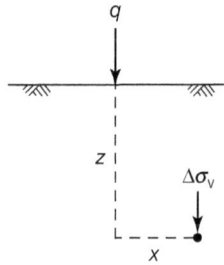

Figure 3.9 Vertical stress induced by a uniform strip load

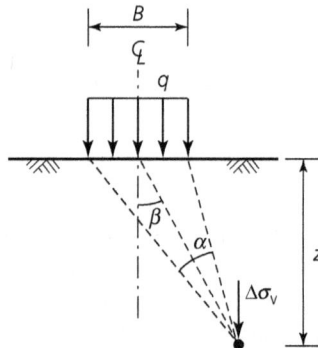

For any point directly below the centre of the strip load, $\beta = 0$; therefore, Equation 3.25 becomes

$$\Delta\sigma_v = \frac{q}{\pi}\left(\alpha + \sin\alpha\right) \tag{3.26}$$

Note that α is the angle in radians subtended by the strip width, B, at the point of interest.

3.7.2.3 Uniform circular load

The increase in vertical stress induced by the uniform circular load (Figure 3.10) is

$$\Delta\sigma_v = qI_c \tag{3.27}$$

where q is the uniform pressure on a circular area of radius, R; I_c is an influence factor given as

$$I_c = 1 - \left[\frac{1}{1 + (R/z)^2}\right]^{3/2} \tag{3.28}$$

where z is the depth of the point of interest below the centre of the circular area.

3.7.2.4 Uniform rectangular load

The increase in vertical stress induced by the uniform rectangular load (Figure 3.11) is

$$\Delta\sigma_v = qI_r \tag{3.29}$$

where q is the uniform pressure on a rectangular area of width, B, and length, L; I_r is an influence factor given as

$$I_r = \frac{1}{4\pi}\left[\left(\frac{2mn\sqrt{m^2 + n^2 + 1}}{m^2 + n^2 + m^2n^2 + 1}\right)\left(\frac{m^2 + n^2 + 2}{m^2 + n^2 + 1}\right) + \tan^{-1}\left(\frac{2mn\sqrt{m^2 + n^2 + 1}}{m^2 + n^2 - m^2n^2 + 1}\right)\right] \tag{3.30}$$

where z is the depth of the point of interest below the corner of the rectangular area; $m = B/z$; $n = L/z$. Note that m and n are interchangeable.

The increase in vertical stress below any point within or outside the rectangular area can be obtained by applying the principle of superposition, as explained in Example 3.6.

Figure 3.10 Vertical stress induced by a uniform circular load

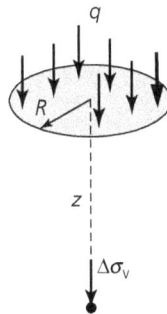

Figure 3.11 Vertical stress induced by a uniform rectangular load

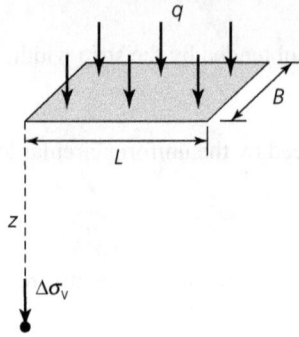

Figure 3.11 Vertical stress induced by a uniform rectangular load

Note that in Equation 3.30, the inverse of the tangent term must be a positive angle in radians. Hence, when $(m^2 + n^2 + 1) < m^2 n^2$, $\pi (= 180°)$ should be added to the negative angle in radians.

Example 3.6

A horizontal rectangular area, $3\,\text{m} \times 5\,\text{m}$, is loaded uniformly with a pressure of $150\,\text{kPa}$. Determine the increase in vertical stress at a depth of $4\,\text{m}$ below its centre.

Solution

Divide the rectangular area, $3\,\text{m} \times 5\,\text{m}$, into four rectangles, each of area $1.5\,\text{m} \times 2.5\,\text{m}$, as shown in Figure E3.6. The common corner of the rectangles is the centre of the complete area.

Figure E3.6

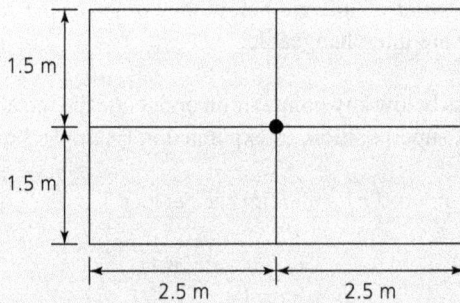

For any one of the rectangles in Figure E3.6, refer to Figure 3.11. Here, $B = 1.5\,\text{m}$, $L = 2.5\,\text{m}$, $z = 4\,\text{m}$ and $q = 150\,\text{kPa}$.

For $m = B / z = 1.5\,\text{m} / 4\,\text{m} = 0.375$ and $n = L / z = 2.5\,\text{m} / 4\,\text{m} = 0.625$, using Equation 3.30, the influence factor is calculated as

$$I_r = \frac{1}{4\pi}\left[\left[\frac{(0.469)\sqrt{1.531}}{1.531+0.055}\right]\left(\frac{2.531}{1.531}\right)+\tan^{-1}\left[\frac{(0.469)\sqrt{1.531}}{1.531-0.055}\right]\right] = \frac{1}{4\pi}\left[0.605+\tan^{-1}(0.393)\right]$$

or

$$I_r = \frac{1}{4\pi}\left[0.605+\left(\frac{\pi}{180}\right)(21.42)\right] = 0.078$$

From Equation 3.29, the increase in vertical stress

$$\Delta\sigma_v = q(4I_r) = (150)(4\times0.078) = 46.8\ \text{kPa}$$

3.7.3   2V:1H stress distribution method

The increase in vertical stress induced by a rectangular load can also be determined by the 2V:1H stress distribution method, which is a simple empirical method based on the assumption that the load spreads in a 2 vertical (V) to 1 horizontal (H) manner on each side of the loaded rectangular area. Thus the load over a rectangular area $(B\times L)$ at the ground surface acts on the rectangular area $(B+z)(L+z)$ at a depth, z, below the ground surface because at this depth, the area carrying the load increases by $z\,/\,2$ on each side, as shown in Figure 3.12. Thus, the increase in vertical stress at a depth, z, induced by the uniform pressure, q, over a rectangular area, BL, at the ground surface is

$$\Delta\sigma_v = \frac{qBL}{(B+z)(L+z)} \tag{3.31}$$

For a strip load $(L>>B$ i.e. for a given B as $L\to\infty)$, Equation 3.31 reduces to

$$\Delta\sigma_v = \frac{qB}{(B+z)} \tag{3.32}$$

Figure 3.12 Determination of vertical stress increase by the 2V:1H stress distribution method

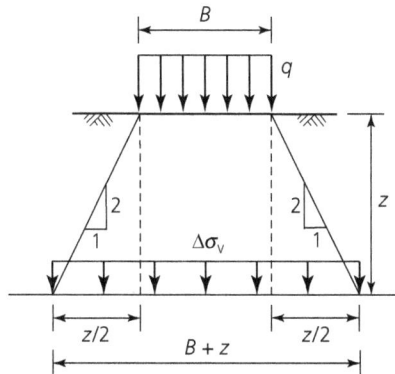

Example 3.7

A horizontal rectangular area, $3\,\text{m} \times 5\,\text{m}$, is loaded uniformly with a pressure of $150\,\text{kPa}$. Determine the increase in vertical stress at a depth of $4\,\text{m}$ using the 2V:1H stress distribution method. Compare this value with the one obtained in Example 3.6.

Solution

Refer to Figure 3.12. Here, $q = 150\,\text{kPa}$, $B = 3\,\text{m}$, $L = 5\,\text{m}$ and $z = 4\,\text{m}$. From Equation 3.31

$$\Delta\sigma_z = \frac{qBL}{(B+z)(L+z)} = \frac{(150)(3)(5)}{(3+4)(5+4)} = 35.71\ \text{kPa}$$

This increase in vertical stress is less than the value calculated in Example 3.6.

The following points regarding stresses within soil induced by applied loads are noteworthy.

- The assumptions considered in the elastic theory are rarely fulfilled by the soils. Therefore the computed values of increase in stresses may be in error by as much as $\pm 25\%$ or more, although a few comparisons of computed values with measured values indicate surprisingly good agreement, especially in the case of increase in vertical stresses (Lambe and Whitman, 1979). However, due to a lack of better techniques, geotechnical engineers often make these assumptions.
- The 2V:1H stress distribution method typically gives lower $\Delta\sigma_v$ at shallow depths ($z/B < 2$) and higher $\Delta\sigma_v$ at large depths ($z/B > 2$) when compared to the maximum stress increase calculated by using Boussinesq's equation. At a depth of $z/B = 2$, the two values are approximately equal.
- A line that connects all points of equal vertical stress increase within a soil under an applied load is called the stress contour or pressure isobar. Stress contours can be generated for any load system, for a uniform strip load using Boussinesq's equation, as illustrated in Figure 3.13.
- The zone within soil under a loaded area wherein the vertical stresses are significant is frequently termed the bulb of stresses, or simply the stress bulb or pressure bulb. Usually, the stress bulb is considered to be the volume of soil within the stress contour for $0.01q$, but this choice is strictly arbitrary (Lambe and Whitman, 1979).
- The increase in vertical stress is less than $0.01q$ at a depth of $4R$ for a uniformly loaded circular area of radius, R, and $2.5B$ for a uniformly loaded square area of side length, B. For a uniformly loaded strip area of width, B, the stress contour of $0.01q$ extends to a depth of more than $5B$. Thus, the stress contours for a uniform strip load extend significantly deeper than those for uniform circular and square loads.
- In computing the change in vertical stresses within a soil, it is generally assumed that the footing carrying the applied loads is flexible; this causes the pressure to be uniform over the loaded areas, as considered in the core principles of stress distribution presented in this chapter.
- These days the available computer software that is based on Boussinesq's equation is widely and routinely used to determine the stress distribution induced by applied loads within the soils.

Figure 3.13 Stress contours (or pressure isobars) for a uniform strip load

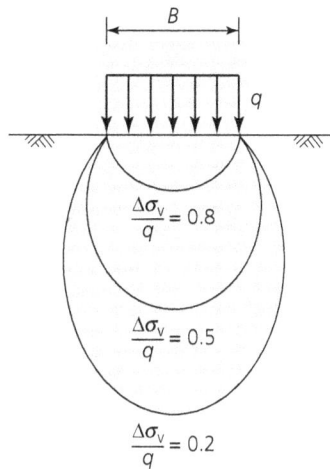

Chapter summary

1. For the geostatic state of stress within soil, which generally occurs in sedimentary soil deposits, when the surface of a soil deposit is level and the unit weight is constant with depth, the vertical and horizontal stresses increase linearly with depth. The total vertical normal stress at a point within the soil deposit with a level surface is simply the sum of the products of the thickness of the soil layer and its total unit weight for all the layers lying above the point.

2. Effective stress may be transmitted from particle to particle through contact and/or through net electrical stress between them. In general, the effective stress is essentially the force carried by the soil skeleton divided by the total sectional area. Within saturated soil, the effective stress at a point is equal to the total stress minus the pore water pressure at that point.

3. For the water table below the ground surface, a rise in the water table causes a reduction in the effective stress at any point within the soil, while a fall in the water table results in an increase in effective stress. For the water table above the ground surface, a fluctuation in the water table does not alter the effective stress at any point within the soil.

4. Within the capillary zone of continuous moisture, the water pressure is negative; therefore, the capillary water does not contribute to hydrostatic pressure below the water table. The only effect of capillary rise above the water table is to increase the total unit weight of the soil; therefore, both the total and the effective vertical stresses at any point at or below the water table are increased by the same amount.

5. The stresses within a soil are not necessarily hydrostatic – that is, the horizontal stress is not exactly equal to the vertical stress. The ratio of effective horizontal stress to effective vertical stress is the coefficient of lateral stress. With no lateral strain within the ground, this ratio is called the coefficient of lateral stress at rest (K_0), which typically varies

between 0.4 and 0.5 for sand deposits when they have not been preloaded earlier. If a soil deposit has been heavily preloaded in the past, K_0 may be greater than 1, and may even reach a value of 3.

6. Boussinesq's equation (Equation 3.21) provides a reasonably suitable value for the vertical stress increase, although soils are not even approximately elastic, homogeneous and isotropic. However, the computed values may be in error by as much as $\pm 25\%$ or more for some cases.

7. The 2V:1H stress distribution method gives lower vertical stress increase at shallow depths and higher values at large depths when compared to the maximum stress increase calculated by using Boussinesq's equation.

8. The stress contours or pressure isobars within the soil for a uniform strip load applied to the ground surface extends significantly deeper than those for uniform circular and square loads.

Questions for practice

Select the most appropriate answer for each of the multiple choice questions from Q3.1 to Q3.10. Questions Q3.11 to Q3.31 require more detailed answers.

3.1 The total number of independent stress components at a point within a soil loaded at its boundary is
(a) 3
(b) 6
(c) 9
(d) 12.

3.2 If a 4 m-deep river consists of the sandy bed with a saturated unit weight of $20\,kN/m^3$, the effective vertical stress at a depth of 6 m below the sand bed will be approximately
(a) 60 kPa
(b) 80 kPa
(c) 120 kPa
(d) 200 kPa.

3.3 Within a soil, at a point, the rise of the water table above the ground surface causes
(a) an increase in total vertical stress but a decrease in pore water pressure
(b) a decrease in total vertical stress but an increase in pore water pressure
(c) an equal decrease in both total vertical stress and pore water pressure
(d) an equal increase in both total vertical stress and pore water pressure.

3.4 The capillary rise in a soil deposit is 1 m. What will be the suction?
(a) 1 kPa
(b) −1 kPa
(c) 10 kPa
(d) −10 kPa.

3.5 Within a soil, at a point below the water table, the capillary rise above the water table causes
    (a) an equal increase in both total vertical stress and pore water pressure
    (b) an equal increase in both total vertical stress and effective vertical stress
    (c) an increase in total vertical stress but a decrease in effective vertical stress
    (d) a decrease in total vertical stress but an increase in pore water pressure.

3.6 If K is the coefficient of lateral stress, the hydrostatic stress condition within a soil refers to
    (a) $K=0$
    (b) $K=0.5$
    (c) $K=1$
    (d) $K=\infty$.

3.7 If a vertical point load of 10 kN acts on the surface of a linearly elastic half-space, the increase in vertical stress induced by this load within the elastic half-space at a depth of 1 m below the point of application of the load will be
    (a) 0.04775 kPa
    (b) 0.4775 kPa
    (c) 4.775 kPa
    (d) 47.75 kPa.

3.8 The vertical stress increase within the foundation soil at some depth below the corner of a 3 m × 4 m rectangular surface footing induced by a certain applied uniform pressure on the footing is 150 kPa. What will be the vertical stress increase below the centre of a 6 m × 8 m rectangular surface footing under the same applied pressure and at the same depth?
    (a) 75 kPa
    (b) 150 kPa
    (c) 300 kPa
    (d) 600 kPa.

3.9 A surface footing, 1 m × 1 m, exerts a uniform pressure of 100 kPa on the ground. Assuming a load dispersion of 2 vertical to 1 horizontal, the average vertical stress increase within the soil at a depth of 1 m below the footing is determined as
    (a) 25 kPa
    (b) 50 kPa
    (c) 75 kPa
    (d) 100 kPa.

3.10 For a uniformly loaded square level ground area of side length B, the increase in vertical stress within the soil is less than 10% of the applied pressure at a depth below the ground level of
    (a) $0.5B$
    (b) B
    (c) $1.5B$
    (d) $2.5B$

3.11 At a building site, the foundation soil consists of the following three layers from top to bottom: 2.5 m-thick silty sand layer with total unit weight $= 17.1 \, kN/m^3$, 4 m-thick clay layer with total unit weight $= 19 \, kN/m^3$, and 1.5 m-thick sandy layer with total unit weight $= 20.8 \, kN/m^3$. Determine the total vertical stress at the bottom of the third layer.

3.12 At a construction site, the vertical stress, σ_v (kPa), within a soil is related to its total unit weight, γ, (kN/m³) as $\gamma = 15.08 + 0.0027 \, \sigma_v$. Determine the vertical stress at a depth of 25 m for a geostatic stress condition.

3.13 The heights of saturated zones by capillarity in fine sand, medium silt and clay are 0.6 m, 6 m and 60 m, respectively. Determine the (matric) suction. Consider the unit weight of water, $\gamma_w = 10 \, kN/m^3$.

3.14 For a coarse sand, effective particle size, $D_{10} = 0.11$ mm and void ratio, $e = 0.27$. Determine the approximate range of height of capillary rise.

3.15 The water table is 2 m above a deep deposit of clay. The average water content of clay is 50% and the specific gravity of clay particles is 2.75. Determine the effective vertical stress at a depth of 12 m below the water table.

3.16 The water table in a very deep deposit of fine sand is 0.5 m below the ground surface. Above the water table, the sand is saturated by capillarity. The total unit weight of the saturated sand is 20.5 kN/m³. Determine the effective vertical stress at a depth of 5 m below the ground surface.

3.17 A 5 m-thick sand stratum rests over a deep clay stratum. The water table is 2 m above the top surface of the sand stratum. The saturated unit weights of the sand and clay are 20 kN/m³ and 19 kN/m³, respectively. Determine the total vertical stress, pore water pressure and effective vertical stress at elevations of 5 m and 10 m below the top surface of the sand stratum and plot their values against the elevation.

3.18 A point load of 100 kN is applied vertically on a horizontal ground surface. Determine the increase in vertical stress at
   (a) a depth of 2 m below the point of application of the load
   (b) a depth of 2 m below the point of application of the load and at a radial distance of 1 m from the line of action of the load.

3.19 A pressure of 100 kPa is applied over a 5 m-wide strip area of a ground surface. The total unit weight of the soil below the ground surface is 20 kN/m³. Determine the total vertical stress at a depth of 2.5 m below the centre of the loaded area.

3.20 A foundation soil with total unit weight of 17.3 kN/m³ is loaded by a pressure of 200 kPa over a circular area, 5 m in diameter. Determine the total vertical stress at a depth of 2 m below the centre of the loaded area.

3.21 A very wide embankment 5 m high is constructed from a fill with a unit weight of 16 kN/m³ over a saturated sand deposit with the presence of the water table coinciding with its top surface. At a depth of 3 m below the top surface of the sand deposit, determine
   (a) total and effective vertical stresses before construction of the embankment
   (b) increase in vertical stress induced by the construction of the embankment
   (c) total and effective vertical stresses after construction of the embankment.

   Compare the values of the total and effective vertical stresses before and after the construction of the embankment in (a) and (c). What do you observe?

3.22 The plan of a uniformly loaded square area is shown in Figure Q3.22. Determine the increase in vertical stress at a depth of 4 m below point A.

Figure Q3.22

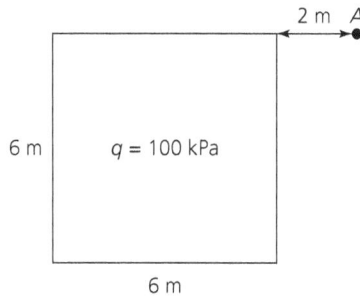

3.23 A vertical point load of 100 kN acts at the horizontal ground surface. Construct a stress contour (isobars) for a vertical stress increase of 10 kPa.

3.24 What are geostatic stresses and how can they be determined?

3.25 Explain the importance of effective stress in geotechnical engineering. Can you measure experimentally the effective stress within a soil? Justify your answer.

3.26 Present Equation 3.13(b) in a simplified form by considering the typical values of the surface tension and unit weight of water.

3.27 Explain the effect of capillary rise within a soil on the distribution of total vertical stress, pore water pressure and effective stress against the depth.

3.28 What are the factors that affect soil suction?

3.29 What is the effect of surcharge loading on the effective stress?

3.30 Where can you walk more easily: on a wet sandy beach or a dry sandy beach? Justify your answer.

3.31 A loose sand bed is submerged in a small tank. If you tap the side of the tank, the sand bed may become densified. Will densification cause any change in the total and effective vertical stresses at the base of the tank? Justify your answer.

REFERENCES

Bishop AW (1960) The principle of effective stress. *Norwegian Geotechnical Institute Publication No.* 32, pp. 1–5.

Bishop AW, Alpan I, Blight GE and Donald IB (1960) Factors controlling the strength of partially saturated cohesive soils. *Proceedings of the Research Conference on Shear Strength of Cohesive Soils*. ASCE, Boulder, CO, USA, pp. 502–532.

Boussinesq J (1885) *Application des potentiels à l'étude de l'équilibre et du mouvement des solides élastiques*. Gauthier-Villars, Paris, France. (In French.)

Lambe TW and Whitman RV (1979) *Soil Mechanics, SI Version*. Wiley, New York, NY, USA.

Mitchell JK and Soga K (2005) *Fundamentals of Soil Behaviour*, 3rd edn. Wiley, Hoboken, NJ, USA.

Nelson JD and Miller DJ (1992) *Expansive Soils*. Wiley, New York, NY, USA.

Poulos HG and Davis EH (1974) *Elastic Solutions for Soil and Rock Mechanics*. Wiley, New York, NY, USA.

Terzaghi K (1943) *Theoretical Soil Mechanics*. Wiley, New York, NY, USA.

Terzaghi K, Peck RB and Mesri G (1996) *Soil Mechanics in Engineering Practice*, 3rd edn. Wiley, New York, NY, USA.

FURTHER READING

Atkinson J (2007) *The Mechanics of Soils and Foundations*, 2nd edn. Taylor and Francis, London, UK.

Das BM (2022) *Principles of Geotechnical Engineering*, 10th edn. Cengage, Boston, MA, USA.

Holtz RD, Kovacs WD and Sheahan TC (2011) *An Introduction to Geotechnical Engineering*, 2nd edn. Pearson, Upper Saddle River, NJ, USA.

emerald
PUBLISHING

ice
Publishing

Sanjay Kumar Shukla
ISBN 978-1-83608-519-5
https://doi.org/10.1108/978-1-83608-516-420252007
Emerald Publishing Limited: All rights reserved

Chapter 4
Fluid flow through soil

Learning aims

This chapter explores the core concepts of the following topics

- basic concepts of heads and fluid flow
- Darcy's law and determination of permeability of soil
- seepage force and its effect on effective stress
- basic flow equation
- construction of flow net and its application
- filter and its design requirements.

4.1.   Introduction

In most geotechnical and other civil engineering applications, water is the primary fluid that flows through soil and rock. The flow of water through soil significantly influences its volume change, deformation or settlement, and stability behaviour. The flow of water can also cause removal and/or introduction of chemicals, colloids and microorganisms. Therefore, the core principles of fluid flow through a soil are required to solve several problems in civil engineering, such as water leakage through an earth dam and its foundation, flow of leachate through a liner of the landfill, settlement of a foundation, and slope stability. This chapter introduces the basic concept of flow of water through soil and focuses on the property of soil, called permeability, which is its ability to conduct fluid. The emphasis is also placed on explaining how water influences the effective stress within the soil through which it flows. The basic flow equation is derived, and its graphical solution is presented using flow nets along with their applications. Basic details of granular filters are also described.

4.2.   Nature and concept of fluid flow

A soil mass generally contains continuous voids in the form of irregular passages through which fluid flows. Thus, soil is a permeable material like other non-metallic materials such as rock and concrete. The fluid that flows through the soil is called the permeant. On a microscopic scale, the fluid flows from one point to another in a winding path, and its velocity may vary from point to point within a soil mass. However, on a macroscopic or an engineering scale, the fluid can be considered to flow along a straight line between any two points within soil at an average velocity.

For fluid flow analysis, the energy of the fluid is commonly expressed in terms of head, which is the energy per unit weight of fluid. Thus, the SI unit of head is metre (m). As a fluid can have potential energy, kinetic energy and pressure energy, they correspond, respectively, to the following three heads.

Elevation head, also referred to as position head or potential head

$$h_e = z \tag{4.1a}$$

Velocity head

$$h_v = \frac{v^2}{2g} \tag{4.1b}$$

Pressure head

$$h_p = \frac{u}{\gamma_w} \tag{4.1c}$$

where z is the distance of the point of interest above or below a reference elevation or datum plane; v is the velocity of water at the point of interest; u is the water pressure at the point of interest; γ_w is the unit weight of water.

The pressure head is the height to which the water rises in a standpipe, known as the piezometer or piezometric tube, with its inlet installed at the point of interest.

For most flow problems in soils, the velocity head is too small to be of any practical importance because of low flow velocity through soils and it, therefore, may be neglected. Thus, the total (or hydraulic) head at any point is

$$h = h_e + h_p \tag{4.2a}$$

or

$$h = z + \frac{u}{\gamma_w} \tag{4.2b}$$

In fluid mechanics applications, the total head is the sum of elevation, velocity and pressure heads, and the sum of elevation and pressure heads is called the piezometric head. In soil mechanics applications, because of negligible velocity head, the total head and piezometric head are equal. Note that in soil mechanics, h_p only is also called the piezometric head by some authors because it is determined by a piezometer; therefore, care must be taken when dealing with this term.

If the total head at any two points within a soil mass is known, the flow through the soil can be determined as it depends solely on the difference in the total head between these points. When water flows through a soil, energy or head losses occur. Therefore, considering flow between points A and B, as shown in Figure 4.1

$$h_A = h_B + \Delta h \tag{4.3a}$$

or

$$z_A + \frac{u_A}{\gamma_w} = z_B + \frac{u_B}{\gamma_w} + \Delta h \tag{4.3b}$$

where subscripts A and B with h, u and z refer to their values at points A and B, respectively; Δh is the loss in total head, commonly referred to as head loss, between points A and B.

Equation 4.3 for the total heads and head loss associated with flow between two points is a special case of the generalised one-dimensional (1D) energy equation, called Bernoulli's equation, which also contains velocity head as considered in fluid mechanics. This equation is the basis for the computation of pore water pressure and analysis of flow through soil and other porous media.

Figure 4.1 Flow of water through a soil with the definition of total head (h) in terms of elevation head (z) and pressure or piezometric head (u / γ_w)

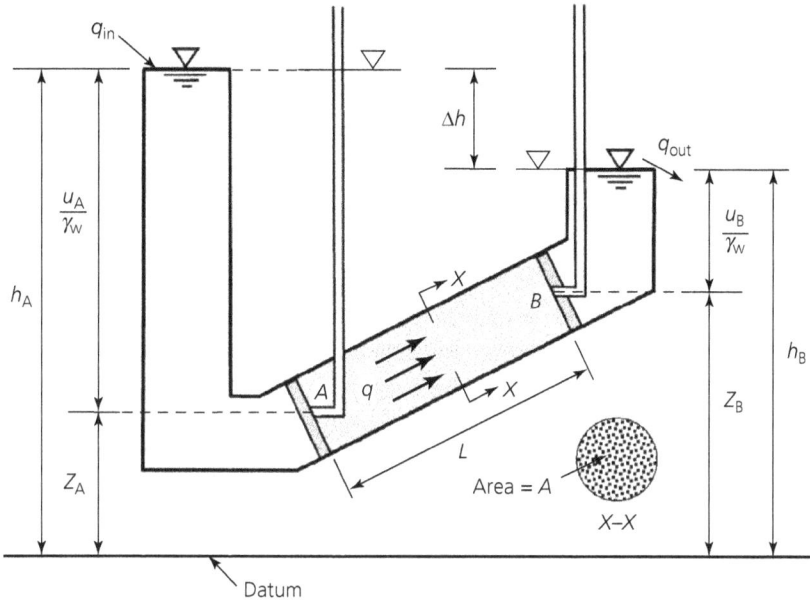

Note: q is the rate of flow or discharge; q_{in} is the inflow discharge; q_{out} is the outflow discharge; L is the flow path length. Section line X–X corresponds to the cross-sectional view presented. In general, $q = q_{in} = q_{out}$.

From Equation 4.3, the head loss

$$\Delta h = \left(z_A + \frac{u_A}{\gamma_w} \right) - \left(z_B + \frac{u_B}{\gamma_w} \right) \tag{4.4}$$

The ratio of the head loss, Δh, between any two points A and B to the flow path length, L, between them is called the hydraulic gradient, i, which is one of the core concepts of soil mechanics. Thus,

$$i = \frac{\Delta h}{L} \tag{4.5}$$

Note that the hydraulic gradient is dimensionless. For no head loss ($\Delta h = 0$) between any two points A and B, $i = 0$; therefore, the total heads at A and B are equal, resulting in no flow. In other words, if there is no flow between two points within the soil mass, there is no head loss – that is, $\Delta h = 0$ and $i = 0$.

The following points regarding the nature and core concepts of fluid flow are noteworthy.

- The absolute value of elevation head depends on the location of the datum, but the difference in elevation head, which has importance in flow problems, remains unaffected by the location of the datum.
- In soil flow problems, it is assumed that all head is lost only in soil, while water flows through a soil mass (Lambe and Whitman, 1979).
- Negative water pressure can exist during flow through soil. A typical example is the existence of negative water pressure in the capillary flow zone.

89

Example 4.1

The high flow velocity of water in a soil can be 1 m/min. What will be the velocity head? Is this value significantly low for neglecting the velocity head in analysis of flow through soils?

Solution

The velocity of water

$$v = 1 \text{ m/min} = 0.017 \text{ m/s}$$

From Equation 4.1b, the velocity head

$$h_v = \frac{v^2}{2g} = \frac{(0.017)^2}{(2)(9.81)} = 1.473 \times 10^{-5} \text{ m} \approx 0.015 \text{mm}$$

This value of the velocity head is far less than the accuracy with which the elevation and pressure heads are commonly measured. Thus, the velocity head can be neglected without any significant error in the analysis of flow through soil.

Example 4.2

Figure E4.1 shows the downward flow of water through the soil in a tube. Atmospheric pressure is maintained at the top of the reservoir water (elevation (El.) 5.0 m) and at the bottom of the tail water (El. 0 m, i.e. the datum). Determine the elevation head, total head and pressure head at El. 5.0 m, El. 4.0 m, El. 2.5 m and El. 0 m. What is the pore water pressure at the point P (El. 2.5 m)? Assume that the soil is homogeneous.

Figure E4.1

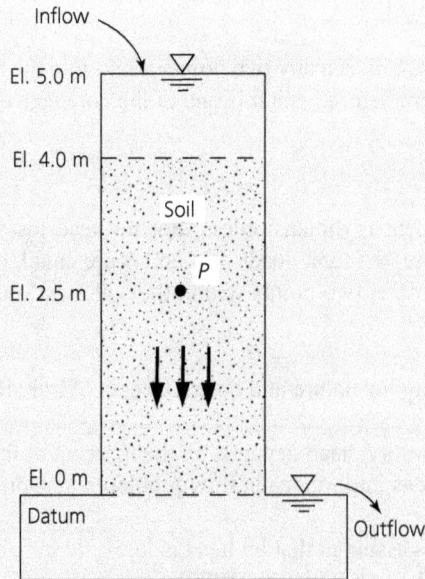

Solution

For convenience, first determine the elevation head (h_e) and total head (h) and then determine the pressure head (h_e) by subtracting the elevation head from the total head.

At El. 5.0 m

$h_e = 5.0$ m

$h = 5.0$ m, because $h_p = 0$ m.

At El. 4.0 m

$h_e = 4.0$ m

$h = 5.0$ m, because in flowing from El. 5.0 m to El. 4.0 m, no total head is lost.

$h_p = h - h_e = 5.0 - 4.0 = 1.0$ m

At El. 0 m

$h_e = 0$ m

$h = 0$ m, because $h_p = 0$ m

Thus, in flowing through the soil from El. 4.0 m to El. 0 m, loss in total head

$\Delta h = 5$ m

The length of soil column, $L = 4 - 0 = 0$ m

As the soil is homogeneous, the heads will vary linearly, and there will be a uniform hydraulic gradient, which can be calculated using Equation 4.5 as

$$i = \frac{\Delta h}{L} = \frac{5}{4} = 1.25$$

At the point P (El. 2.5 m)

$h_e = 2.5$ m

$h = 5 - i(4 - 2.5) = 5 - (1.25)(1.5) = 3.125$ m

$h_p = h - h_e = 3.125 - 2.5 = 0.625$ m

This implies that if a standpipe is inserted at point P, the water will rise to a height of 0.625 m in it. Hence, the pore water pressure at point P is

$u = 0.625\gamma_w = (0.625)(9.81) = 6.13$ kPa

4.3. Darcy's law and flow velocity

In 1856, Henry Darcy, while working in Paris, investigated the flow properties of water through a vertical homogeneous sand filter bed. His investigations are also applicable to the set-up shown in Figure 4.1 where the flow takes place through an inclined soil bed. According to Darcy's study, the rate of flow (also known as the rate of discharge, or simply discharge) or the volume of water flowing per unit time, is

$q \propto A$

$q \propto \Delta h$

and

$q \propto \dfrac{1}{L}$

where A = total cross-sectional area of the soil through which the flow takes place. When combined

$q \propto \left(\dfrac{\Delta h}{L}\right) A$

or

$q = k \left(\dfrac{\Delta h}{L}\right) A$ (4.6)

where k is a constant of proportionality, called the coefficient of permeability, or simply permeability or hydraulic conductivity.

Permeability is one of the important properties of soil, indicating the relative ease with which a fluid flows through it. Note that permeability is also one of the important properties of other porous media. The SI unit of k is m/s, which is the unit of velocity.

Using Equation 4.5, Equation 4.6 becomes

$q = kiA$ (4.7)

or

$v = ki$ (4.8)

where

$v = \dfrac{q}{A}$ (4.9)

is the flow velocity.

Any one of Equations 4.6, 4.7 or 4.8 individually defines Darcy's law, which is one of the core principles of soil mechanics. As A is the total cross-sectional area of the soil specimen, including the solid area, A_s, through which water cannot flow (Figure 4.2), v is the superficial velocity of water flowing through the soil. The flow velocity, v, is basically the approach velocity at the point A or the discharge velocity at point B of the soil bed (Figure 4.1). As water can only flow through the voids in the cross-section, the average effective velocity of flow, called the seepage velocity, through the soil is

$$v_s = \frac{q}{A_v} \qquad (4.10)$$

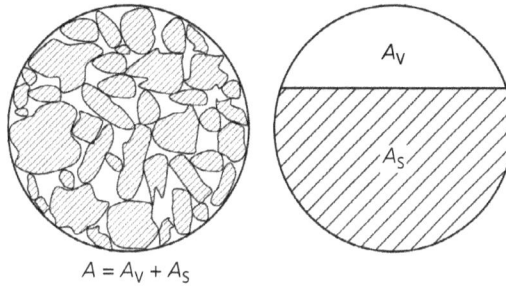

Figure 4.2 Phase diagram for the cross-sectional area of the soil bed

$A = A_V + A_S$

Note that the term seepage refers to the slow movement of water through the soil or rock caused by the hydraulic gradient.

By applying the continuity (or conservation of mass) principle of water flow, by comparing Equations 4.9 and 4.10,

$$q = vA = v_s A_v \qquad (4.11)$$

or

$$v = \left(\frac{A_v}{A}\right) v_s = \left(\frac{A_v L}{AL}\right) v_s = \left(\frac{V_v}{V}\right) v_s = n v_s \qquad (4.12a)$$

or

$$v_s = \frac{v}{n} \qquad (4.12b)$$

where $n (= V_v / V)$ is the porosity of the soil, as defined in Equation 1.2.

As n varies between 0 and 1, $v_s > v$. Using Equation 4.8, Equation 4.12b reduces to

$$v_s = \frac{ki}{n} \qquad (4.12c)$$

Darcy's law holds good for all types of soil if the fluid flows under laminar flow conditions, where a fluid particle follows a definite path and does not cross the paths of other fluid particles. In other words, Darcy's law is valid in situations where the fluid flows in parallel layers without mixing and with no internal energy dissipation, resulting in a linear plot of hydraulic gradient, i, against flow velocity, v.

From Equation 4.8, for $i = 1$, $v = k$; therefore, the hydraulic conductivity, k, can be defined as the approach velocity for a hydraulic gradient of unity. Table 4.1 lists the typical values of coefficient of permeability and degree of permeability, defined as high to practically impermeable, for some common soils.

Table 4.1 Typical values of the coefficient of permeability and degree of permeability for some common soils

Soil type	Coefficient of permeability, k: m/s	Degree of permeability
Clean gravel	$\geq 10^{-2}$	High
Clean coarse sand	$10^{-2} - 10^{-4}$	Medium
Clean fine sand	$10^{-4} - 10^{-5}$	Low
Silty sand	$10^{-5} - 10^{-6}$	Very low
Silt	$10^{-6} - 10^{-9}$	Extremely low
Clay	$\leq 10^{-9}$	Practically impermeable

The k values depend on the characteristics of both soil and fluid (commonly referred to as the permeant), as listed in Table 4.2. For clean sands with less than 5% passing the 75 µm sieve size, and with D_{10} between 0.1 mm and 3.0 mm (Hazen, 1911)

$$k = cD_{10}^2 \tag{4.13}$$

where k is in cm/s, D_{10} is in mm, and c is a constant that varies from 0.4 to 1.2. In the absence of a specific value, c is commonly assumed to be unity.

Equation 4.13 is valid for $k \geq 10^{-5}$ m/s. Equation 4.13 is known as the Hazen equation. Carrier (2003) reports that the Hazen equation is valid at 10°C, and not at 20°C as commonly assumed, and suggests that c in Equation 4.13 should be multiplied by 1.3.

The coefficient of permeability of sands can also be computed using the Kozeny–Carman equation (Carman, 1938, 1956; Kozeny, 1927; Lambe and Whitman, 1979), as given below.

$$k = \frac{1}{k_0 S_s^2} \frac{\gamma_w}{\eta} \left(\frac{e^3}{1+e} \right) \tag{4.14}$$

where k_0 is a factor depending on the void or pore shape and ratio of length of actual flow path to soil bed length; S_s is the specific surface area; e is the void ratio of soil; γ_w is the unit weight of water (or permeant); η is the viscosity of water (or permeant).

Table 4.2 Factors affecting permeability of soils

Soil characteristics	Fluid (or permeant) characteristics
Particle size	Unit weight
Void ratio/state of denseness	Viscosity
Composition	Polarity
Soil structure/fabric	
Degree of saturation	

The polarity of fluid, which is not included in Equation 4.14, can also affect the permeability because the mobility of fluid immediately adjacent to the soil particles depends on the polarity of the pore fluid.

Equation 4.13 suggests that smaller soil particles result in smaller voids, which act as flow channels, thereby reducing permeability. Equation 4.14 suggests that as e decreases, which may be caused by an increase in effective stress, k also decreases.

For clays, when compacted at a constant compactive effort, the permeability decreases with increasing water content and reaches a minimum value at about the optimum water or moisture content; this is explained in more detail in Chapter 7. A clay with a flocculated structure, which generally has some large channels available for flow, has relatively higher permeability compared to the permeability of a clay with dispersed structure, especially when the flow is normal to the particles.

The influence of soil composition is of great importance with clay. Among the common exchangeable ions, sodium is the one that gives the lowest permeability to clay. Sodium montmorillonite is one of the least permeable soil minerals and is, therefore, widely used as an additive to other soils to reduce their permeability.

Air present in the voids in the form of bubbles can block some of the voids, thereby reducing permeability. Therefore, the permeability generally increases with an increase in the degree of saturation of soil.

Equation 4.14 may be expressed as

$$k = \left(\frac{\gamma_w}{\eta}\right) K \tag{4.15a}$$

where

$$K = \frac{1}{k_0 S_s^2}\left(\frac{e^3}{1+e}\right) \tag{4.15b}$$

The value of K depends solely on the properties of the soil and is, therefore, referred to as the specific or absolute permeability, primarily used in hydrogeology and petroleum engineering applications. The SI unit of K is m², and is also expressed in terms of darcys where 1 darcy $= 0.987 \times 10^{-12}$ m².

The following points regarding Darcy's law and flow velocity are noteworthy.

▨ Fluid flow in most soils is sufficiently slow to be considered laminar; therefore, Darcy's law applies to flow through them. For liquid flow at very high velocities, and for gas flow at either very low or very high velocities, Darcy's law becomes invalid.

▨ Since a drop of water moving through the soil follows a winding path with varying velocity, the seepage velocity, v_s, is a fictitious velocity for an assumed drop of water that moves in a straight line joining any two points within the soil at a constant velocity. Although both v and v_s are fictitious quantities, they are often used to determine the time required for water to flow through a given distance within a soil mass.

▨ In general, e against $\log k$ is close to a straight line for nearly all soils (Lambe and Whitman, 1979).

Example 4.3

The coefficient of permeability of a beach sand at a void ratio of 0.8 is 0.002 m/s. What will be the coefficient of permeability if this sand has a void ratio of 0.6?

Solution

From Equation 4.14

$$k \propto \frac{e^3}{1+e}$$

Therefore, for two different void ratios, e_1 and e_2 of the beach sand with its corresponding coefficients of permeability, k_1 and k_2

$$\frac{k_1}{k_2} = \frac{\dfrac{e_1^3}{1+e_1}}{\dfrac{e_2^3}{1+e_2}}$$

or

$$\frac{0.002}{k_2} = \frac{\dfrac{(0.8)^3}{1+0.8}}{\dfrac{(0.6)^3}{1+0.6}} = 2.107$$

or

$$k_2 = 9.49 \times 10^{-4} \text{ m/s} \approx 0.001 \text{ m/s}$$

4.4.   Laboratory determination of permeability

The two test methods commonly used in the laboratory to determine the permeability of soils are

- constant head permeability test (Figure 4.3), suitable for coarse-grained and highly permeable soils having coefficient of permeability typically in the range of 10^{-2}–10^{-7} m/s (ASTM D2434-22 (ASTM, 2022); BS 1377-2:2022 (BSI, 2022); IS 2720 (Part 17)-1986 (Bureau of Indian Standards, 2021); AS 1289.6.7.1 (Standards Australia, 2001a))
- falling head (or variable head) permeability test (Figure 4.4), suitable for fine-grained and less permeable soils having coefficient of permeability typically smaller than 10^{-7} m/s (ASTM D5084-16a (ASTM, 2024); IS 2720 (Part 17)-1986 (Bureau of Indian Standards, 2021); AS 1289.6.7.2 (Standards Australia, 2001b)).

Both methods use a cylindrical device, called the permeameter, which consists essentially of a cylinder containing the soil specimen between two filters.

The constant head permeability test is performed by allowing de-aired water to flow through the soil specimen at a head controlled and kept constant by the overflow in a reservoir. When the steady flow conditions are obtained, the time, Δt, taken to collect a certain discharge volume, Q, of water in a measuring cylinder is recorded. The discharge is calculated as

$$q = \frac{Q}{\Delta t} \tag{4.16}$$

From Darcy's law (Equation 4.6) with the replacement of head loss, Δh, by h, which is constant as applicable to the present case, and Equation 4.16

$$\frac{Q}{\Delta t} = k \left(\frac{h}{L} \right) A$$

or

$$k = \frac{QL}{Ah\Delta t} \tag{4.17}$$

In the case of the falling head permeability test, the de-aired water is allowed to flow through the soil specimen at a head that keeps falling with time in a standpipe fitted at the top of the permeameter. When the steady flow conditions are obtained, the time for the water to fall from one level to another in the standpipe is noted. If h is the water level in the standpipe at any time, t, measured from the constant tail water level (Figure 4.4), and if dh is the head loss in time interval, dt, the rate of flow or discharge, q, at time, t, can be given by

$$q = -a \frac{dh}{dt} \tag{4.18}$$

where a is the cross-sectional area of the standpipe.

From Darcy's law (Equation 4.6) with the replacement of head loss, Δh, by h, which is a function of time, t, as applicable to the present case, and Equation 4.18

Figure 4.3 Constant head laboratory permeability test

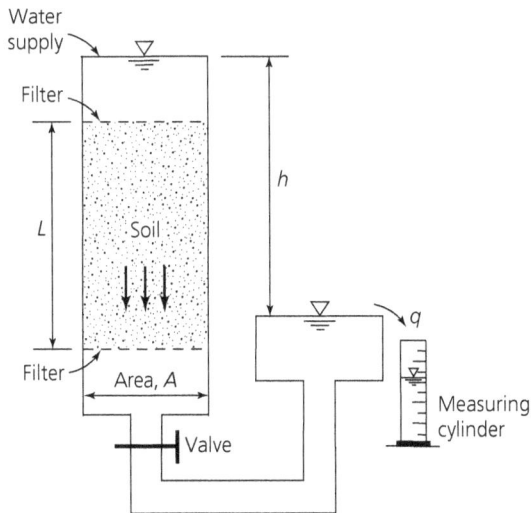

$$-a\frac{dh}{dt} = k\left(\frac{h}{L}\right)A$$

or

$$-\frac{dh}{h} = k\left(\frac{A}{aL}\right)dt$$

If h decreases from h_1 at $t = t_1$ to h_2 at $t = t_2$, then integrating over these limits

$$-\int_{h_2}^{h_1}\frac{dh}{h} = k\left(\frac{A}{aL}\right)\int_{t_2}^{t_1}dt$$

or

$$k = \frac{aL}{A\Delta t}\ln\left(\frac{h_1}{h_2}\right) = \frac{2.3aL}{A\Delta t}\log_{10}\left(\frac{h_1}{h_2}\right) \qquad (4.19a)$$

where $\Delta t(= t_2 - t_1)$ is the time interval, which can be measured directly using a stopwatch.

If d and D are the diameters of the standpipe and the cylindrical permeameter, respectively, Equation 4.19a can be expressed as

Figure 4.4 Falling head laboratory permeability test

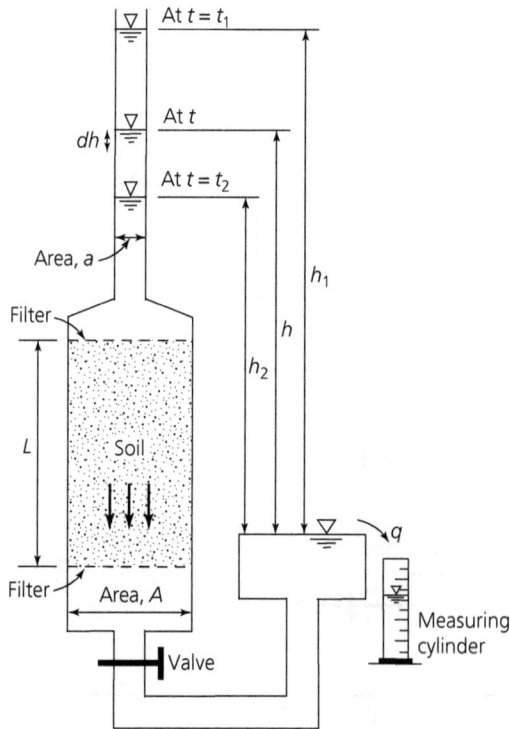

$$k = \frac{d^2 L}{D^2 \Delta t} \ln\left(\frac{h_1}{h_2}\right) = \frac{2.3 d^2 L}{D^2 \Delta t} \log_{10}\left(\frac{h_1}{h_2}\right) \qquad (4.19b)$$

Example 4.4

In a falling head permeability test, the permeameter has a diameter of 100 mm, and the length of the soil specimen is 150 mm. The diameter of the standpipe is 10 mm. During the test, the head loss decreased from 1500 mm to 900 mm in 150 s at room temperature. Determine the coefficient of permeability of soil in m/s at room temperature.

Solution

Given $D = 0.1$ m, $L = 0.15$ m, $d = 0.01$ m, $h_1 = 1.5$ m, $h_2 = 0.9$ m and $\Delta t = 150$ s

From Equation 4.19b

$$k = \frac{d^2 L}{D^2 \Delta t} \ln\left(\frac{h_1}{h_2}\right) = \frac{(0.01)^2 (0.15)}{(0.1)^2 (150)} \ln\left(\frac{1.5}{0.9}\right) = 5.108 \times 10^{-6} \text{ m/s}$$

4.5.   Field determination of permeability

The permeability values obtained from laboratory tests may not truly represent the field or in situ values. The major reasons are the difficulty in simulating the in situ soil structure and the small size of the test specimen. Additionally, the soil samples for the laboratory test are sampled vertically and the tests generally use the vertical flow, while permeability may be higher in the horizontal direction in several field conditions.

There are several in situ tests for determining the in situ permeability of soil; however, only the well-pumping test is described here. This test is the most suitable field test for determining the in situ permeability of homogeneous coarse-grained soil strata, which are both highly porous and permeable, called aquifers. In this test, water is pumped from a well, generally at least 30 cm in diameter, until steady-state conditions are reached. Unless the soil stratum is too thick, the well should penetrate to the full thickness of the soil stratum, as shown in Figure 4.5 in the case of an unconfined aquifer (highly porous and permeable soil stratum over an impermeable soil or rock stratum) and Figure 4.6 in the case of a confined aquifer (highly porous and permeable soil stratum sandwiched between two impermeable soil or rock strata). The pumping in the well generates a radial flow of water towards the well, resulting in lowering of the original groundwater table in the form of a cone of depression, which in the cross-section is called the drawdown curve. Water levels are observed in at least two small-diameter boreholes, called observation wells, spaced radially in different directions away from the pumping well.

In Figure 4.5 for the case of an unconfined aquifer, r_1 and r_2 are the radial distances of observation wells from the centre of the pumping well; h_1 and h_2 are the heights of water levels in respective observation wells above the bottom of the unconfined aquifer. At any radial distance, r, from the centre of pumping well, the area through which seepage takes place, $A = 2\pi r h$, where h is the height of drawdown curve at a radial distance, r, from the centre of the pumping well.

It is assumed that the hydraulic gradient at any distance, r, from the centre of the pumping well is constant with depth and is equal to the slope of the drawdown curve – that is, $i = dh/dr$, which is Dupuit's assumption, reasonably accurate except at points close to the pumping well. Using Darcy's law (Equation 4.7)

$$q = kiA = k\left(\frac{dh}{dr}\right)(2\pi rh)$$

or

$$q\frac{dr}{r} = 2\pi khdh$$

Integrating between the limits of r from r_1 to r_2 and the corresponding limits of h from h_1 to h_2

$$q\int_{r_1}^{r_2}\frac{dr}{r} = 2\pi k\int_{h_1}^{h_2} hdh$$

or

$$q\ln\left(\frac{r_2}{r_1}\right) = \pi k\left(h_2^2 - h_1^2\right)$$

or

$$k = \frac{q}{\pi}\frac{\ln\left(\dfrac{r_2}{r_1}\right)}{\left(h_2^2 - h_1^2\right)} = \frac{2.3q}{\pi}\frac{\log_{10}\left(\dfrac{r_2}{r_1}\right)}{\left(h_2^2 - h_1^2\right)} \qquad (4.20)$$

Figure 4.5 Well-pumping test in an unconfined aquifer

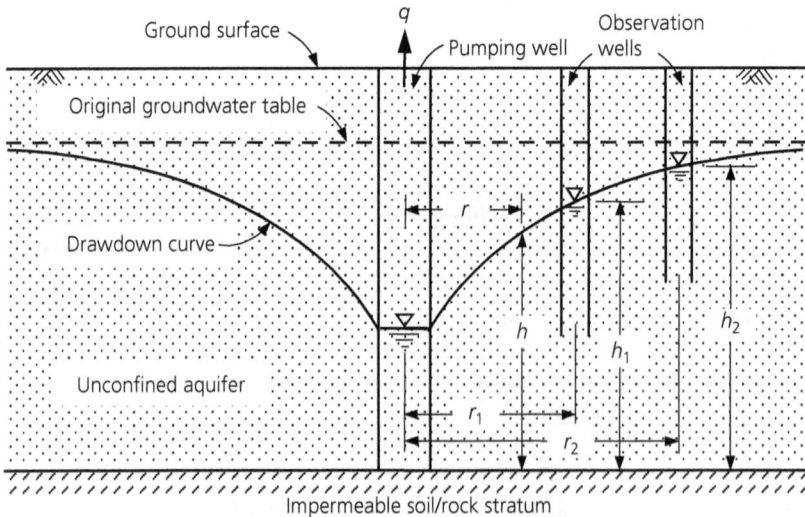

In Figure 4.6 for the case of a confined aquifer, the area through which seepage takes place, $A = 2\pi rH$, where H is the thickness of the confined soil stratum. Since the groundwater flow is confined between two impermeable strata and is under pressure greater than atmospheric pressure (similar to the groundwater table of an unconfined aquifer), the original water table (called original piezometric surface) and the drawdown curve will be above the confined stratum. Because the flow is everywhere horizontal, Dupuit's assumption applies without error (Todd and Mays, 2005). Using Darcy's law (Equation 4.7)

$$q = kiA = k\left(\frac{dh}{dr}\right)(2\pi rH)$$

or

$$q\frac{dr}{r} = 2\pi kHdh$$

Integrating between the limits of r from r_1 to r_2 and the corresponding limits of h from h_1 to h_2

$$q\int_{r_1}^{r_2}\frac{dr}{r} = 2\pi kH\int_{h_1}^{h_2}dh$$

or

$$q\ln\left(\frac{r_2}{r_1}\right) = 2\pi kH\left(h_2 - h_1\right)$$

or

$$k = \frac{q}{2\pi H}\frac{\ln\left(\dfrac{r_2}{r_1}\right)}{\left(h_2 - h_1\right)} = \frac{2.3q}{2\pi H}\frac{\log_{10}\left(\dfrac{r_2}{r_1}\right)}{\left(h_2 - h_1\right)} \qquad (4.21)$$

Figure 4.6 Well-pumping test in a confined aquifer

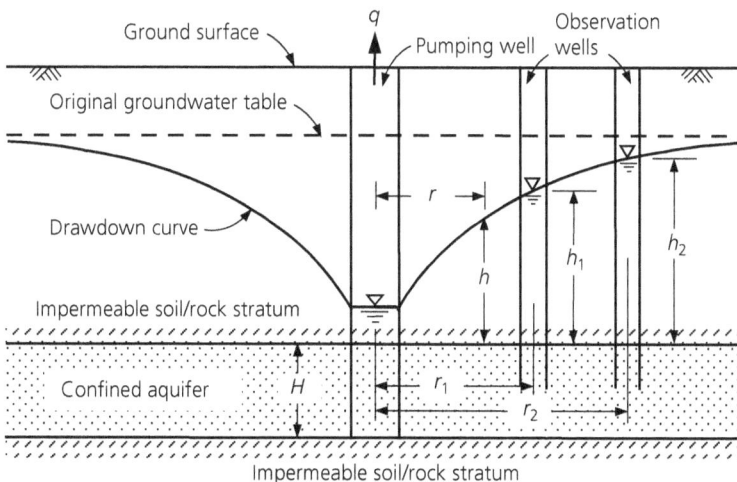

Example 4.5

A 12 m-thick sand stratum with a horizontal ground surface overlies an impermeable horizontal rock bed. A pumping well is driven to the bottom of the sand stratum, and two observation wells are made at distances of 20 m and 40 m from the centre of the pumping well. Water is pumped from the well at a constant rate of 0.11 m³/s. The water levels in the two observations wells are found to be at 2.3 m and 3.0 m below the ground surface. Determine the coefficient of permeability of the sand deposit.

Solution

Refer to Figure 4.5. Given $q = 0.11 \text{ m}^3/\text{s}$, $r_1 = 20$ m, $r_2 = 40$ m, $h_1 = 12 - 3 = 9$ m, and $h_2 = 12 - 2.3 = 9.7$ m. Using Equation 4.20, the coefficient of permeability of sand deposit

$$k = \frac{q}{\pi} \frac{\ln\left(\dfrac{r_2}{r_1}\right)}{\left(h_2^2 - h_1^2\right)} = \frac{0.11}{\pi} \frac{\ln\left(\dfrac{40}{20}\right)}{(9.7)^2 - (9)^2} = 1.85 \times 10^{-3} \text{ m/s}$$

4.6. Permeability of stratified soil

Figure 4.7 shows a vertical section through a stratified or layered soil mass of three homogeneous and isotropic strata with horizontal bedding planes. Note that stratification commonly occurs as a sedimentary soil deposit. The coefficients of permeability of the individual strata are k_1, k_2, k_3 and their respective thicknesses are H_1, H_2, H_3. This soil mass can be assumed to be equivalent to a single homogeneous and anisotropic stratum of thickness $H(= H_1 + H_2 + H_3)$ with an average/equivalent coefficient of permeability, k_{II}, parallel to the bedding planes (usually horizontal), and an average/equivalent coefficient of permeability, $k_\perp$, perpendicular to the bedding planes, as shown in Figure 4.7.

For water flow parallel to the bedding planes, the hydraulic gradient, i, remains the same for all the strata as well as for the equivalent single stratum. Therefore, the total discharge, q, through the equivalent single stratum is equal to the sum of the discharges q_1, q_2 and q_3 through the respective strata – that is,

$$q = q_1 + q_2 + q_3$$

Figure 4.7 A stratified/layered soil deposit and its equivalent single stratum

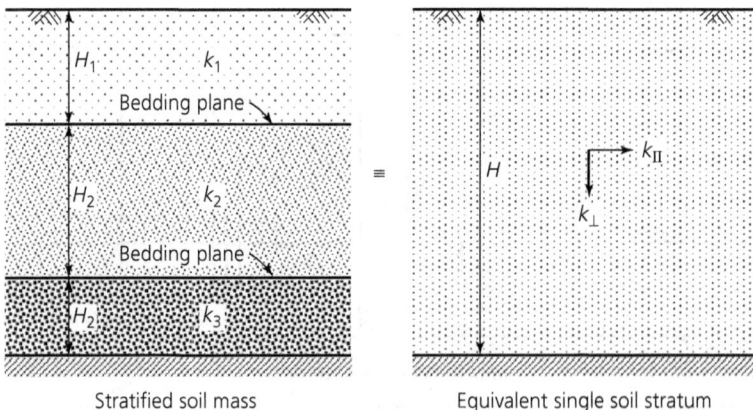

Stratified soil mass   Equivalent single soil stratum

or, using Darcy's law (Equation 4.7) with discharge per unit length

$$k_{II}iH = k_1iH_1 + k_2iH_2 + k_3iH_3$$

or

$$k_{II} = \frac{k_1H_1 + k_2H_2 + k_3H_3}{H} \tag{4.22}$$

For water flow perpendicular to the bedding planes, by the principle of continuity of flow, the discharge per unit area (q/A) – that is, the discharge velocity, v, through the equivalent single stratum – is equal to each individual discharge velocity v_1, v_2 and v_3 through the respective strata – that is,

$$v = v_1 = v_2 = v_3$$

or, using Darcy's law (Equation 4.6) with $A = 1$

$$v = k_\perp \frac{\Delta h}{H} = k_1 \frac{\Delta h_1}{H_1} = k_2 \frac{\Delta h_2}{H_2} = k_3 \frac{\Delta h_3}{H_3}$$

where Δh is the head loss through the equivalent single stratum; Δh_1, Δh_2 and Δh_3 are the head losses through individual strata.

Also

$$\Delta h = \Delta h_1 + \Delta h_2 + \Delta h_3$$

Combining these equations

$$\frac{vH}{k_\perp} = \frac{vH_1}{k_1} + \frac{vH_2}{k_2} + \frac{vH_3}{k_3}$$

or

$$k_\perp = \frac{H}{\dfrac{H_1}{k_1} + \dfrac{H_2}{k_2} + \dfrac{H_3}{k_3}} \tag{4.23}$$

Note that for a soil deposit that consists of n strata of thicknesses H_1, H_2,..., H_n with respective coefficients of permeability k_1, k_2,..., k_n, Equations 4.22 and 4.23 can be generalised as

$$k_{II} = \frac{\displaystyle\sum_{j=1}^{n} k_j H_j}{H} \tag{4.24}$$

and

$$k_\perp = \frac{H}{\displaystyle\sum_{j=1}^{n} \dfrac{H_j}{k_j}} \tag{4.25}$$

where $H = \displaystyle\sum_{j=1}^{n} H_j$

Example 4.6

A horizontally bedded sedimentary soil deposit consists of a clay stratum sandwiched between a silty sand stratum at the top and a coarse sand stratum at the bottom. All the strata have equal thickness. The coefficient of permeability of clay, silty sand and coarse sand are 1.5×10^{-9} m/s, 5.6×10^{-6} m/s and 2.4×10^{-3} m/s, respectively. Calculate the average coefficient of permeability in directions parallel and perpendicular to the bedding planes.

Solution

Refer to Figure 4.7. Given $H_1 = H_2 = H_3 = H$, $k_1 = 5.6 \times 10^{-6}$ m/s, $k_2 = 1.5 \times 10^{-9}$ m/s and $k_3 = 2.4 \times 10^{-3}$ m/s. From Equation 4.22, the average coefficient of permeability in a direction parallel to the bedding planes

$$k_{\mathrm{II}} = \frac{k_1 H_1 + k_2 H_2 + k_3 H_3}{H} = \frac{5.6 \times 10^{-6} H + 1.5 \times 10^{-9} H + 2.4 \times 10^{-3} H}{3H} = 8.0 \times 10^{-4} \text{ m/s}$$

From Equation 4.23, the average coefficient of permeability in a direction perpendicular to the bedding planes

$$k_{\perp} = \frac{3H}{\dfrac{H}{5.6 \times 10^{-6}} + \dfrac{H}{1.5 \times 10^{-9}} + \dfrac{H}{2.4 \times 10^{-3}}} = 4.5 \times 10^{-9} \text{ m/s}$$

4.7. Seepage force and its effect on effective stress

As explained earlier, the term seepage refers to the slow movement of water or any other fluid through soil or rock caused by the hydraulic gradient. When water flows or moves through a soil, it exerts a force, called the seepage force, on the soil particles through friction drag. In an isotropic soil, the seepage force always acts in the direction of fluid flow. The application of seepage force by the flowing fluid on the soil particles causes a gradual loss in the total head of the flowing fluid – that is, a transfer of energy takes place from the flowing fluid to the soil particles. If h is the loss in total head through frictional drag over a length, L, of the soil and A is the area of cross-section through which fluid flows, then the seepage force is

$$J = \gamma_w h A \tag{4.26}$$

or

$$J = \frac{h}{L} \gamma_w (AL) = i \gamma_w (AL)$$

or

$$j = i \gamma_w \tag{4.27}$$

where $j = J / (AL)$ is the seepage force per unit volume of soil, called the seepage pressure, which is a convenient expression for description of the seepage force. Note that the SI unit of j is N/m^3, which is different from the SI unit of pressure (N/m^2, i.e. pascal (Pa)).

Figure 4.8(a) shows a vessel containing a soil stratum submerged in water and connected to a water reservoir. Note that when the water level is the same in both the soil vessel and the reservoir, there is no flow of water through the soil. At the base X-X of the soil stratum total stress

$$\sigma_v = \gamma_w H + \gamma_{sat} L \qquad\qquad (4.28a)$$

pore water pressure

$$u = \gamma_w (H + L) \qquad\qquad (4.28b)$$

effective stress

$$\sigma_v' = \sigma_v - u = (\gamma_{sat} - \gamma_w)L = \gamma' L \qquad\qquad (4.28c)$$

where γ_w, γ_{sat} and γ' are unit weight of water, saturated unit weight of soil, and submerged unit weight of soil, respectively.

Figure 4.8(b) shows an equilibrium situation of the downward flow of water through soil caused by the lower level of water reservoir by a constant height, h, which is basically the head loss in flow across length, L, of the soil stratum. At the base X-X of the soil stratum

total vertical stress

$$\sigma_v = \gamma_w H + \gamma_{sat} L \qquad\qquad (4.29a)$$

pore water pressure

$$u = \gamma_w (H + L - h) \qquad\qquad (4.29b)$$

effective vertical stress

$$\sigma_v' = \sigma_v - u = (\gamma_{sat} - \gamma_w)L + \gamma_w h = \gamma' L + \gamma_w h = \gamma' L + i\gamma_w L = \gamma' L + jL \qquad\qquad (4.29c)$$

From Equation 4.29c, the downward flow of water through a soil causes an increase in effective stress by $\gamma_w h (= jL)$.

Figure 4.8(c) shows an equilibrium situation of the upward flow of water through the soil caused by the higher level of water reservoir by a constant height, h, which is basically the head loss in flow across length, L, of the soil stratum.

At the base X-X of the soil stratum

total vertical stress

$$\sigma_v = \gamma_w H + \gamma_{sat} L \qquad\qquad (4.30a)$$

pore water pressure

$$u = \gamma_w (H + L + h) \qquad\qquad (4.30b)$$

effective vertical stress

$$\sigma_v' = \sigma_v - u = (\gamma_{sat} - \gamma_w)L - \gamma_w h = \gamma' L - \gamma_w h = \gamma' L - i\gamma_w L = \gamma' L - jL \qquad\qquad (4.30c)$$

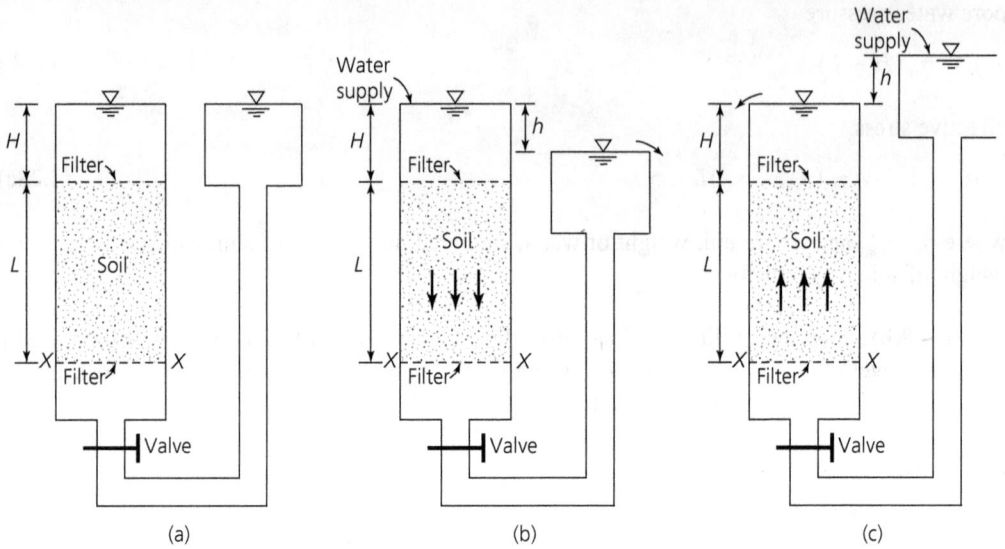

Figure 4.8 Different equilibrium flow conditions within a soil stratum: (a) no flow; (b) downward flow; (c) upward flow

From Equation 4.30c, the upward flow of water through a soil causes a decrease in effective stress by $\gamma_w h (= jL)$.

As an upward flow of water through a soil mass causes a decrease in effective stress, there is a possibility of zero effective stress within a soil mass. For the effective vertical stress, $\sigma_v' = 0$, the hydraulic gradient, $i = i_c$, where i_c is the critical hydraulic gradient. Therefore, for $\sigma_v' = 0$, from Equation 4.30c

$$\gamma' L - i_c \gamma_w L = 0 \tag{4.31}$$

or

$$i_c = \frac{\gamma'}{\gamma_w} \tag{4.32}$$

From Equation 1.24

$$\gamma' = \left(\frac{G-1}{1+e}\right)\gamma_w$$

Therefore,

$$i_c = \left(\frac{G-1}{1+e}\right) \tag{4.33}$$

where G is the specific gravity of soil particles and e is the void ratio of soil.

The following points regarding seepage force are noteworthy.

- Typically, the saturated soil unit weight, $\gamma_{sat} \approx 20 \text{ kN/m}^3$, unit weight of water, $\gamma_w \approx 10 \text{ kN/m}^3$, and the submerged unit weight, $\gamma' = \gamma_{sat} - \gamma_w \approx 10 \text{ kN/m}^3$; therefore, the critical hydraulic gradient, $i_c \approx 1$.
- The shear strength of a cohesionless soil is directly proportional to the effective stress (see Chapter 6). Hence, when the critical hydraulic gradient is reached, resulting in $\sigma'_v = 0$, the shear strength of the cohesionless soil becomes zero, and the soil behaves like a dense fluid in which a person can easily float because the unit weight of the dense fluid is about twice that of water. This hydraulic condition where the shear strength or resistance of a soil is zero due to the zero effective stress is called the quick condition or quicksand. In nature, the quick condition is usually observed in coarse silt and fine sand deposits under artesian flow conditions. Cohesive soils generally do not become quick because they can have some shear strength caused by cohesion even at zero effective stress. Note that quicksand is not a type of soil, but simply a hydraulic condition.
- Equation 4.31 can be expressed as $\gamma'LA = i_c\gamma_w LA$, which means that under the quick condition of upward water flow, the submerged weight of unloaded soil is equal to the seepage force or, in other words, the total weight $(\gamma_{sat}LA)$ of unloaded saturated soil is equal to the total upward water force (seepage force + buoyant force $= i_c\gamma_w LA + \gamma_w LA$).
- If the critical hydraulic gradient is exceeded, the soil surface appears to boil or liquefy because the soil particles move with the water flow, and this may create a hole or pipe in the soil. This phenomenon is called piping (or internal erosion), which is a serious problem for the stability of water-retaining structures such as an earth or rock-fill dam, a concrete dam or weir, and a sheet pile wall. The factor of safety against the piping (FS_{piping}) is defined as the ratio of the critical hydraulic gradient, i_c, to the maximum hydraulic gradient, i, caused by the loss in total head near the soil surface or discharge boundary surface. The maximum value of i near the ground or discharge boundary surface on the downstream side of a water-retaining structure is generally called the exit hydraulic gradient, i_e. Thus,

$$FS_{piping} = i_c / i_e \tag{4.34}$$

For the stability of water-retaining structures, FS_{piping} should be greater than unity.

- A shock, defined as a load of very short duration, applied by earthquake, blasting, pile driving or traffic movement to a certain loose, saturated cohesionless soil, generally coarse silt or fine sand, may also result in zero effective stress within the soil mass. This occurs because the shock causes an increase in density of the soil due to its initially loose condition, which in turn raises the pore water pressure relative to the total stress. As a result, the soil behaves like a dense fluid, which has zero shear strength. This phenomenon is called liquefaction, which has been the major cause of many slope failures and foundation settlements during past earthquakes. A brief description of soil liquefaction is provided in Section 9.6 of Chapter 9.

Example 4.7

A sand deposit at a construction site has a void ratio of 0.75. The specific gravity of sand particles is 2.65. The ground water table exists at a depth of 1.5 m below the ground surface. A long trench is excavated up to a depth of 4.2 m below the ground surface, and the sides of the trench are supported by the sheet pile walls driven up to 1.8 m below the bottom of the trench.

The water accumulated in the trench is pumped out continuously in order to have a dry working area. Check whether a quick condition can occur. If so, what specific remedial measures would you recommend to avoid the quick condition and mitigate the risk of piping?

Solution

From Equation 4.33, the critical hydraulic gradient

$$i_c = \frac{G-1}{1+e} = \frac{2.65-1}{1+0.75} = 0.94$$

Figure E4.7 shows the given site details as described in the question. The water will flow through the 1.8 m-thick soil within the trench under a hydraulic head, $\Delta h = 4.2 - 1.5 = 2.7$ m. Therefore, the hydraulic gradient, which causes the flow, can be calculated using Equation 4.5 as

$$i = \frac{\Delta h}{L} = \frac{2.7}{1.8} = 1.5 > i_c$$

Figure E4.7

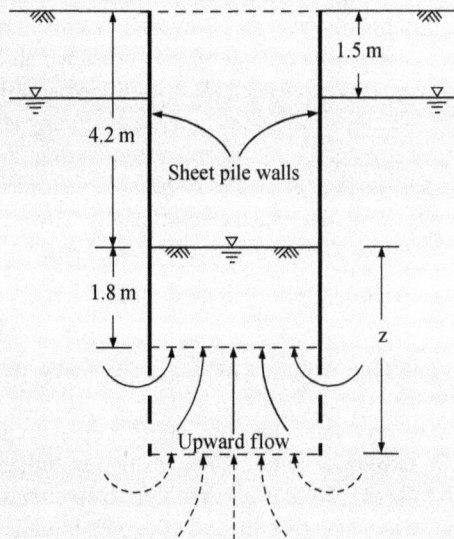

Since $i > i_c$, the quick condition will occur. One of the following two solutions can be adopted to avoid the quick condition and mitigate the risk of piping.

Solution (a)

Reduce Δh significantly by lowering the water table by a suitable dewatering method so that $i < i_c$. Hausmann (1990) provides detailed information on several dewatering methods, while some basic details are also presented in Shukla (2025).

Solution (b)

Increase the depth of embedment of the sheet pile wall below the bottom of the trench signifi-cantly. If z is the required depth of embedment to achieve a factor of safety of say 1.3 against piping, then

Hydraulic gradient (Equation 4.5)

$$i = \frac{\Delta h}{L} = \frac{2.7}{z}$$

and

Factor of safety against the piping (Equation 4.34)

$$\frac{i_c}{i} = 1.3$$

These two relationships give

$$\frac{0.94}{\frac{2.7}{z}} = 1.3$$

or

$$z = \frac{1.3 \times 2.7}{0.94} = 3.73 \text{ m}$$

4.8. Basic flow equation

The previous sections presented the concepts of one-dimensional (1D) flow of fluid/water through soil. Flows around sheet pile walls, through foundation of dams and through earth dams are generally two-dimensional (2D). Flow near the abutments of an earth dam is an example of three-dimensional (3D) flow. Although 3D flow is the most general flow situation, its analysis is too complex to be practical, hence most flow problems are solved on the assumption that the flow is 2D, which is reasonably correct, and is introduced here.

The basic equation of fluid flow through soil is based on the principle of conservation of mass, which forms the principle of continuity of flow. Figure 4.9 shows an element of soil through which water is flowing under laminar flow conditions in the x–z plane at a flow rate of q (water volume per unit time) with components q_x and q_z in x (horizontal) and z (vertical) directions, respectively. Note that the component of flow rate in the y direction, q_y, is zero for the 2D flow condition. Using Darcy's law (Equation 4.7), flow into the bottom of the soil element

$$q_{zin} = k_z \left(-\frac{\partial h}{\partial z} \right) dxdy$$

and flow out of the top of soil element

$$q_{zout} = k_z \left(-\frac{\partial h}{\partial z} - \frac{\partial^2 h}{\partial z^2} dz \right) dxdy$$

where k_z is the coefficient of permeability of soil in z direction at point (x, y, z); h is the total head causing the flow, which is a function of x and z. Note that k_z is assumed to be constant in the z direction, considering the soil is homogeneous.

The net flow into the soil element from the vertical flow in z direction

$$\Delta q_z = q_{zin} - q_{zout} = k_z \left(-\frac{\partial h}{\partial z} \right) dx\, dy - k_z \left(-\frac{\partial h}{\partial z} - \frac{\partial^2 h}{\partial z^2} dz \right) dx\, dy$$

or

$$\Delta q_z = k_z \frac{\partial^2 h}{\partial z^2} dx\, dy\, dz$$

In a similar way, the net flow into the soil element can be derived from the horizontal flow in x direction as

$$\Delta q_x = k_x \frac{\partial^2 h}{\partial x^2} dx\, dy\, dz$$

Thus, for 2D flow in the x–z plane, the net flow into the soil element is

$$\Delta q = \Delta q_x + \Delta q_z = \left(k_x \frac{\partial^2 h}{\partial x^2} + k_z \frac{\partial^2 h}{\partial z^2} \right) dx\, dy\, dz \tag{4.35a}$$

If S is the degree of saturation and n is the porosity of soil, the volume of water in the soil element is

$$V_w = S n\, dx\, dy\, dz$$

or, using Equation 1.16a, where $n = e/(1+e)$

$$V_w = \frac{Se}{1+e} dx\, dy\, dz$$

Figure 4.9 Two-dimensional flow of water through a soil element in the x–z plane

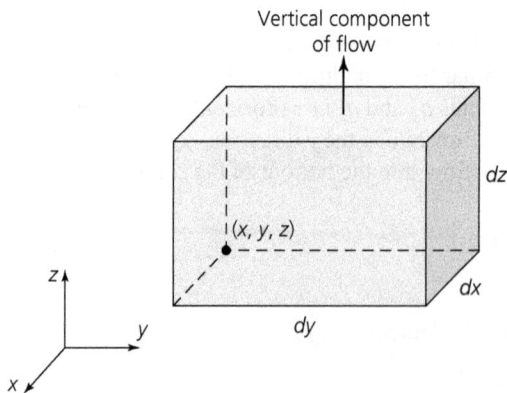

The net flow into the soil element, which is basically the rate of change of the water volume in the soil element, can be given as

$$\Delta q = \frac{\partial V_w}{\partial t} = \frac{\partial}{\partial t}\left(\frac{Se}{1+e}\,dx\,dy\,dz\right)$$

As $dx\,dy\,dz\,/\,(1+e)$ is the volume of soil solids in the element and is a constant

$$\Delta q = \frac{\partial V_w}{\partial t} = \frac{dx\,dy\,dz}{1+e}\frac{\partial(Se)}{\partial t} \tag{4.35b}$$

From Equations 4.35a and 4.35b

$$\left(k_x\frac{\partial^2 h}{\partial x^2} + k_z\frac{\partial^2 h}{\partial z^2}\right)dx\,dy\,dz = \frac{dx\,dy\,dz}{1+e}\frac{\partial(Se)}{\partial t}$$

or

$$k_x\frac{\partial^2 h}{\partial x^2} + k_z\frac{\partial^2 h}{\partial z^2} = \frac{1}{1+e}\left(e\frac{\partial S}{\partial t} + S\frac{\partial e}{\partial t}\right) \tag{4.36}$$

For steady flow (velocity, pressure, density etc. at any point within the flow domain do not change with time), e and S are both constant; therefore, Equation 4.36 reduces to

$$k_x\frac{\partial^2 h}{\partial x^2} + k_z\frac{\partial^2 h}{\partial z^2} = 0 \tag{4.37}$$

Depending on the variation in e and S, there can be three other types of unsteady flow (e varies and is S constant; e is constant and S varies; e and S both vary) with corresponding flow equations. If e decreases, the flow condition or phenomenon is described as consolidation; if e increases, flow condition causes an expansion of soil. Chapter 5 explains consolidation and expansion of a soil mass in detail. The other two unsteady flow types are complex, which may be considered as research problems.

For the isotropic condition of permeability, $k_x = k_z$; hence, Equation 4.37 becomes

$$\frac{\partial^2 h}{\partial x^2} + \frac{\partial^2 h}{\partial z^2} = 0 \tag{4.38}$$

Equation 4.38 is a partial differential equation known as Laplace's equation, which also describes the flow of heat through a thermal medium as well as the flow of electric charge through a conducting medium with replacement of variable, h, by relevant parameters. Note that this equation is valid only for 2D steady-state flow in homogeneous, isotropic soils that follow Darcy's law of fluid flow.

The solution of Laplace's equation (Equation 4.38) for a given set of boundary conditions describes how the total head, h, varies with x and z directions – that is, how the total head causing flow is dissipated in the x–z flow space. With this information, along with the value of coefficient of permeability, the quantity of water flowing through the soil mass within known boundaries can be determined. The general solution of Laplace's equation in an analytical closed form is too complicated for engineering use (Harr, 1962). Its graphical solution, despite being approximate with specific boundary conditions, which results in a flow net as described in the following section, is relatively easily presented. The graphical solution is also more useful for most flow problems such as water flow through the foundation of a water-retaining structure (sheet-pile wall or dam), called the confined flow, and flow through an earth dam embankment, called the unconfined flow. In a confined flow, all boundaries of the flow domain are completely defined, whereas in an unconfined flow, at least one boundary is not clearly defined unless the flow net is drawn.

4.9. Concept of flow nets and their applications

The graphical solution of Laplace's equation is presented by carefully sketching a flow net that consists of a system of two sets of orthogonal curves, called the flow lines and the equipotential lines, on a cross-section of the flow problem. A flow line is the path within the soil along which water flows; this means that each drop of water that moves through the soil follows a flow line. An equipotential line is a line drawn through points of equal total head (not equal pressure head). The space between two adjacent flow lines is called the flow path or flow channel.

The guidelines for the construction/drawing of flow nets are presented here. To construct a flow net, a cross-section of the site and the structure is drawn to a scale, and the boundary conditions are observed. Being orthogonal curves, flow lines and equipotential lines are drawn crossing each other at right angles. The component of flow normal to a flow line is zero. Two flow lines can never meet or intersect because in that case the water flowing in the flow path between them disappears, which is physically not correct. Two equipotential lines also do not meet or cross each other because in that case there will be two total heads at the same point, which is not possible physically. The impermeable boundary is taken as a flow line, with the permeable boundary as an equipotential line. As water flows, its total head is dissipated by friction drag with soil particles, and equipotential lines in a flow net act as contours to show how the total head is lost. All flow paths must carry equal quantity of flow, Δq. The difference in total head, Δh, between adjacent equipotential lines is kept constant throughout the flow net. The area formed by intersecting flow lines and equipotential lines, called the flow element, must be a square or close to a square; this means that a circle can be placed inside the enclosed area touching all four sides. The datum to which the total head refers may be any level but, in most flow problems, it may be convenient to select the downstream water level as the datum.

Figure 4.10 shows 1D flow in a constant head permeameter, similar to that shown in Figure 4.3, with the flow net constructed following the guidelines described above. All flow lines and equipotential lines including the boundary ones are clearly indicated. Here, h is the loss in total head in flow – that is, the difference in total head between the top and bottom boundary equipotential lines. The rate of flow, q, can be determined using Darcy's law (Equation 4.6), but can also be determined using the flow net. In the flow net shown in Figure 4.10, the rate of flow per unit length perpendicular to the page through the shaded area with length, l, and breadth, b, is

$$\Delta q = k\left(\frac{\Delta h}{l}\right)(b \times 1) = k\Delta h\left(\frac{b}{l}\right)$$

where Δh is the head loss over length, l, of the soil.

Figure 4.10 One-dimensional confined flow of water through a soil in the constant head permeameter and the corresponding flow net

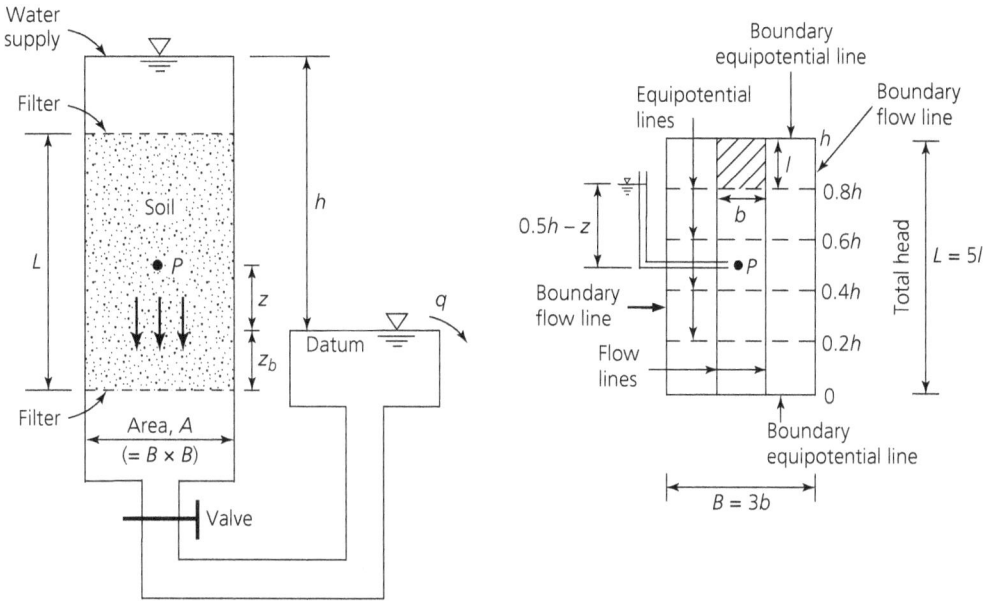

Figure 4.10 One-dimensional confined flow of water through a soil in the constant head permeameter and the corresponding flow net

If n_d is the number of head drops in the flow net, each representing the same total head loss Δh, then

$$\Delta h = \frac{h}{n_d}$$

and Δq becomes

$$\Delta q = k \left(\frac{h}{n_d} \right) \left(\frac{b}{l} \right)$$

If n_f is the number of flow channels, each carrying the same flow, Δq, the total rate of flow per unit length perpendicular to the page through the entire soil is

$$q = n_f \Delta q = kh \left(\frac{n_f}{n_d} \right) \left(\frac{b}{l} \right)$$

Since the shapes of the flow elements are square, $b = l$

$$q = kh \left(\frac{n_f}{n_d} \right) = khS_f \tag{4.39}$$

where

$$S_f = \frac{n_f}{n_d} \tag{4.40}$$

113

is called the shape factor of the flow net, which is independent of the coefficient of permeability, k, and total head loss, h. Note that n_f and n_d may not be integers if areas enclosed by flow and equipotential lines differ significantly from the square shape – for example, if the ratio $b/l = 0.5$ in a specific flow channel, it should be counted as 0.5, rather than as 1.

Equation 4.39, although developed giving a simple example of a flow net, is also applicable to most complex flow nets. If the permeability, k, and total head loss, h, for a flow problem are known, the rate of flow, q, can be determined using Equation 4.39.

As indicated in the flow net shown in Figure 4.10, when the datum is at the lower water reservoir level, the total head at the top boundary equipotential line is h, while the total head at the bottom boundary equipotential line is zero. From the flow net, with the total head and elevation head at any point, the pressure head can be determined, which is the height above the point to which water will rise in a piezometer installed at the point – for example, consider a point, P, in the soil at a height, z, above the datum where the total head is $0.5h$. Since elevation head is z, the pressure head is therefore $0.5h - z$. Similarly, the pressure head at any other point can be calculated, say a point at the bottom of soil where the total head is zero and the elevation head is $-z_b$, resulting in a pressure head equal to $0 - (-z_b)$ – that is, z_b.

The flow net can also be used to determine the hydraulic gradient i ($= \Delta h / l$) at any point in the soil, using the head loss, Δh, over the potential drop length, l, of the soil at the point. In Figure 4.10, all the squares are of the same size; therefore, the hydraulic gradient remains constant within the soil.

Flow nets for 2D flows can be constructed and the rate of flow through the soil can be determined; the heads and the hydraulic gradient can then be determined from the flow net, as explained in the example below.

Example 4.8

A sheet pile wall is driven to a depth of 5 m into a permeable soil which extends to a depth of 10.3 m below the ground level where there is an impermeable soil/rock stratum. The depths of water over the ground level are 8.6 m and 1.1 m on the two sides of the wall. Draw a neat sketch of the flow net and determine the following

(a) loss in total head
(b) pore water pressure at the lower end of the sheet pile
(c) rate of flow – that is, quantity of seepage through the soil under the sheet pile wall – taking the coefficient of permeability of soil, $k = 2.3 \times 10^{-7}$ m/s
(d) exit gradient – that is, the maximum hydraulic gradient near the ground surface on the downstream side.

Solution

The flow net for the flow under the sheet pile wall is drawn by following the guidelines for the flow net construction and is illustrated in Figure E4.8. Here, the boundary conditions are AB, upstream equipotential line; DE, downstream equipotential line; BCD, boundary flow line; and FG, boundary flow line. The number of flow channels, $n_f = 3.5$ because the flow elements in the lowermost flow channel resemble rectangles with a scaled width-to-length ratio of approximately 0.5, and the number of potential drops, $n_d = 8$.

(a) Loss in total head, $h = h_1 - h_2 = 8.6 - 1.7 = 6.9$ m.
(b) At the lower end, C, of the sheet pile wall, $h - 4\Delta h = 6.9 - 4 \times (6.9/8) = 3.45$ m, and elevation head $= -(h_2 + d) = -(1.7 + 5) = -6.7$ m. Therefore, pressure head $= 3.45 - (-6.7) = 10.15$ m, which means that water will rise in a piezometer, if inserted at C, to a height of 10.15 m, and the pore water pressure $= 10.15 \times \gamma_w = 10.15 \times 9.81 = 99.57$ kPa.
(c) From Equation 4.39, the rate of flow – that is, quantity of seepage

$$q = kh\left(\frac{n_f}{n_d}\right) = (2.3 \times 10^{-7})(6.9)\left(\frac{3.5}{8}\right) = 6.94 \times 10^{-7} \text{ m}^3 \text{ /s /m}$$

(d) Scaled average length of flow in the top element adjacent to the sheet pile wall on the downstream side, $l \approx 2$ m.

$$\text{Exitgradient, } i = \frac{\Delta h}{l} = \frac{6.9/8}{2} = 0.431$$

Figure E4.8 Flow net for the confined flow under a sheet file wall

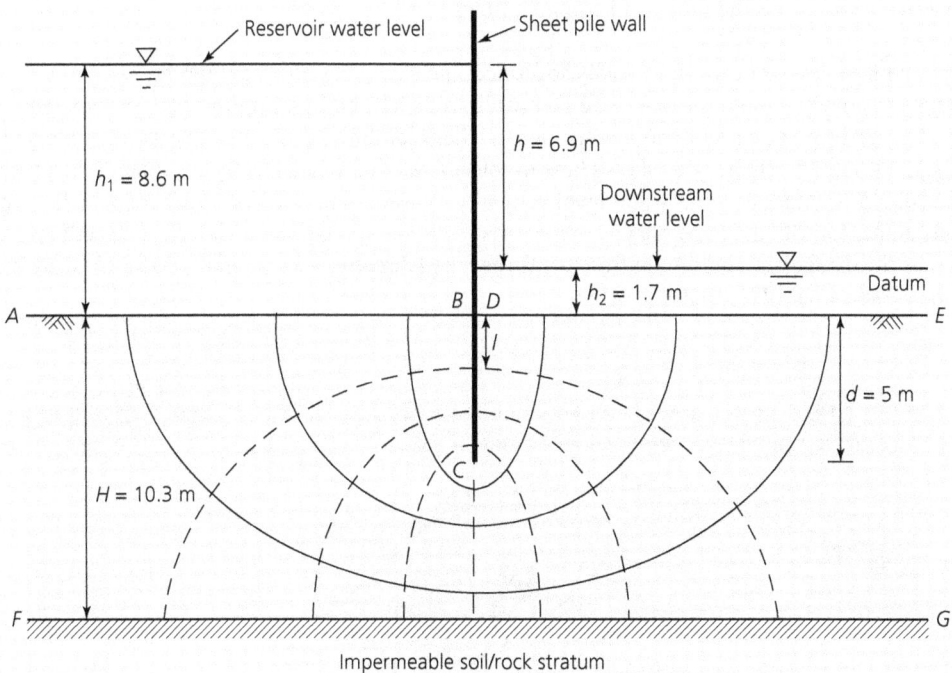

In Figure E4.8, note that a prism of soil in front of the sheet pile on the downstream side of width equal to about half of the embedment depth ($d/2$) and depth equal to the embedment depth (d) of the sheet pile itself, is most susceptible to piping due to heave. In the flow net, the flow lines in this region are closely spaced and emerge vertically upward (a dangerous flow situation). The factor of safety against heave piping ($FS_{\text{heave piping}}$) is determined as a ratio of the submerged or effective weight (W') of the soil prism to the seepage force (J) acting upward at the base of the prism (Terzaghi et al., 1996). Thus,

$$FS_{\text{heave piping}} = \frac{W'}{J} \tag{4.41}$$

The flow nets discussed so far are for confined flows through soils in the permeameter and under the sheet pile wall where all boundaries of the flow domain are completely defined. Figure 4.11 shows the flow net for steady-state seepage through a homogeneous earth dam with a horizontal blanket drain resting on an impermeable or impervious foundation. The three boundary conditions: AB, the upstream equipotential line (total head $= H$ at every point on AB with datum at the downstream water level); BF, the boundary flow line; and FC, the boundary equipotential line (total head $= 0$ at every point of FC) are well defined. The line AC is the fourth boundary as a boundary flow line with the zero-pressure head – that is, atmospheric pressure at its all points; thus AC is a phreatic line. This characteristic feature of AC makes the difference in total head between two equipotential lines equal to the change in elevation between the points where these equipotential lines intersect AC. In other words, AC is a flow line along which the total head is equal to the elevation head. Thus, the location of AC as the fourth boundary, which is not known until the flow net is constructed, describes the flow as unconfined flow.

Since the flow line AC cannot be easily identified, the sketching of a flow net for unconfined flow requires relatively more effort and time compared to the sketching of a flow net for confined flow. This task can be made easier by following the guidelines for drawing the flow line AC as explained by Casagrande (1937). Figure 4.12 shows how to select a relatively satisfactory location of the phreatic line. Point E is at the reservoir water level vertically above B, and $AD = 0.3b$ where $b = AE$. A part of AC coincides with the parabola CD, which has focus and directrix as F and IG, respectively. Any point on the parabola is equidistant from the focus F and directrix IG; therefore, if $P(x, y)$ is any point on the parabola with $F(0, 0)$ as the origin

Figure 4.11 Flow net for unconfined flow through a homogeneous earth dam with a horizontal blanket drain resting on an impermeable stratum

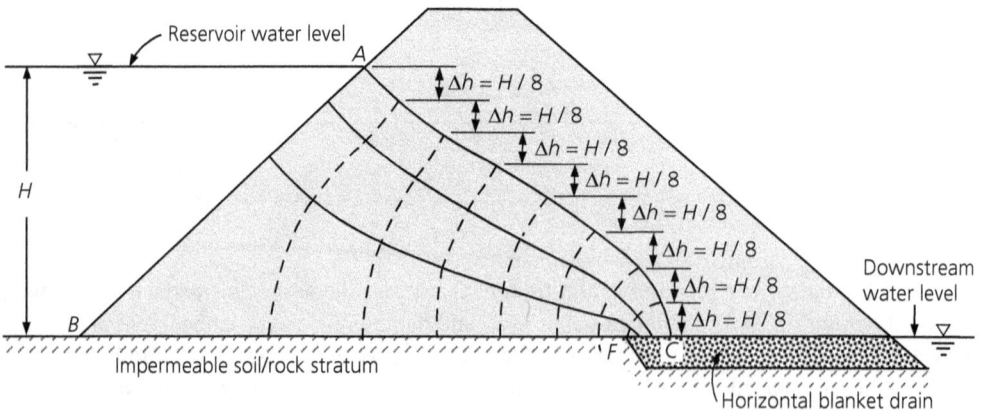

$$\sqrt{x^2 + y^2} = x + s \qquad (4.42a)$$

or

$$y^2 = 2sx + s^2 \qquad (4.42b)$$

where $s = FG$, called the focal distance.

Equation 4.42b is the equation of the parabola. Since the coordinates of point D on the parabola are (d, H), Equation 4.42a reduces to

$$s = \sqrt{d^2 + H^2} - d \qquad (4.43)$$

Equation 4.43 also comes directly from $DF = DI$.

A simple expression can be derived for the seepage through the dam per unit length perpendicular to the page in terms of s, considering the vertical section PQ with application of Darcy's law as

$$q = kiA = k\left(\frac{dy}{dx}\right)(y \times 1) = k\left(\frac{s}{\sqrt{2sx + s^2}}\right)\left(\sqrt{2sx + s^2}\right) = ks \qquad (4.44)$$

Note that a horizontal blanket drain or any other similar drainage system lowers the position of the top flow line; otherwise, the top flow line would exit at the downstream slope of the dam and, therefore, the flowing water may carry soil particles, gradually leading to the failure of the dam.

In the case of an anisotropic soil with $k_x \neq k_z$, to which Equation 4.37 is applicable, it can be rewritten as

$$\frac{\partial^2 h}{\partial x_T^2} + \frac{\partial^2 h}{\partial z^2} = 0 \qquad (4.45)$$

Figure 4.12 Casagrande's method of determining the phreatic line for unconfined flow through a homogeneous earth dam with a horizontal blanket drain resting on an impermeable stratum (Casagrande, 1937)

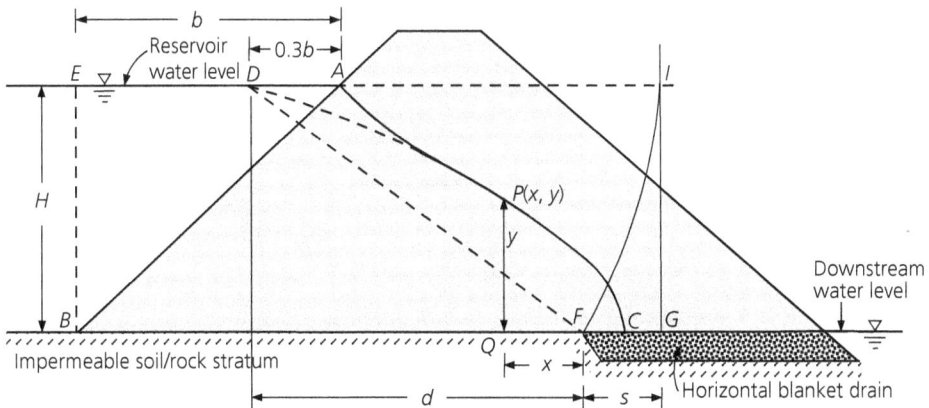

where

$$x_T = \left(\sqrt{\frac{k_z}{k_x}} \right) x \qquad (4.46)$$

Therefore, a flow net can be constructed for the anisotropic subsoil profile on the cross-section drawn to a transformed scale using all x dimensions according to Equation 4.46, following the same method used for the isotropic subsoil profile, as Equation 4.45 represents Laplace's equation. However, the permeability to be used with the transformed section is (Lambe and Whitman, 1979)

$$k_e = \sqrt{k_x k_z} \qquad (4.47)$$

where k_e is the effective (or average) coefficient of permeability in both x and z directions.

If the flow occurs through the boundary of two soils with different permeability values, both the direction of flow and the geometry of flow elements in the flow net changes at the boundary. Figure 4.13 illustrates how a flow path deviates from its initial direction and shape as it passes through the boundary of soils 1 and 2 with coefficients of permeability k_1 and k_2, respectively. With $k_1 > k_2$, the square shape of the flow element enclosed by flow and equipotential lines becomes rectangular because of the flow path being wider ($b_2 > b_1$) in soil 2. From the principle of continuity of flow, the rates of flow in soil 1 and soil 2 remain the same; therefore,

$$q_1 = q_2$$

or, using Darcy's law,

$$k_1 \frac{\Delta h}{l_1} b_1 = k_2 \frac{\Delta h}{l_2} b_2$$

or

$$\frac{k_1}{k_2} = \frac{\dfrac{l_1}{b_1}}{\dfrac{l_2}{b_2}} = \frac{\tan \alpha_1}{\tan \alpha_2} \qquad (4.48)$$

Figure 4.13 Flow between soils with different coefficients of permeability

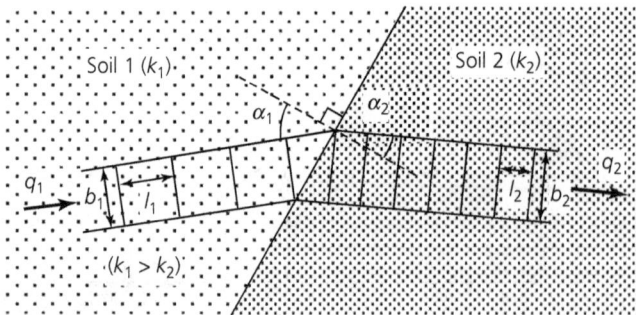

The following points regarding the flow net are noteworthy.

- For any flow problem, there can be an infinite number of flow lines available, but only a few (3 to 5) are sufficient for engineering solutions to flow problems. Considering many flow lines can complicate the result.
- The graphical method for drawing or sketching the flow nets has been discussed. Other methods such as the analytical method and the electrical analogy method are available, but the graphical method is the quickest and the most widely used. The skill of drawing flow nets graphically can be developed by practice; commercial software can also be used.
- A flow net depends solely on the boundary conditions and changes only when the dimensions of the flow region are modified. There is no change with any variation in coefficient of permeability, k, of soil and/or total head loss, h, that causes the flow through the soil. In other words, for a given set of boundary conditions, there is only one correct flow net, resulting in a unique solution to the flow problem.

4.10. Filters and their design requirements

A filter allows for an adequate flow of fluid across its plane while preventing the migration of soil particles along with fluid flow during the projected service period of its application under consideration. Filters are used extensively to prevent internal erosion by flowing water and to control seepage in geotechnical structures such as earth dams. A soil layer used as the filter is commonly called a granular filter. For satisfactory filtration function, a granular filter may be constructed as a series of soil layers of different particle size distribution. In this case, it is called a graded granular filter.

Geosynthetics, especially nonwoven geotextiles, which are planar, permeable, polymeric products in the form of flexible sheets, are also used as filters in dams and other geotechnical applications where flow of water through soil can cause erosion problems. Geosynthetic filters have been described in detail by Shukla (2016, 2022).

If a proper filter is used, sufficient head is lost in flow through the filter, and no significant amount of soil being protected moves through the filter. The selection of the dimensions and material of a filter depends on the type of soil and its particle size distribution, and also on the type of flow pattern in the structure under consideration. Based on the experiment, Terzaghi (Terzaghi, 1922; Terzaghi *et al.*, 1996) provided the following criteria.

To ensure that the soil particles do not migrate through the filter (retention criterion)

$$\frac{D_{15\,\text{filter}}}{D_{85\,\text{soil}}} < (4 \text{ to } 5) \tag{4.49}$$

and to ensure a high permeability of the filter (permeability criterion)

$$\frac{D_{15\,\text{filter}}}{D_{15\,\text{soil}}} > (4 \text{ to } 5) \tag{4.50}$$

where $D_{15\,\text{filter}}$ is the sieve size through which 15% of filter material passes; $D_{15\,\text{soil}}$ is the sieve size through which 15% of soil to be protected passes; $D_{85\,\text{soil}}$ is the sieve size through which 85% of soil to be protected passes. For the satisfactory performance of a filter, it is also important that the particle size distribution curves of the filter material and the soil to be protected should be almost similar in general shape.

Example 4.9

For a fine-grained soil, $D_{15\,soil} = 0.005$ mm and $D_{85\,soil} = 0.03$ mm. Is it possible to protect this soil against erosion caused by flow of water using a soil with $D_{15\,filter} = 0.08$ mm as a filter material?

Solution

From Equation 4.49

$$\frac{D_{15\,filter}}{D_{85\,soil}} = \frac{0.08}{0.03} = 2.67 < (4 \text{ to } 5) \text{ is true.}$$

From Equation 4.50

$$\frac{D_{15\,filter}}{D_{15\,soil}} = \frac{0.08}{0.005} = 16 > (4 \text{ to } 5) \text{ is also true.}$$

Thus, the soil with $D_{15\,filter} = 0.08$ mm meets both the retention and the permeability criteria. Therefore, it can be a suitable filter material provided it has a similar general shape of the particle size distribution curve to that of the soil to be protected.

Chapter summary

1. The direction of flow through soil is determined by the difference in total head, rather than elevation head or pressure head alone. In other words, if the total heads at two points A and B are equal, there can be no flow between them, even if $z_A \neq z_B$ and $u_A \neq u_B$.
2. Permeability is one of the important properties of soil. It indicates the ability of a soil to conduct fluid and depends on the characteristics of both the permeant and the soil. The range of permeability is extremely large, ranging from above 10^{-2} m/s for gravel to below 10^{-9} m/s for clay.
3. Darcy's law ($v = ki$) is applicable to most soils, especially those finer than coarse sands.
4. The seepage force per unit volume of soil, called seepage pressure, is $i\gamma_w$. In isotropic soil, it acts in the direction of flow. In the upward flow through a soil, the seepage force causes a decrease in effective stress at any point within the soil.
5. Quicksand refers to a hydraulic condition wherein a cohesionless soil loses its strength when the upward flow of water takes place through it. This happens because the effective stress reduces to zero.
6. Laplace's equation is valid for 2D steady-state flow in homogeneous, isotropic soils that follow Darcy's law of fluid flow. It states that for flow in the x (horizontal)–z (vertical) plane, the rate of change of hydraulic gradient in the x direction plus the rate of change of hydraulic gradient in the z direction is zero.
7. A flow net, which is a graphical solution of Laplace's equation, is a system of two sets of orthogonal curves, called the flow lines and the equipotential lines. It can be used to determine the rate of flow, heads and hydraulic gradient at any point within the flow domain for which the flow net is constructed. The rate of flow, $q = khS_f$, where S_f is the shape factor, which is obtained from the flow net.
8. Filters are used extensively to prevent internal erosion by flowing water and to control seepage in geotechnical structures such as earth dams.

Questions for practice

Select the most appropriate answer for each of the multiple choice questions from Q4.1 to Q4.10. Questions Q4.11 to Q4.28 require more detailed answers.

4.1 The flow of water through soil between any two points is governed only by the difference in
(a) elevation head
(b) pressure head
(c) total head
(d) velocity head.

4.2 The approach velocity of water through a soil is 1 m/min. If the void ratio of the soil is 0.4, the seepage velocity of water through the soil will be
(a) 0.28 m/min
(b) 0.40 m/min
(c) 2.50 m/min
(d) 3.50 m/min.

4.3 If the coefficient of permeability of a soil is 1 mm/s, the soil is considered to be
(a) gravel
(b) sand
(c) silt
(d) clay.

4.4 The coefficient of permeability of a soil is 3×10^{-6} m/s for a certain pore fluid. If the viscosity of the pore fluid is reduced to half without any change in its other properties, the coefficient of permeability is estimated to be
(a) 0.75×10^{-6} m/s
(b) 1.5×10^{-6} m/s
(c) 6×10^{-6} m/s
(d) 1.2×10^{-5} m/s.

4.5 Which of the following laboratory methods is most suitable for determination of hydraulic conductivity of a silty clay?
(a) falling head permeability test
(b) constant head permeability test
(c) both (b) and (c)
(d) none of the above.

4.6 Along a phreatic line in a homogeneous earth dam
(a) total head is zero everywhere
(b) total head is equal to the elevation head
(c) pressure head is zero everywhere
(d) both (b) and (c).

4.7 Excavation is carried out in a soil with porosity, $n = 0.43$, and specific gravity of particles, $G = 2.66$. What is the critical hydraulic gradient for this soil?
(a) 0.95
(b) 1.16
(c) 1.28
(d) 2.09.

4.8 Piping in soil occurs when
(a) soil is highly porous
(b) sudden change in permeability occurs
(c) soil is highly stratified
(d) effective stress becomes zero.

4.9 In 2D flow, coefficients of permeability of an anisotropic soil in x and z directions are k_x and k_z, respectively. The effective (or average permeability) of the soil is

(a) $k_x k_z$

(b) $\sqrt{k_x k_z}$

(c) $k_x + k_z$

(d) $\sqrt{k_x^2 + k_z^2}$.

4.10 Select the incorrect statement.
(a) In all seepage problems, the velocity head is generally disregarded.
(b) Piping in soil occurs when the effective stress becomes zero.
(c) Water pressure is atmospheric at all points on a phreatic line. The phreatic line is, therefore, an equipotential line.
(d) The higher the temperature of the permeant, the greater the permeability of the soil.

4.11 Figure Q4.11 shows the downward flow of water through the soil in a tube. Atmospheric pressure is maintained at the top of the reservoir water (El. 5.5 m) and at the bottom of the tail water (El. 0). Determine the elevation head, pressure head and total head at elevations 0 m, 1.0 m, 4.5 m and 5.5 m. Assume the soil is homogeneous.

(Hint: For convenience, first determine the elevation and total heads and then determine the pressure head by subtracting the elevation head from the total head. Since the soil is homogeneous, the heads will vary linearly.)

Figure Q4.11

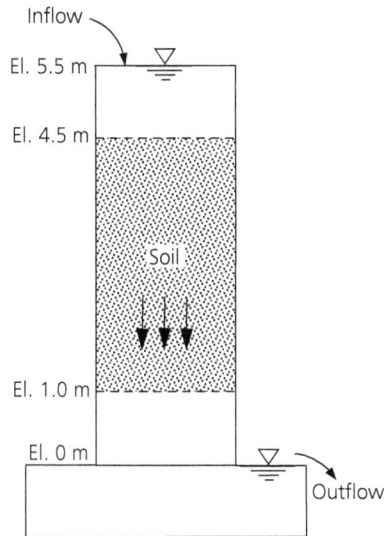

4.12 A cylindrical specimen of soil, 10 cm in diameter and 25 cm long, was tested in a constant head permeability apparatus by maintaining a constant head of 60 cm at 20°C. The amount of water collected in 30 s was 435 cm³. Determine the coefficient of permeability of soil. Also determine the flow velocity (approach or discharge velocity) and the seepage velocity, given that the void ratio of the soil is 0.56.

4.13 Refer to Figure 4.10 and consider the area of cross-section of the permeameter in plan, $A = B \times B = 0.9$ m $\times 0.9$ m, flow length through soil, $L = 1.5$ m, coefficient of permeability of soil, $k = 0.15 \times 10^{-2}$ m/s, and loss in total head in flow, $h = 2.2$ m. Determine the rate of flow using the flow net shown in the figure. Also determine the rate of flow using Darcy's law directly and compare the two answers. If $z = 0.4$ m, determine the pressure head at point P. What is the hydraulic gradient at point P?

4.14 A 1.5 m-thick soil layer is subjected to upward seepage under a head of 2.5 m. The porosity of soil and the specific gravity of soil particles are 38% and 2.65, respectively. What depth of coarse sand is required above the existing soil layer to provide a factor of safety of 2 against piping? Assume that the coarse sand has the same porosity and specific gravity as the soil, and that there is negligible head loss in the sand – in other words, the placement of coarse sand does not alter seepage force but increases the effective stress.

4.15 At a construction site, a 13.5 m-thick clay layer rests over the sand bed. The saturated unit weight of the clay is 19.2 kN/m³. When the depth of an open trench excavated in the clay reached 8.3 m below the ground surface, the bottom cracked and the water started entering the trench from below. Determine the height to which water would have risen from the top of a sand bed in a borehole if it were drilled into the sand bed prior to excavation.

4.16   If, in a flow net for the confined flow under a sheet pile wall the number of flow channels, $n_f = 4$, and the number of equipotential drops, $n_d = 12$, determine the quantity of seepage under the wall in litres/day, given the coefficient of permeability of soil, $k = 5 \times 10^{-5}$ cm/s. The difference in water level between upstream and downstream is 5.5 m. The length of the wall is 80 m.

4.17   A homogeneous anisotropic earth dam, which is 25 m high, is constructed on an impermeable foundation. The coefficients of permeability of the soil used for the construction of the dam in the horizontal and vertical directions are 4.8×10^{-8} m/s and 1.6×10^{-8} m/s, respectively. The depths of water above the base of the dam on upstream and downstream sides are 20 m and 2 m, respectively. A flow net with square areas enclosed by the flow lines and the equipotential lines is constructed on the dam cross-section drawn to a transformed scale in the same way as a flow net is drawn for an isotropic soil. The number of flow channels and equipotential drops are 4 and 18, respectively. Determine the quantity of seepage per unit length in m³/s through the dam.

4.18   Refer to Figure 4.11. Given $H = 20$ m and the coefficient of permeability of soil used for the construction of dam, $k = 1.7 \times 10^{-8}$ m/s, determine the quantity of seepage through the dam per metre run along the dam axis. Is it necessary to check the factor of safety against the piping?

4.19   Refer to Figure 4.13. Given $k_1 = 1.5 \times 10^{-7}$ m/s, $k_2 = 6.0 \times 10^{-7}$ m/s and $\alpha_1 = 30^0$. Determine α_2.

4.20   For a fine-grained soil, $D_{15\,soil} = 0.005$ mm and $D_{85\,soil} = 0.03$ mm. Is it possible to protect this soil against erosion caused by flow of water using a soil with $D_{15\,filter} = 0.008$ mm as a filter material?

4.21   During a falling head laboratory permeability test on a soil specimen, equal time intervals are noted for the head to drop from h_1 to h_2 and h_2 to h_3. Can you draw a relationship involving h_1, h_2 and h_3? If yes, are h_1, h_2 and h_3 in a geometric progression?

4.22   Consider a bucket full of water. Do you observe any flow between two points aligned vertically, one above the other, within the water? Provide an explanation by plotting the elevation head, total head and pressure head against elevation.

4.23   Consider a capillary tube inserted in the bucket full of water. Do you observe any rise of water in the capillary tube? Is there any total head gradient between two points within water inside the capillary tube? Plot the elevation head, total head and pressure head against elevation for water in the capillary tube.

4.24   A stratified soil deposit consists of two strata of equal thickness. The coefficient of permeability of the top stratum is half of the coefficient of permeability of the bottom stratum. Determine the ratio of the average coefficients of permeability in directions parallel and perpendicular to the bedding planes.

4.25   In the upward flow of water through the soil, the effective stress at any point within the soil decreases, and it may reduce to zero. Is it possible that the effective stress can be negative? Justify your answer.

4.26 How many boundary conditions can be defined clearly for the confined flow through the permeable foundation of a dam? Is the flow through an earth dam resting on an impermeable stratum confined?

4.27 For a given flow problem, under what conditions does the flow net change?

4.28 Derive the following expression for effective permeability: $k_e = \sqrt{k_x k_z}$ (Equation 4.47).

REFERENCES

ASTM (2022) ASTM D2434-22: Standard test method for measurement of hydraulic conductivity of coarse-grained soils. ASTM International, West Conshohocken, PA, USA.

ASTM (2024) ASTM D5084-16a: Standard test methods for measurement of hydraulic conductivity of saturated porous materials using a flexible wall permeameter. ASTM International, West Conshohocken, PA, USA.

BSI (2022) BS 1377-2:2022: Method of test for soils for civil engineering purposes – classification tests and determination of geotechnical properties. BSI, London, UK.

Bureau of Indian Standards (2021) IS 2720 (Part 17)-1986 (Reaffirmed 2021): Laboratory determination of permeability. Bureau of Indian Standards, New Delhi, India.

Carman PC (1938) The determination of the specific surface of powders. *Journal of the Chemical Society, Transactions* **57**: 225.

Carman PC (1956) *Flow of Gases Through Porous Media*. Butterworths Scientific Publications, London, UK.

Carrier III WD (2003) Goodbye, Hazen; hello, Kozeny-Carman. *Journal of Geotechnical and Geoenvironmental Engineering*, ASCE **129(11)**: 1054–1056.

Casagrande A (1937) Seepage through dams. *Journal of the New England Water Works Association* **51(2)**: 131–172, reprinted in *Contributions to Soil Mechanics, 1925–1940*, Boston Society of Civil Engineers, pp. 295–336.

Harr ME (1962) *Groundwater and Seepage*. McGraw-Hill, New York, NY, USA.

Hausmann MR (1990) *Engineering Principles of Ground Modification*. McGraw-Hill, New York, NY, USA.

Hazen A (1911) Discussion of "Dams on sand foundations" by AC Koenig. *Transactions, ASCE* **43**: 199–203.

Kozeny J (1927) Ueber kapillare Leitung des Wassers im Boden. *Akademie der Wissenschaften in Wien* **136(2a)**: 271. (In German.)

Lambe TW and Whitman RV (1979) *Soil Mechanics, SI Version*. Wiley, New York, NY, USA.

Shukla SK (2016) *An Introduction to Geosynthetic Engineering*. CRC Press, Taylor and Francis, London, UK.

Shukla SK (2022) *ICE Handbook of Geosynthetic Engineering*, 3rd edn. ICE Publishing, London, UK.

Shukla SK (2025) *ICE Core Concepts: Geotechnical Engineering*, 2nd edn. Emerald/ICE Publishing, London, UK.

Standards Australia (2001a) AS 1289.6.7.1 (2001): Determination of permeability of a soil – constant head method for a remoulded specimen. Standards Australia, Strathfield, NSW, Australia.

Standards Australia (2001b) AS 1289.6.7.2 (2001): Determination of permeability of a soil – falling head method for a remoulded specimen. Standards Australia, Strathfield, NSW, Australia.

Terzaghi K (1922) Der Grundbruch an Stauwerken und seine Verhutung (The failure of dams by piping and its prevention). *Die Wasserkraft* **17**: 445–449, reprinted in *From Theory to Practice in Soil Mechanics*, Wiley, New York, NY, USA, 1960, pp. 114–118.

Terzaghi K, Peck RB and Mesri G (1996) *Soil Mechanics in Engineering Practice*, 3rd edn. Wiley, New York, NY, USA.

Todd DK and Mays LW (2005) *Groundwater Hydrology*, 3rd edn. Wiley, New York, NY, USA.

FURTHER READING

Atkinson J (2007) *The Mechanics of Soils and Foundations*, 2nd edn. CRC Press, Taylor and Francis, London, UK.

Das BM (2022). *Principles of Geotechnical Engineering*, 10th edn. Cengage, Boston, MA, USA.

Holtz RD, Kovacs WD and Sheahan TC (2023) *An Introduction to Geotechnical Engineering*, 3rd edn. Pearson Education, Hoboken, NJ, USA.

Sanjay Kumar Shukla
ISBN 978-1-83608-519-5
https://doi.org/10.1108/978-1-83608-516-420252008

Chapter 5
Consolidation and compressibility of soil

Learning aims

This chapter explores the core concepts of the following topics

- basic stress–strain relationships
- consolidation test and compressibility parameters
- mechanics of primary consolidation
- consolidation theory and rate of settlement
- concept of secondary consolidation
- computation of settlements.

5.1.   Introduction

When a soil deposit is subjected to loading by a structure, an earth fill, a heavy machine or any other human activity, a part of the soil mass is stressed significantly, resulting in deformations or strains in the form of changes in volume and/or shape. The deformations also occur when the soil deposit is unloaded by excavation, removal of structure or erosion. A change in volume of the soil by loading is noticed mainly through its compression that generally appears as the downward verti-cal deformation of the ground surface, called settlement. A change in volume of the soil by unload-ing can be seen in the form of swelling or heaving of the ground. Note that there is a possibility of both compression and heaving of the ground occurring simultaneously in different locations during loading of the ground; this aspect is discussed in Chapter 8. Deformation in the form of a change in shape, called distortion, can also result in settlement or heaving of the ground surface.

The property of a soil pertaining to its susceptibility to decrease in volume when subjected to loading is called compressibility. In civil engineering applications, the decrease in volume of individual soil particles under applied stress is very small and generally considered negligible. As a result, any reduction in the volume of a soil mass primarily occurs due to a decrease in void volume, accompanied by the expulsion of air and/or water, depending on whether the soil is dry, unsaturated or saturated.

The decrease in volume of a soil mass due to expulsion of air from its voids or pores caused by compressive load or stress application is almost instantaneous; thus, this process of volume reduc-tion is time-independent and is commonly used as one of the ground improvement methods, called compaction, to improve the strength characteristics of soils. The details of soil compaction are presented in Chapter 7.

The process of volume reduction of a saturated soil mass under compressive loading or stress application requires the expulsion of water from its voids or pores. This process of volume reduction, known as primary consolidation, or simply consolidation, is time-dependent, especially in low-permeability soils such as silts and clays. For consolidation to proceed effectively, the applied load or stress must remain active over an extended period.

The volume reduction in some soils may continue even after the completion of the primary consolidation process due to the plastic adjustment of the soil structure. This phenomenon is known as secondary consolidation, also referred to as secondary compression or creep.

There is a possibility that the processes of primary and secondary consolidations may occur simultaneously for some period, especially during the later stages of the primary consolidation process. When the compressive load applied to the soil is removed, the soil may expand in volume. This process is known as swelling or heaving.

When designing the foundation of engineered structures or checking the stability of an engineered fill, the following must be estimated for the ground surface supporting the foundation or the fill

- magnitude of settlement
- rate of both the total and the differential settlements.

To meet the structural safety and the serviceability requirements of structures and fills, the settlements caused by the applied load or stress must be within some permissible limits over their design lives, which may be as much as 100 years. An understanding of the stress–strain–time relationships, called the constitutive relationships, is necessary to estimate the settlement of soils. Since for a given load or stress application, the amount of deformation or strain in a soil depends on several factors, namely composition, void ratio, past stress history and the way the stress is applied, the constitutive relationships for soils are complex. For simplicity, engineers often use the stress–strain relationships given for elastic materials for some soil problems, although soils behave differently from most elastic materials.

As the consolidation process in soil increases its strength, reduces its compressibility and lowers its permeability, it is important for geotechnical engineers to thoroughly understand this process. This chapter primarily focuses on the process of consolidation and compressibility characteristics of soils. The fundamental concepts of stress–strain relationships for elastic materials are also presented. Additionally, detailed explanations are provided on analysing the rate of settlement and estimating the magnitude of ground surface settlement due to loading.

5.2. Basic stress–strain relationships

The concept of stress within soil was presented in Chapter 3. Stress at a point within a material is defined as the force or load acting on a unit area of the desired orientation centred at the point of interest. When a soil mass is loaded – for instance, by the construction of a structure – the soil particles move closer to each other by interacting through sliding, rolling, bending and/or crushing, resulting in a decrease in its volume, but when unloaded – for instance, by excavation – they may move away from each other, resulting in an increase in volume to some extent. The ratio of the change in volume, ΔV, to the original volume, V, of the soil element – that is, $\Delta V / V$ – is called the volumetric strain, ε_v. If the volume of a soil element changes, the linear dimensions of the soil element also change. If the length, L, of a soil element in the z (vertical) direction changes by ΔL, the longitudinal or vertical strain, ε_z, which is generally compressive strain in most soil problems, is $\Delta L / L$. There is also a possibility of shear deformation – that is, a change in shape of the soil element. The shear strain is defined as the angle, say γ_{zx} referenced to x and z axes, by which any length of the soil element rotates.

Note that throughout this book, when discussing strain, z is used as the subscript for vertical, x as the subscript for horizontal and v as the subscript for volume; when discussing stress, v (in place of z) will generally be used as the subscript for vertical and h (in place of x) as the subscript for horizontal. Also note that when talking about stress and strain within a soil, the area and length of the soil element for the computation of stress and strain at the point under consideration must be large enough to include a representative number of soil particles. Because soils, except some cemented soils, cannot sustain tensile stresses, compressive stresses and strains are considered positive. Thus, a positive compressive strain causes the shortening of the linear dimension of the soil element, and a positive volumetric strain causes a volume decrease of the soil element.

Chapter 3 explained that the state of stress at a point within soil is completely described by six independent stresses as: $\sigma_x, \sigma_y, \sigma_z, \tau_{xy}, \tau_{yz}, \tau_{zx}$, which are the components of the stress tensor at the point of interest with respect to the Cartesian coordinate directions x, y and z. From the theory of elasticity, the stress–strain relationships for uniaxial loading, simple shear, isotropic (volumetric) compression and confined compression are given below.

Uniaxial loading on an elastic cylinder ($\sigma_x = 0, \sigma_y = 0, \varepsilon_x \neq 0, \varepsilon_y \neq 0$) (Figure 5.1(a))

$$\varepsilon_z = \frac{\sigma_z}{E} \tag{5.1}$$

$$\varepsilon_x = \varepsilon_y = -\mu\varepsilon_z \tag{5.2}$$

$$\varepsilon_v = \varepsilon_x + \varepsilon_y + \varepsilon_z = (1-2\mu)\varepsilon_z = (1-2\mu)\frac{\sigma_z}{E} \tag{5.3}$$

where σ_z is the uniaxial stress in the z direction, $\varepsilon_x, \varepsilon_y, \varepsilon_z$ are the strains in the x, y and z directions, respectively (they are positive when compressive), E is Young's modulus of elasticity and μ is Poisson's ratio.

Simple shear loading on an elastic cube (Figure 5.1(b))

$$\gamma_{zx} = \frac{\tau_{zx}}{G} \tag{5.4}$$

where τ_{zx} is the shear stress and G is the shear modulus of elasticity.

Isotropic or volumetric loading on an elastic cube ($\varepsilon_x = \varepsilon_y = \varepsilon_z$) (Figure 5.1(c))

$$\varepsilon_v = 3\varepsilon_z = \frac{\sigma_0}{K} \tag{5.5}$$

where $\sigma_0 = \sigma_x = \sigma_y = \sigma_z =$ isotropic stress and K is the bulk modulus of elasticity.

Uniaxial compression loading on an elastic cylinder or disc confined laterally ($\sigma_x \neq 0, \sigma_y \neq 0$, $\varepsilon_x = \varepsilon_y = 0$) (Figure 5.1(d))

$$\varepsilon_z = \varepsilon_v = \frac{\sigma_z}{D} \tag{5.6}$$

$$\sigma_x = \sigma_y = \frac{\mu}{1-\mu}\sigma_z \tag{5.7}$$

where σ_z is the uniaxial compressive stress and D is the constrained modulus of elasticity.

The elastic constants E, G, K, μ and D have the following relationships.

$$G = \frac{E}{2(1+\mu)} \tag{5.8}$$

$$K = \frac{E}{3(1-2\mu)} \tag{5.9}$$

$$D = \frac{E(1-\mu)}{(1+\mu)(1-2\mu)} \tag{5.10}$$

Figure 5.1 Some common types of loading on elastic bodies and their deformation: (a) uniaxial loading; (b) simple shear; (c) isotropic or volumetric loading; (d) confined compression loading

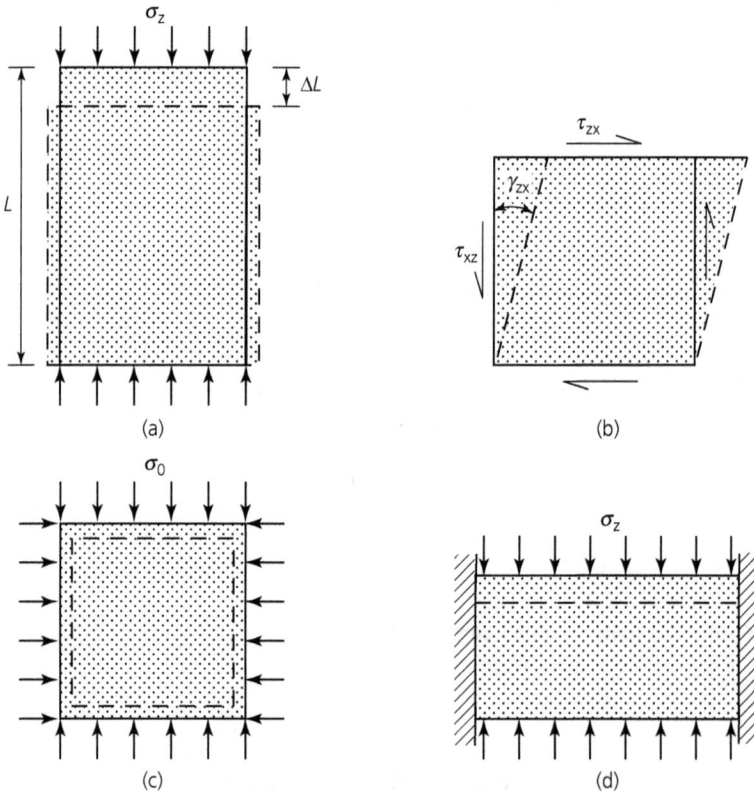

(a)

(b)

(c)

(d)

Figure 5.2(a) shows a typical stress (σ) against strain (ε) curve for a soil. This curve is similar to the stress–strain curves for common engineering materials such as metals and plastics. At low stress levels, the relationship between stress and strain is linear; therefore, the soil can be assumed to be a linear elastic material. However, this does not mean that it has the ability to recover its original size and shape after removal of stress as with truly or purely elastic materials. The gradient of the linear portion of the curve gives the modulus of elasticity, which describes the stiffness of the soil. The modulus of elasticity can be determined at any stress level where the stress–strain curve may even be nonlinear as the tangent modulus or secant modulus.

Note that in Figure 5.2(a), the gradient changes noticeably when stress, σ, reaches a value, σ_y, called the yield stress. For $\sigma > \sigma_y$, the stress–strain relationship is highly nonlinear, and it can be idealised as elastic-perfectly plastic (Figure 5.2(b)), elastic-strain hardening or softening plastic (Figure 5.2(c)) or rigid-perfectly plastic (Figure 5.2(d)), depending on the level of stress, as well as the failure condition under consideration. The stress–strain models shown in Figure 5.2 can be considered with their specific features applicable to soils for solving any specific load deformation problem. The peak or ultimate stress that the soil can sustain gives the strength of the soil.

Figure 5.2 (a) Typical stress–strain curve for a soil; (b) elastic-perfectly plastic model; (c) elastic-strain hardening or softening model; (d) rigid-perfectly plastic model

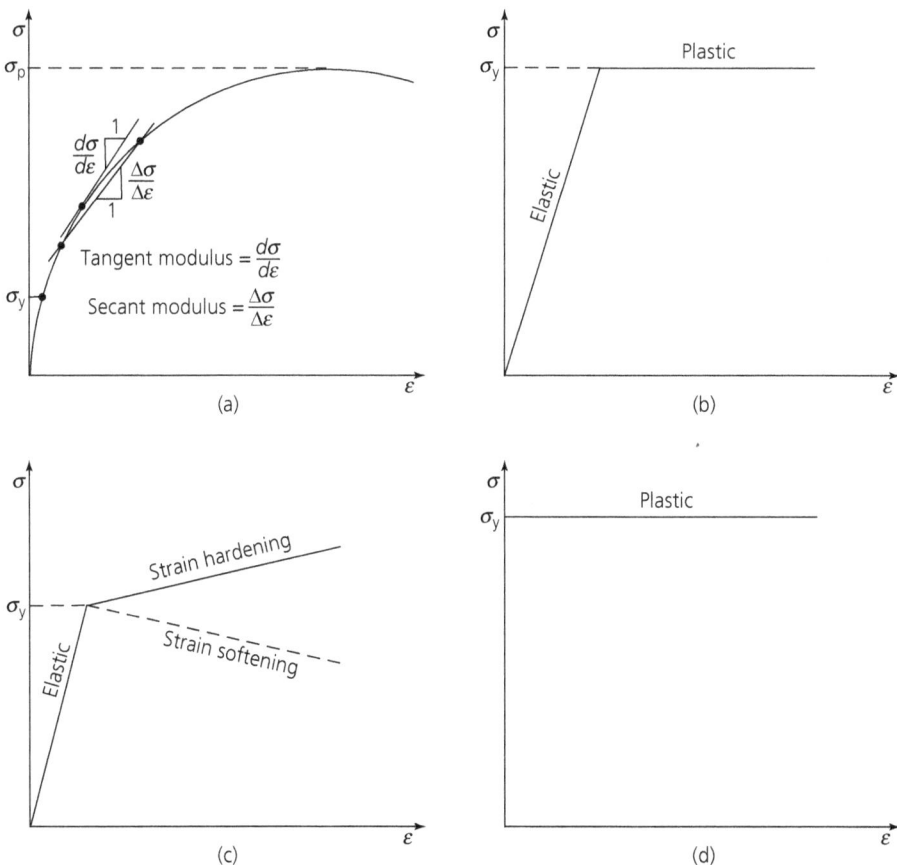

Although engineers talk about tensile strength, compressive strength, shear strength and so on, the maximum shear stress that the soil or any other material can sustain is the link between these different strengths (Atkinson, 2007). In geotechnical engineering applications, by the strength of a soil, engineers generally mean its shear strength – that is, the maximum shear stress that the soil can sustain. The details of shear strength of soil are presented in Chapter 6.

For isotropic and linearly elastic materials such as many metals, the elastic constants, E (Young's modulus) and μ (Poisson's ratio), can be determined by a single simple test. These constants can then be used to solve several actual problems involving G, K and D under different loading conditions. However, a similar simple approach is generally not possible with a soil because, being a multiphase system and having a particulate nature of soil solid phase, the soil behaves in a complex manner. Hence, the type of test to study the stress–strain behaviour of a soil varies with the type of loading observed in actual problems. The most common tests used to study the stress–strain behaviour of soils in geotechnical engineering are

- confined compression test
- isotropic compression test
- triaxial compression test
- unconfined compression test
- direct shear test.

This chapter provides details of the confined compression test; most others will be described in Chapter 6.

The following points regarding the basic stress–strain behaviour are noteworthy.

- The volumetric strain of a soil element of linearly elastic material subjected to all stress components is

$$\varepsilon_v = \varepsilon_x + \varepsilon_y + \varepsilon_z = (1 - 2\mu)\frac{(\sigma_x + \sigma_y + \sigma_z)}{E} \tag{5.11}$$

- For the majority of materials, including steel, aluminium, concrete and many drained soils, the Poisson's ratio μ ranges from 0.1 to 0.3. If, for a linearly elastic material, $\mu = 0.5$, its volumetric strain, $\varepsilon_v = 0$. Rubber is an example of such a material. For saturated clays subjected to undrained loading, $\mu = 0.5$; hence $\varepsilon_v = 0$.
- Equations 5.1 to 5.11 apply for increments of stress starting from some initial stress, as well as for increments of stress starting from zero stress (Lambe and Whitman, 1979).
- The stiffness of soil determines the strain or displacement at working load in the ground when it is loaded or unloaded, whereas the strength governs the maximum load that the ground can sustain. A soil can be stiff, meaning it has high stiffness, or soft, meaning it has low stiffness. Additionally, soil can also be classified as strong or weak based on its strength characteristics (Atkinson, 2007).
- More generalised constitutive relationships involve time effect with stress and strain. In soils, there are a number of time effects mainly due to drainage of water from the soil and, to a limited extent, due to creep (continued deformation or strain under constant load or stress) of soil.

Example 5.1

For an elastic material, the Poisson's ratio, μ, is 0.5. What will be the ratio of the Young's modulus of elasticity, E, to the shear modulus of elasticity, G? Can you compute the values of the bulk modulus of elasticity, K, and constrained modulus of elasticity, D, for this material? What will be the volumetric strain, ε_v, during loading? Can you name the material?

Solution

Given $\mu = 0.5$. From Equation 5.8

$$G = \frac{E}{2(1+\mu)} = \frac{E}{2(1+0.5)} = \frac{E}{3}$$

or

$$\frac{E}{G} = 3$$

From Equation 5.9, the bulk modulus of elasticity

$$K = \frac{E}{3(1-2\mu)}$$

For $\mu = 0.5$, $K \rightarrow \infty$

From Equation 5.10, the constrained modulus of elasticity

$$D = \frac{E(1-\mu)}{(1+\mu)(1-2\mu)}$$

For $\mu = 0.5$, $D \rightarrow \infty$

From Equation 5.11, the volumetric strain

$$\varepsilon_v = \varepsilon_x + \varepsilon_y + \varepsilon_z = (1-2\mu)\frac{(\sigma_x + \sigma_y + \sigma_z)}{E}$$

For $\mu = 0.5$, $\varepsilon_v = 0$, that is, $\varepsilon_z = -(\varepsilon_x + \varepsilon_y)$.

The material can be saturated clay subjected to undrained loading – that is, with no change in volume during loading.

5.3.   Confined compression test

The confined compression test, also known as the oedometer test or the one-dimensional (1D) compression/consolidation test, is relatively simple to perform and widely used by geotechnical engineers. Figure 5.3 illustrates the basic test condition and the type of deformation of the soil specimen in the test. Note that there is no horizontal deformation – that is, the horizontal strain, $\varepsilon_h = 0$, but there is some finite vertical deformation because the surface AB can settle to a new position $A'B'$; thus, the vertical strain, $\varepsilon_z \neq 0$, which can be positive for loading case or negative for unloading case. Since the horizontal strain, $\varepsilon_h = 0$, the vertical strain, ε_z, is exactly equal to the

Figure 5.3 Confined compression test – basic test condition and type of deformation

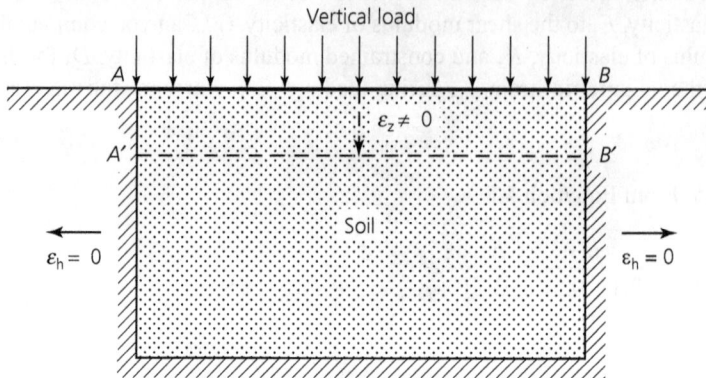

volumetric strain, ε_v. Thus, deformation is primarily volumetric, mainly compression as the dominant source of strain, with some distortion. The ratio of effective horizontal stress, σ_h', to effective vertical stress, σ_v', within the soil is the coefficient of lateral stress at rest, K_0.

The best feature of the confined compression test is that the strain or deformation condition of the soil specimen in the test approximates certain field problems. Some examples of field situations simulated by this test are application of the surface load from the raft and other wide footings, placement of a structural fill or an embankment over a large areal extent, and natural/geological deposition (loading) and erosion (unloading) of the earth materials.

In this test, a soil specimen (minimum diameter = 50 mm, minimum initial specimen height = 12 mm, but not less than ten times the maximum soil particle diameter) is restrained laterally in a ring that is either fixed to the base or floating with porous discs on each side of the specimen. The specimen is submerged in water and is loaded axially with total stress increments. The device used for holding the specimen and performing the measurement task is called the consolidometer. Figure 5.4(a) shows a schematic view of the floating ring consolidometer and Figure 5.4(b) shows the 1D consolidation test set-up in the laboratory. Each stress increment is maintained until the excess pore water pressure is completely dissipated; this duration is traditionally 24 h for each stress increment. The loads are applied in steps to obtain the successive stress on the soil to be twice the preceding one. The stresses commonly used are approximately 12, 25, 50, 100, 200, 400 kPa etc. During the consolidation process, the changes in the specimen height are measured for each stress increment at time intervals of approximately 0.1, 0.25, 0.5, 1, 2, 4, 8, 15, 30 min and 1, 2, 4, 8 and 24 h.

After the consolidation under the last stress increment is over, the specimen is unloaded in two or three stages and the soil is allowed to swell. Only the final changes in the specimen height are measured at each unloading stage. At the end of the final unloading stage, the soil specimen in the ring is dried in an oven to determine the dry weight of the soil specimen – that is, the weight of solids, W_s. The complete test details are given in several test standards (ASTM D2435-04 (ASTM, 2001); BS 1377-2:2022 (BSI, 2022); IS 2720 (Part 15) (Bureau of Indian Standards, 2021); AS 1289.6.6.1:2020 (Standards Australia, 2020)).

There are two methods, namely the height of solids method and the change in void ratio method, to determine the void ratio at the end of each stress increment period.

Figure 5.4 One-dimensional consolidation test: (a) schematic view of the floating ring consolidometer; (b) test set-up in the laboratory

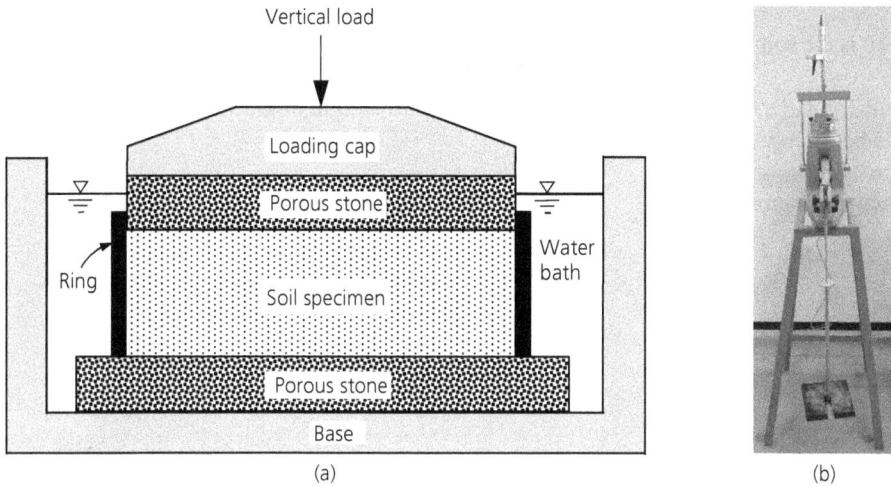

Figure 5.4 One-dimensional consolidation test: (a) schematic view of the floating ring consolidometer; (b) test set-up in the laboratory

5.3.1 Height of solids method

Considering the solid phase of the soil separated from liquid and/or gas present in the voids of the soil, the height of solids, H_s, of the soil specimen can be calculated as

$$H_s = \frac{V_s}{A} = \frac{\dfrac{W_s}{G\gamma_w}}{A} = \frac{W_s}{G\gamma_w A} \qquad (5.12)$$

where V_s is the volume of solids in the soil specimen, A is the cross-sectional area of the soil specimen, which remains constant throughout the test, and G is the specific gravity of soil solids.

The void ratio before test

$$e_0 = \frac{H_0 - H_s}{H_s} \qquad (5.13)$$

where H_0 is the initial soil specimen height, which is the sum of the height of solids and the height of voids occupied by air/water before the test.

The void ratio after test

$$e_f = \frac{H_f - H_s}{H_s} \qquad (5.14)$$

where H_f is the final soil specimen height, which is the sum of the height of solids and the height of voids occupied by air/water after the test.

The void ratio at the end of any stress increment period can be calculated as

$$e = \frac{H - H_s}{H_s}$$

(5.15)

where H is the soil specimen height at the end of stress increment period under consideration.

If W_0 is the initial moist weight and W_f is the final moist weight of the soil specimen, the following can also be determined.

Initial water content

$$w_0 = \frac{W_0 - W_s}{W_s}$$

(5.16)

Final water content

$$w_f = \frac{W_f - W_s}{W_s}$$

(5.17)

Initial degree of saturation

$$S_0 = \frac{W_0 - W_s}{\gamma_w A(H_0 - H_s)}$$

(5.18)

Final degree of saturation

$$S_0 = \frac{W_f - W_s}{\gamma_w A(H_f - H_s)}$$

(5.19)

Note that the height of solids method is also applicable to unsaturated soils.

5.3.2 Change in void ratio method

This method is applicable only to saturated soils for which the degree of saturation, $S = 100\%$. Here the final water content, w_f, determined at the end of the test is used to compute the void ratio, e_f, at the end of the test from Equation 1.18 (see Chapter 1) as

$$e_f = w_f G$$

(5.20)

If ΔH is the total decrease in thickness or settlement of the soil specimen during the test due to a total increase in effective vertical stress of $\Delta\sigma_v'$, the void ratio of the soil before test

$$e_0 = e_f + \Delta e$$

(5.21)

where the total decrease in void ratio, Δe, is calculated from the following relationship in which during the test, the total decrease in volume, ΔV, is equal to the total decrease in void volume, ΔV_v, because the change in volume of solids $\Delta V_s = 0$.

$$\frac{\Delta H}{H_0} = \frac{A(\Delta H)}{AH_0} = \frac{\Delta V}{V_0} = \frac{\Delta V_v}{V_0} = \frac{-\Delta e}{1+e_0} \tag{5.22}$$

This procedure can be used to determine the void ratio at the end of any stress increment period, and it is generally carried out backwards from the known values of soil specimen thickness, H_f, and the void ratio, e_f, at the end of the test. In Equation 5.22, a negative sign is inserted to make Δe positive. Note that in Equation 5.22, $\Delta H / H_0 = \varepsilon_z$ and $\Delta V / V_0 = \varepsilon_v$, where ε_z and ε_v are vertical and volumetric strains, respectively, and both are equal. Equation 5.22 is also applicable to the situation when ΔH is an incremental change in thickness of the soil specimen due to an incremental change in effective vertical stress, $\Delta \sigma_v'$, but in that situation, $\Delta H / H_0 = \Delta \varepsilon_z$ and $\Delta V / V_0 = \Delta \varepsilon_v$, where $\Delta \varepsilon_z$ and $\Delta \varepsilon_v$ are incremental changes in vertical and volumetric strains, respectively, and both are equal. When calculating the incremental strains, it does not usually make a significant difference if H_0 is taken as the soil specimen thickness at the beginning of the loading process.

5.3.3 Compressibility parameters

Figure 5.5 shows a typical stress–strain relationship between the volumetric strain expressed as the void ratio, e, and the effective vertical stress (or effective consolidation stress), σ_v', during loading, unloading and then reloading of a saturated clay in the confined compression test. The slope of the e against σ_v' curve is defined as the coefficient of compressibility, a_v. Thus

$$a_v = -\frac{\partial e}{\partial \sigma_v'} \tag{5.23a}$$

or

$$a_v = -\frac{\Delta e}{\Delta \sigma_v'} \tag{5.23b}$$

Figure 5.5 A typical relationship between the void ratio, e, and the effective vertical stress, σ_v', based on the confined compression test

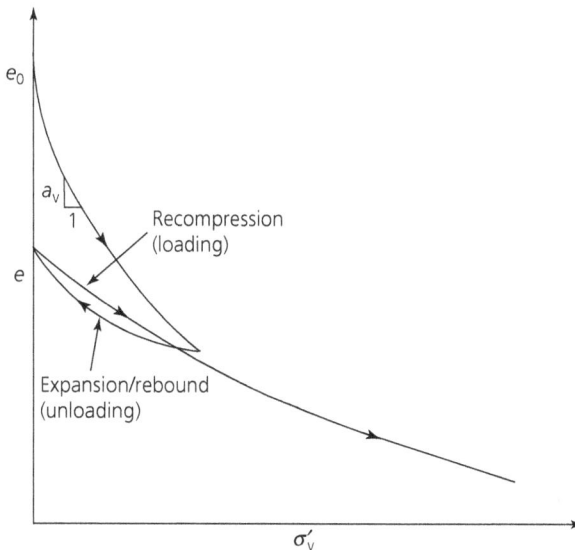

Since e decreases as σ'_v increases, the negative sign is introduced to make a_v a positive parameter.

It is common practice to plot the confined compression test data as e against $\log_{10}\sigma'_v$ as shown in Figure 5.6. This plot is convenient for presenting the stress–strain behaviour over a wide range of stresses. Additionally, the curve usually becomes more-or-less a straight line, called the virgin compression curve, at large stresses, which are being applied for the first time on the soil in its geological history. The slope of the virgin compression curve is defined as the compression index, C_c. Thus,

$$C_c = -\frac{\partial e}{\partial(\log_{10}\sigma'_v)} \tag{5.24a}$$

or

$$C_c = -\frac{\Delta e}{\Delta(\log_{10}\sigma'_v)} = -\frac{e-e_0}{\log_{10}\sigma'_v - \log_{10}\sigma'_{v0}} \tag{5.24b}$$

or

$$e = e_0 - C_c \log_{10}\left(\frac{\sigma'_v}{\sigma'_{v0}}\right) \tag{5.24c}$$

where e_0 is the initial void ratio corresponding to the initial effective vertical stress, σ'_{v0}. Equation 5.24c is the equation of the virgin compression curve.

Figure 5.6 A typical relationship between the void ratio, e, and the logarithm of effective vertical stress, $\log_{10}\sigma'_v$, based on the confined compression test

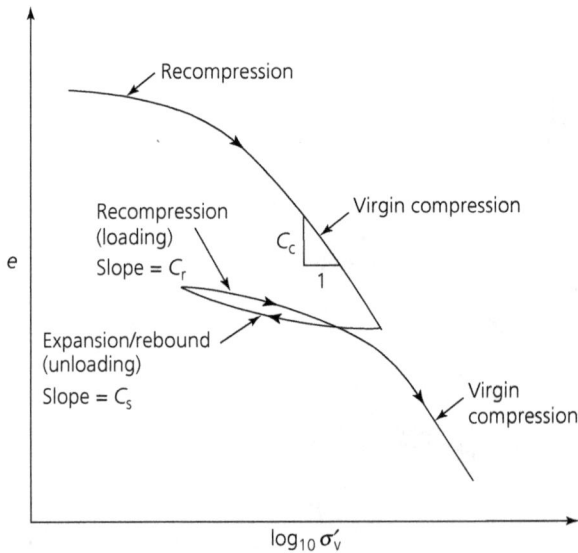

The coefficient of volume change, also known as coefficient of volume compressibility, m_v, is another soil compressibility parameter for the confined compression or 1D consolidation defined in terms of an incremental change in vertical strain, $\Delta\varepsilon_z$, due to an incremental change in effective vertical stress, $\Delta\sigma_v'$, as

$$m_v = \frac{\partial\varepsilon_z}{\partial\sigma_v'} \tag{5.25a}$$

or by using Equation 5.22 and 5.23b

$$m_v = \frac{\Delta\varepsilon_z}{\Delta\sigma_v'} = \left(-\frac{\Delta\varepsilon_z}{\Delta e}\right) \times \left(-\frac{\Delta e}{\Delta\sigma_v'}\right) = \frac{a_v}{1+e_0} \tag{5.25b}$$

Comparing Equation 5.25 with Equation 5.6 shows that m_v is simply the reciprocal of the constrained modulus. Thus, in the confined compression test for an incremental change in vertical strain, $\Delta\varepsilon_z$, due to an incremental change in effective vertical stress, $\Delta\sigma_v'$

$$D = \frac{1}{m_v} = \frac{\Delta\sigma_v'}{\Delta\varepsilon_z} = \frac{1+e_0}{a_v} \tag{5.26}$$

Note that if $\Delta\sigma_v'$ is the total change in effective vertical stress and ΔH is the total change in thickness of the soil specimen during the entire confined compression test, Equations 5.25b and 5.26 can be presented as

$$m_v = \frac{\varepsilon_z}{\Delta\sigma_v'} = \frac{\varepsilon_v}{\Delta\sigma_v'} = \frac{\frac{\Delta H}{H_0}}{\Delta\sigma_v'} = -\frac{\Delta e}{1+e_0}\frac{1}{\Delta\sigma_v'} = \frac{a_v}{1+e_0} \tag{5.27}$$

and

$$D = \frac{1}{m_v} = \frac{\Delta\sigma_v'}{\varepsilon_z} = \frac{\Delta\sigma_v'}{\varepsilon_v} = \frac{\Delta\sigma_v'}{\frac{\Delta H}{H_0}} = -\frac{1+e_0}{\Delta e}\Delta\sigma_v' = \frac{1+e_0}{a_v} \tag{5.28}$$

Example 5.2

In a laboratory 1D consolidation test, the void ratio of a clay decreases from 1.012 to 0.870 during the virgin compression when the effective vertical stress increases from 400 to 800 kPa. Determine the following

(a) coefficient of compressibility
(b) compression index
(c) coefficient of volume change.

Solution

Given $e_1 = 1.012$, $e_2 = 0.870$, $\sigma'_{v1} = 400$ kPa and $\sigma'_{v2} = 800$ kPa.

(a) From Equation 5.23b, the coefficient of compressibility

$$a_v = -\frac{\Delta e}{\Delta \sigma'_v} = -\frac{e_2 - e_1}{\sigma'_{v2} - \sigma'_{v1}} = -\frac{0.870 - 1.012}{800 - 400} = 3.55 \times 10^{-4} \text{ m}^2 / \text{kN}$$

(b) From Equation 5.24b, the compression index

$$C_c = -\frac{\Delta e}{\Delta (\log_{10} \sigma'_v)} = -\frac{e_2 - e_1}{\log_{10} \sigma'_{v2} - \log_{10} \sigma'_{v02}} = -\frac{0.870 - 1.012}{\log_{10}(800) - \log_{10}(400)} = 0.47$$

(c) From Equation 5.27 with $e_0 = e_1$, the coefficient of volume change

$$m_v = \frac{a_v}{1 + e_1} = \frac{3.55 \times 10^{-4}}{1 + 1.012} = 1.76 \times 10^{-4} \text{ m}^2 / \text{kN}$$

Note that in Figure 5.6, the virgin compression curve is connected to the initial flat portion, called the recompression curve, by a smooth transition curve. The flat portion indicates that the soil has been subjected to stresses larger than the existing stresses in its geological history due to factors such as overburden deposition followed by subsequent erosion, advancing and receding glaciers, construction followed by demolition of structures, fall and rise of water table, desiccation due to surface drying, and changes in the chemical environment and soil structure. When the soil is stressed to a level greater than the maximum past effective vertical stress, $\sigma'_{v\max}$ (or σ'_p), called the preconsolidation pressure (or maximum past consolidation stress), to which it has ever been subjected in the past, the soil structure probably breaks, resulting in a much higher compressibility. The maximum past effective vertical stress, $\sigma'_{v\max}$, corresponds to a point somewhere on the transition curve and is analogous to the yield stress of a metal. Casagrande (1936) proposed a technique, explained below and shown in Figure 5.7, to estimate $\sigma'_{v\max}$.

- Locate the point A on the e against $\log_{10} \sigma'_v$ curve as the point of maximum curvature – that is, the point of minimum radius of curvature.
- Through A, draw a horizontal line AB and a tangent line AC to the curve.
- Bisect the angle formed by lines AB and AC and locate the bisector line AD.
- Extend the virgin part of the compression curve backward as line EF to meet AD at point P.
- Point P gives the maximum past effective vertical stress $\sigma'_{v\max}$ (or preconsolidation pressure, σ'_p) that the soil at P has experienced in its geological history.

A soil mass is defined as normally consolidated if it is at equilibrium under the maximum effective vertical stress (or effective overburden pressure) it has ever experienced in its geological history. In other words, a normally consolidated soil has never been subjected to effective vertical stresses higher than the present effective vertical stress (σ'_{v0}) since its origin.

A soil mass is defined as overconsolidated (or precompressed or preconsolidated) if it is at equilibrium under an effective vertical stress less than that to which it was once consolidated in its

Figure 5.7 Casagrande's technique for estimating the maximum past effective vertical stress (or preconsolidation pressure) (after Casagrande, 1936)

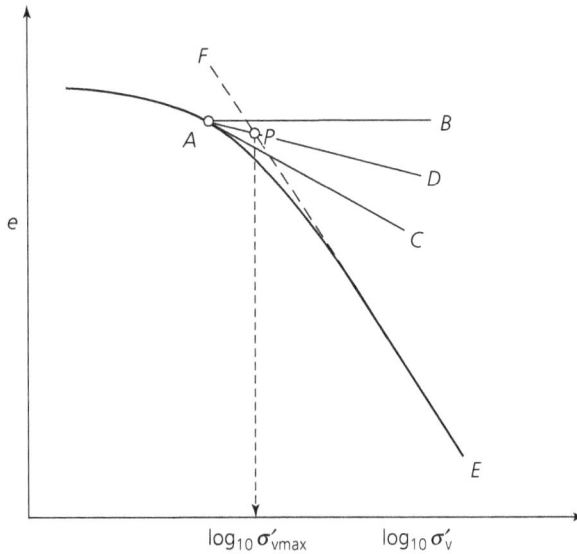

geological history. In other words, an overconsolidated soil has been subjected to effective vertical stresses higher than the present effective vertical stress (σ'_{v0}) since its origin.

The ratio of maximum past effective vertical stress, $\sigma'_{v\,max}$, to the present effective vertical stress, σ'_{v0}, is called the overconsolidation ratio (OCR), which is used to express the degree of overconsolidation of a soil. Thus,

$$OCR = \frac{\sigma'_{v\,max}}{\sigma'_{v0}} \tag{5.29}$$

For normally consolidated soils, OCR = 1 and for overconsolidated soils, OCR > 1. Can soils have OCR < 1? The answer is yes. A soil with an OCR < 1 is called an underconsolidated soil. Such soils are generally recently constructed or geologically deposited saturated fills (mine tailings, fly ash deposits in ash ponds, landslide deposits etc.) which are still consolidating under the present overburden pressure or self-weight. The pore water pressures within the voids of underconsolidated soils will be higher than the static or steady-state pore water pressures.

The following points regarding the confined compression tests and compressibility characteristics of soil are noteworthy.

- The laboratory consolidation curve does not reproduce the field compression curve mainly because of the change of stress and fabric inherent in the soil sampling, and the difference in temperature between the ground and the laboratory. However, for the best estimates of ground surface settlement, realistic values of C_c and $\sigma'_{v\,max}$ are required, which can be determined from the results of the confined compression tests on undisturbed specimens of high quality.
- C_c is a meaningful parameter only for normally consolidated soils and is required to determine the magnitude of the primary settlement.

▨ In Figure 5.6, which is an e against $\log_{10} \sigma_v'$ plot, the slopes $-\Delta e / \Delta(\log_{10} \sigma_v')$ in the expansion and recompression ranges, generally determined as average slopes, are called the rebound, swelling or expansion index, C_s, and recompression index, C_r, respectively. The swelling index is a measure of the increase in volume of the soil caused by unloading.

▨ The compressibility parameters a_v and m_v decrease as σ_v' increases but C_c remains constant. Hence, it can be stated that the compressibility of a soil is a function of the effective vertical stress.

▨ For saturated silt and clay deposits, C_c typically varies between 0.1 and 3, and may increase to 10 for peats and organic soils (Terzaghi et al., 1996). There are some empirical correlations of C_c with simple soil properties such as the liquid limit w_L. For a remoulded soil (a soil with a modified structure) (Skempton, 1944; Terzaghi and Peck, 1967)

$$C_c = 0.007(w_L - 10) \tag{5.30}$$

The value of C_c corresponding to the field consolidation is roughly equal to 1.3 times the value of C_c for the remoulded specimen, hence

$$C_c = 0.009(w_L - 10) \tag{5.31}$$

Note that in Equations 5.30 and 5.31, w_L is in %, and they are applicable for virgin compression. These simple correlations and others available in literature (Bowles, 1996) can be used if an approximate value of settlement is needed, especially during the initial stage of a project.

▨ The measured magnitudes of OCR for a variety of natural soft clays are in the range of 1.2 to 3 except within the desiccated crust, where they may be much higher (Terzaghi et al., 1996).

5.4. Mechanics of 1D consolidation

Consolidation is the process of a gradual reduction in volume of a partially or fully saturated soil mass, resulting from an increase in compressive stress and its sustained application. During this process, water drains out with time from the voids or pores of the soil mass. The process or mechanism involved in the expulsion or drainage of water for the case of 1D consolidation is shown in Figure 5.8. A large surface area of a fully saturated soil stratum, say a clay stratum that has a very low coefficient of permeability, k, is subjected to a compressive stress increment, $\Delta\sigma_v$, by placement or construction of a sand fill. The clay stratum with AB and CD as the top and bottom surfaces, respectively, is thus sandwiched between two highly permeable sand strata. The water table coincides with AB, and therefore the soil below AB is fully saturated. This may be a common construction site situation.

Prior to application of $\Delta\sigma_v$, called the consolidation pressure increment, the pore water pressures are hydrostatic, and the total head is the same at all depths; hence, there is no flow of water. Assuming that the construction time period is very short – that is, the stress increment, $\Delta\sigma_v$, is applied suddenly at AB, say at time $t = 0$ – then, at this time, the pore water pressure, u, will increase equally everywhere below AB within the clay by $\Delta u = \Delta\sigma_v$ above the reference static pore water pressure, u_s, without any stress transfer to the soil skeleton of solid particles – that is, without any increase in effective vertical stress, σ_v'. This happens because the clay has a very low permeability, the water in the voids is almost incompressible and the soil skeleton is compressible. The additional or excess pore water pressure Δu ($= \Delta\sigma_v$) within the clay causes an increase in total head throughout its thickness. Assuming that there is no change in the location of the water table in this problem, the final or steady-state pore water pressure (pore water pressure in the equilibrium condition),

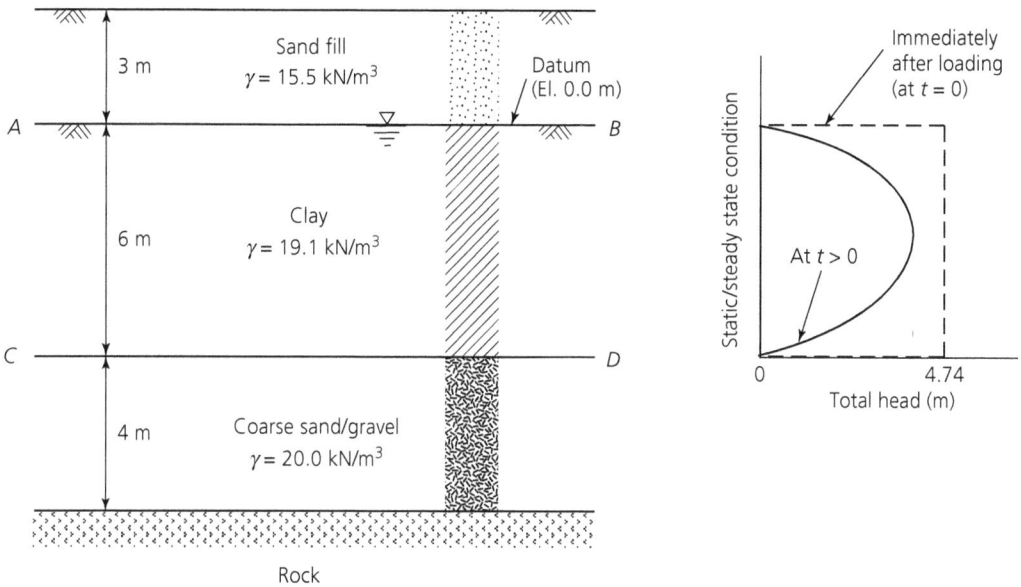

Figure 5.8 Undrained loading by the construction of a sand fill

u_{ss}, at any depth within the clay is just the same as the static pore water pressure, u_s (Lambe and Whitman, 1979).

Since sand is highly permeable, within the sand stratum underlying the clay stratum the increment of pore water pressure $\Delta u\,(= \Delta\sigma_v)$ dissipates quickly; hence no measurable increase in total head occurs. Because of difference in total head at clay stratum boundaries AB and CD with sand, strong hydraulic gradients exist at AB and CD, resulting in flow of water in the vertical direction with a gradual decrease in excess pore water pressure, Δu, within the clay. Since the pore water pressure and the total head within the clay change with time during the fluid flow, this flow is transient (or unsteady). The flow of water out of the clay causes a decrease in void ratio, e, of the soil accompanied by the appropriate increase in effective vertical stress, σ'_v. Since the loaded area is very wide, there is no horizontal strain or deformation as can be observed in the confined compression test. The decrease in void ratio mainly results in the settlement of the top surface AB. This settlement can be estimated from Equation 5.22. The excess pore water pressure at any time $t > 0$ is

$$u_e = u - u_{ss} \tag{5.32a}$$

Since, for this problem, $u_{ss} = u_s$

$$u_e = u - u_s \tag{5.32b}$$

At $t = 0$, $u_e = u_0 = \Delta u = \Delta\sigma_v$, where u_0 is the initial excess pore water pressure. The transient flow stops when u_e becomes zero, which may not occur within a finite time period, as described later in this chapter.

143

The above problem involved an undrained loading followed by dissipation of excess pore water pressure during constant loading. If the construction is slow and/or the permeability is high, there is a possibility of no increase in pore water pressure during loading by construction; this loading is called drained loading. The difference between undrained and drained loadings is explained below.

If, during the loading process, an increment of pore water pressure is developed in a saturated soil mass but the pore water does not flow, the loading is called undrained loading. The most important feature of undrained loading is that there is no change in volume and water content of the soil mass. A relatively quick loading is considered an undrained loading. If, during the loading process, no increment of pore water pressure is developed in a saturated soil mass by allowing pore water to flow out of the voids, the loading is called drained loading. During drained loading, there is a gradual decrease in volume and water content of the soil mass. A relatively slow loading is considered a drained loading.

Note that the transient flow may occur with no change in the total load applied to the soil mass by lowering the water table. With quick lowering at $t = 0$, $u_e = u_0 = u_s - u_{ss}$, and u_e at any time $t > 0$ is given by Equation 5.32a, where u_{ss} is the steady-state pore water pressure long after the lowering (Lambe and Whitman, 1979).

The hydromechanical analogy given in Figure 5.9 illustrates the process or mechanics of 1D consolidation (Taylor, 1948). The spring represents the compressible soil skeleton of a saturated soil mass, and the water in the cylinder represents the water in the voids of the soil. The valve opening in the piston corresponds to the permeability of the soil, and the compressibility of the spring corresponds to the compressibility of the soil. The load applied on the piston corresponds to the structural load. The four cases given below provide an insight into the process of consolidation in a simple way. Note that Q is the initial load on the piston, ΔQ is the increase in load on the piston as an undrained loading at time $t = 0$, A is the area of the piston, V_0 is the initial volume of the saturated soil, V_f is the final volume of the saturated soil at the end of transient flow $(t = t_f)$, and V is the volume of the saturated soil during the transient flow at any time, t.

Case I: load Q on the piston with valve open $(t < 0)$

No flow of water occurring through the valve

$$\sigma_v = \frac{Q}{A}$$

$$u = 0$$

$$\sigma_v' = \frac{Q}{A}$$

$$V = V_0$$

Case II: load $Q + \Delta Q$ on the piston with valve closed $(t = 0)$

No flow of water occurring through the valve

$$\sigma_v = \frac{Q}{A} + \frac{\Delta Q}{A}$$

Figure 5.9 Hydromechanical model for 1D consolidation of saturated soil

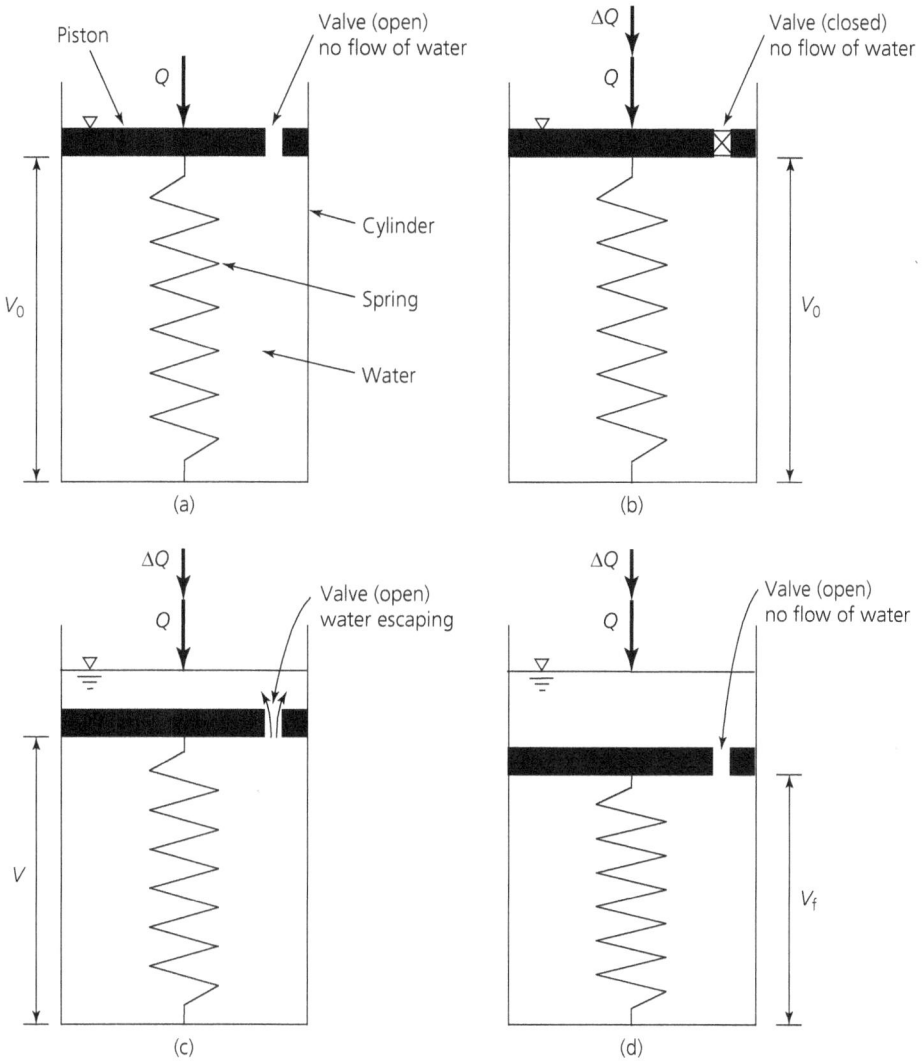

$$u = \frac{\Delta Q}{A}$$

$$\sigma'_v = \frac{Q}{A}$$

$$V = V_0$$

Case III: load $Q + \Delta Q$ on the piston with valve open ($0 < t < t_f$)

Water escaping through the valve under transient flow condition

$$\sigma_v = \frac{Q}{A} + \frac{\Delta Q}{A}$$

$$u < \frac{\Delta Q}{A}$$

$$\sigma_v' > \frac{Q}{A}$$

$$V < V_0$$

Case IV: load $Q + \Delta Q$ on the piston with valve open $(t = t_f)$

No flow occurring through the valve at the end of transient flow

$$\sigma_v = \frac{Q}{A} + \frac{\Delta Q}{A}$$

$$u = 0$$

$$\sigma_v' = \frac{Q}{A} + \frac{\Delta Q}{A}$$

$$V_f < V < V_0$$

Note that the additional load, ΔQ, is initially resisted by the water in the voids of the saturated soil, gradually transferred to the soil skeleton in the form of effective stress as a result of expulsion of pore water from the voids of the soil, and finally at the end of the transient flow – that is, at the end of the consolidation process – completely transferred to the soil skeleton. Additionally, the consolidation process leads to a decrease in both the volume and the water content of the soil.

Example 5.3

Referring to Figure 5.8, determine the total vertical stress (σ_v), pore water pressure (u) and effective vertical stress (σ_v') at the centre of the clay stratum and at the centre of the coarse sand stratum for the following three cases

(a) before placement of the sand fill ($t < 0$)
(b) immediately after completion of the placement of the sand fill ($t = 0$)
(c) after a very long time ($t \to \infty$).

Solution

(a) Before placement of the sand fill ($t < 0$)

In the clay

$$\sigma_v = \gamma z = (19.1)(3) = 57.3 \text{ kPa}$$

$$u = \gamma_w z = (9.81)(3) = 29.4 \text{ kPa}$$

$$\sigma_v' = \sigma_v - u = 57.3 - 29.4 = 27.9 \text{ kPa}$$

In the coarse sand

$$\sigma_v = \Sigma \gamma \Delta z = (19.1)(6) + (20.0)(2) = 154.6 \text{ kPa}$$

$$u = \gamma_w z = (9.81)(8) = 78.5 \text{ kPa}$$

$$\sigma_v' = \sigma_v - u = 154.6 - 78.5 = 76.1 \text{ kPa}$$

(b) Immediately after completion of the placement of the sand fill ($t = 0$)

In the clay

The clay will remain undrained, with no change in effective stress.

$$\sigma_v = \Sigma \gamma \Delta z = (15.5)(3) + (19.1)(3) = 103.8 \text{ kPa}$$

$$\sigma_v' = 27.9 \text{ kPa, as in case (a)}$$

$$u = \sigma_v - \sigma_v' = 103.8 - 27.9 = 75.9 \text{ kPa}$$

In the coarse sand

The sand will remain drained, with no change in pore water pressure.

$$\sigma_v = \Sigma \gamma \Delta z = (15.5)(3) + (19.1)(6) + (20.0)(2) = 201.1 \text{ kPa}$$

$$u = 78.5 \text{ kPa, as in case (a)}$$

$$\sigma_v' = \sigma_v - u = 201.1 - 78.5 = 122.6 \text{ kPa}$$

(c) After a very long time ($t \rightarrow \infty$)

In the clay

With the full dissipation of excess pore water pressure in the clay, there will be static or steady-state pore water pressure with the water table at the original ground level.

$$\sigma_v = 103.8 \text{ kPa, as in case (b)}$$

$$u = 29.4 \text{ kPa, as in case (a)}$$

$$\sigma_v' = \sigma_v - u = 103.8 - 29.43 = 74.4 \text{ kPa}$$

In the coarse sand

There will be no change, and the stresses are the same as those in case (b).

5.5.   Terzaghi's 1D consolidation theory

Consolidation theory refers to the mathematical equation that describes the dissipation of excess pore water pressure and associated deformation or settlement of the soil mass. Terzaghi (1925) presented the consolidation theory for 1D consolidation, which is often regarded as the beginning of modern soil mechanics. The analysis is based on the following assumptions.

- The soil is fully saturated ($S = 100\%$).
- The soil is homogeneous ($k \neq k(z)$, where z is the depth in the vertical direction).
- The flow of pore water and soil compression occur only in the z (vertical) direction $\left(\varepsilon_x = \varepsilon_y = 0 \right)$.
- Darcy's law for flow through porous media is valid $\left(v_z = k\dfrac{\partial h}{\partial z} \right)$.
- The pore water is incompressible ($\gamma_w = $ constant)
- The soil particles are incompressible – that is, the volume of soil solids in any soil element of dimensions dx, dy and dz is constant $\left(\dfrac{dxdydz}{1+e_0} = \text{constant} \right)$.
- Consolidation takes place under small changes in effective vertical stress – that is, the stress–strain relationship ($e - \sigma'_v$ relationship) is linear over the load increment ($a_v = $ constant; $m_v = $ constant).

From the principle of continuity of flow (Equation 4.36), considering the flow of pore water only in the z (vertical) direction with $k_z = k$

$$k\frac{\partial^2 h}{\partial z^2} = \frac{1}{1+e}\frac{\partial e}{\partial t}$$ (5.33)

From the stress–strain relationship (Equation 5.23a)

$$a_v = -\frac{\partial e}{\partial \sigma'_v}$$

or

$$\frac{\partial e}{\partial t} = -a_v \frac{\partial \sigma'_v}{\partial t}$$

By this substitution, Equation 5.33 becomes

$$\frac{k(1+e)}{a_v}\frac{\partial^2 h}{\partial z^2} = -\frac{\partial \sigma'_v}{\partial t}$$ (5.34)

Using Equation 5.32a, the total head, h, from Equation 4.2b (see Chapter 4) becomes

$$h = z + \frac{u}{\gamma_w} = z + \frac{1}{\gamma_w}(u_e + u_{ss})$$

Substituting this value of h into Equation 5.34 with a note that the steady-state pore water pressure u_{ss} varies linearly with depth

$$\frac{k(1+e)}{\gamma_w a_v} \frac{\partial^2 u_e}{\partial z^2} = -\frac{\partial \sigma_v'}{\partial t}$$

or

$$c_v \frac{\partial^2 u_e}{\partial z^2} = -\frac{\partial \sigma_v'}{\partial t} \tag{5.35}$$

where

$$c_v = \frac{k(1+e)}{\gamma_w a_v} = \frac{k}{\gamma_w m_v} \tag{5.36}$$

is called the coefficient of consolidation, a soil property, and m_v is given by Equation 5.25b or 5.27.

As $\sigma_v' = \sigma_v - (u_{ss} + u_e)$, and $\frac{\partial u_{ss}}{\partial t} = 0$, Equation 5.35 reduces to

$$c_v \frac{\partial^2 u_e}{\partial z^2} = \frac{\partial u_e}{\partial t} - \frac{\partial \sigma_v}{\partial t} \tag{5.37}$$

If the applied load remains constant during the consolidation process – that is, the total stress σ_v is constant with time – Equation 5.37 becomes

$$c_v \frac{\partial^2 u_e}{\partial z^2} = \frac{\partial u_e}{\partial t} \tag{5.38}$$

Equation 5.38 is popularly known as Terzaghi's consolidation equation. A similar differential equation is also applicable to other physical problems such as the transient heat flow problem. A solution to this partial differential equation for a given set of initial and boundary conditions provides information on how the excess pore water pressure, u_e, due to loading on a saturated soil stratum dissipates with time, t, and location, z, within the soil stratum. The consolidation process is completed with a stop of the transient flow of water from the voids within the soil stratum when u_e becomes zero at all locations.

The solution to Equation 5.38 uses the Fourier series expansion method, details of which can be found in Taylor (1948). This equation is general and valid for any distribution of initial excess pore water pressure with depth in a consolidating soil stratum of thickness or depth $z = 2H$ with z measured downward from the top surface of the stratum, as shown in Figure 5.10. Various distributions of the initial excess pore water pressure with depth are possible. In a relatively thin (lower value of $2H$) layer of fine-grained soil with drainage at both the top and the bottom, a uniform or constant distribution with depth can be considered (Figure 5.10(b)), which is of major interest for most engineers, being the simplest case of consolidation. This is considered here to present the solution to Equation 5.38. Note that for the total vertical stress at any point to be constant during the consolidation process as well as for a uniform distribution of initial excess pore water pressure with depth, the loading through construction or other activities on the soil stratum should be applied in a time that is very small compared to the consolidation time; in other words, no consolidation should occur during the loading or construction period.

Figure 5.10 (a) A clay stratum with a two-way drainage; (b) a sequence of isochrones, presenting the excess pore water pressure distributions

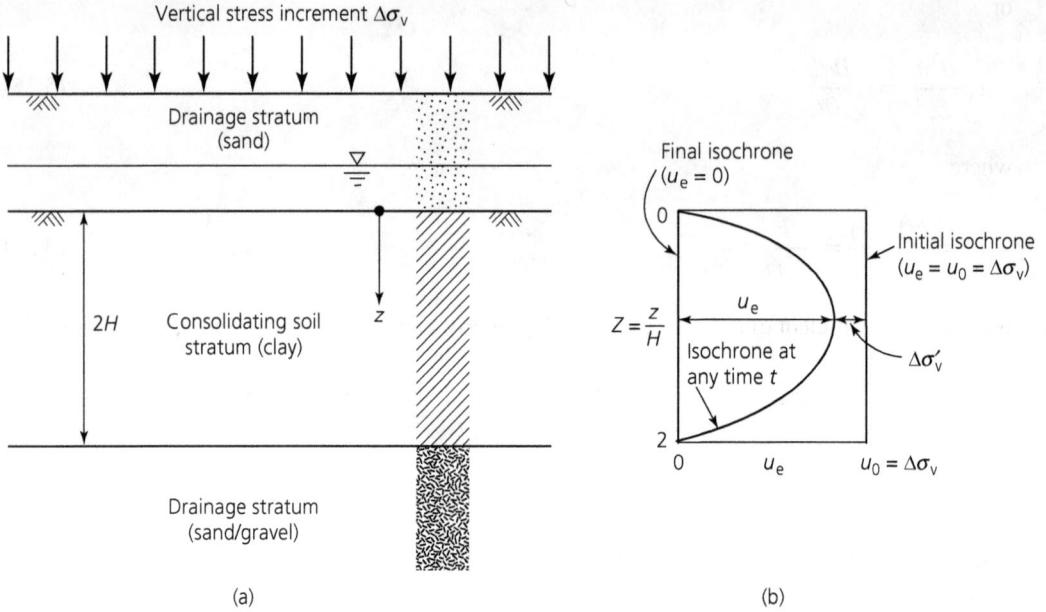

(a)

(b)

For convenience in presenting the solutions, Equation 5.38 is modified as

$$\frac{\partial^2 u_e}{\partial Z^2} = \frac{\partial u_e}{\partial T_v} \tag{5.39}$$

where Z and T_v are nondimensional variables, defined as

$$Z = \frac{z}{H} \tag{5.40}$$

and

$$T_v = \frac{c_v t}{H^2} \tag{5.41}$$

The nondimensional variable T_v is known as the time factor, and H is the maximum drainage path length, which is one-half of the thickness $2H$ of the consolidating stratum in the present case due to the drainage at both its top and bottom. Consider the following conditions.

Initial condition at $t = 0$ (Figure 5.10(b))

$u_e = u_0$, for $0 \leq Z \leq 2$

where u_0 is the initial excess pore water pressure.

Boundary condition at all $t > 0$ (Figure 5.10(b))

$u_e = 0$, for $Z = 0$ and $Z = 2$

In Figure 5.10(b), initial, intermediate and final excess pore water pressure contours, called the isochrones, can be observed. If $\Delta\sigma_v$ is the total vertical stress increment at any point within the fully saturated consolidating soil stratum by loading at a given time, t, during the consolidation process, then the equilibrium requires that $\Delta\sigma_v = \Delta\sigma_v' + u_e$, where $\Delta\sigma_v'$ is the effective vertical stress increment and u_e is the excess pore water pressure at that point at time, t.

The solution to Equation 5.3 is (Lambe and Whitman, 1979; Taylor, 1948; Terzaghi et al., 1996)

$$u_e(z, t) = \sum_{m=0}^{m=\infty} \frac{2u_0}{M} \sin(MZ)e^{-M^2 T_v} \tag{5.42}$$

or

$$u_e(z, t) = \Delta\sigma_v \sum_{m=0}^{m=\infty} \frac{2}{M} \sin(MZ)e^{-M^2 T_v} \tag{5.43}$$

where

$$M = \frac{\pi}{2}(2m+1) \tag{5.44}$$

where m is a positive integer varying from 1 to ∞ – that is, $m = 1, 2, 3,...$

The degree of consolidation, also known as the consolidation ratio or degree of compression, U_z, of a saturated soil stratum of thickness $2H$ is defined as

$$U_z = 1 - \frac{u_e}{u_0} \tag{5.45}$$

Using Equation 5.42, Equation 5.45 reduces to

$$U_z(z, t) = 1 - \sum_{m=0}^{m=\infty} \frac{2}{M} \sin(MZ)e^{-M^2 T_v} \tag{5.46}$$

Note that U_z expresses the amount of compression or settlement at a specific depth, z, and time, t. Equation 5.46 can be plotted to have U_z against z curves for T_v varying from 0 to any value approaching ∞, as shown in Figure 5.11.

As geotechnical engineers are often concerned with the total compression or settlement of the whole soil stratum at a particular time, t, during the consolidation process, it is common practice

Figure 5.11 Plot of degree of consolidation, U_z, against depth or thickness, z, of the consolidating soil stratum for typical values of time factor, T_v, with graphical interpretation of average degree of consolidation, U

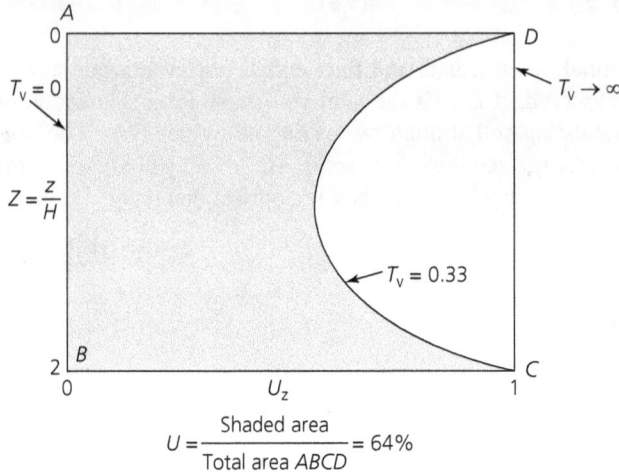

$$U = \frac{\text{Shaded area}}{\text{Total area } ABCD} = 64\%$$

to express the total compression or settlement in terms of the average degree of consolidation, also known as the average consolidation ratio or average degree of compression, defined as

$$U = \frac{\rho}{\rho_f} \tag{5.47}$$

where ρ is the compression or settlement at time, t (or T_v), and ρ_f is the compression or settlement at the end of consolidation when u_e becomes zero throughout the thickness or depth $2H$. U is generally expressed as a percentage.

As explained in Figure 5.11, U can be interpreted as the ratio of shaded area to the total area on a U_z against z plot. Equation 5.42 or Equation 5.46 can be used, by integration to obtain the expression for U as a function of time factor, T_v, as

$$U = 1 - \sum_{m=0}^{m=\infty} \frac{2}{M^2} e^{-M^2 T_v} \tag{5.48}$$

Equation 5.48 is plotted in Figure 5.12 where it can be seen that initially U decreases rapidly, but the rate of decrease gradually slows. As U approaches $T_v = 1$ asymptotically, theoretically the consolidation process never ends. However, $T_v = 1$ is often taken as the end of the consolidation process for most engineering works. Equation 5.48 can be represented with high precision using the following empirical expressions (Taylor, 1948).

$$T_v = \frac{\pi}{4}\left(\frac{U}{100}\right)^2 \quad \text{for } U < 60\% \tag{5.49}$$

and

$$T_v = 1.781 - 0.933 \log_{10}(100 - U) \quad \text{for } U > 60\% \tag{5.50}$$

Figure 5.12 Graphical representation of the U–T_v relationship defined by Equation 5.48

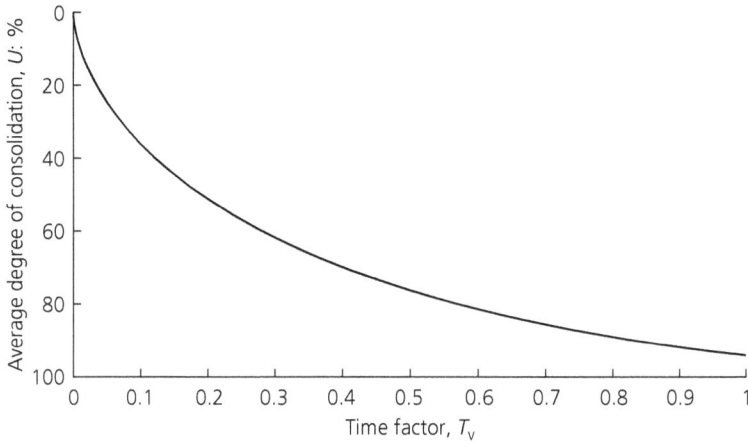

Example 5.4

A 6 m-thick saturated clay stratum has drainage at both its top and bottom surfaces due to the presence of sand strata. The time for 50% consolidation is 2 years. If the coefficient of volume change m_v is 1.46×10^{-3} m^2/kN, determine the following

(a) coefficient of consolidation c_v, in m²/yr
(b) coefficient of permeability, k, in m/yr.

Solution

Given $H = 6/2 = 3$ m because the clay stratum has two drainage surfaces, $U = 50\%$, $t = 2$ years, and $m_v = 1.46\times10^{-3}$ m^2/kN

(a) From Equation 5.49

$$T_v = \frac{\pi}{4}\left(\frac{U}{100}\right)^2 = \frac{\pi}{4}\left(\frac{50}{100}\right)^2 = 0.196$$

From Equation 5.41

$$T_v = \frac{c_v t}{H^2}$$

or

$$c_v = \frac{T_v H^2}{t} = \frac{(0.196)(3)^2}{2} = 0.882 \text{ m}^2 / \text{yr}$$

(b) From Equation 5.36

$$c_v = \frac{k}{\gamma_w m_v}$$

or

$$k = c_v \gamma_w m_v = (0.882)(9.81)(1.46 \times 10^{-3}) = 0.013 \text{ m} / \text{yr}$$

The following points regarding consolidation theory are noteworthy.

- Since k, m_v and γ_w are assumed to be constant during consolidation, c_v is constant during the consolidation process. The SI unit of c_v is m²/s, but it is generally presented in m²/year.
- In a relatively thick layer of fine-grained soil, the initial excess pore water pressure distribution with depth may be triangular or trapezoidal.
- The assumption of a uniform distribution of initial excess pore water pressure applies in the laboratory 1D consolidation test, but it may not be strictly applicable in cases of consolidation in nature (Taylor, 1948).
- If consolidation occurs during the loading or construction period, suitable corrections are required to use the solution to Equation 5.38 as presented in this chapter for a uniform distribution of initial excess pore water pressure.

5.6. Determination of c_v

In view of the importance of the coefficient of consolidation, c_v, in practical applications, several methods, classified as standard and alternative methods, have been proposed in the past for its determination, mainly using results from the laboratory confined compression test with incremental loading. Standard methods, namely Casagrande's logarithm of time fitting method (Casagrande and Fadum, 1940; Taylor, 1948) and Taylor's square root of time fitting method (Taylor, 1948), are the most widely used methods in practical designs as well as in research since their developments. Shukla et al. (2009) presented a critical review of all the available methods.

5.6.1 Casagrande's logarithm of time fitting method

In this method, determining the coefficient of consolidation typically requires compression readings to be taken for at least 24 hours, allowing accurate evaluation of the slope of the compression curve attributed to secondary compression on a compression against logarithm of time plot. The procedure for determining c_v is as follows (Figure 5.13).

- Plot the compression dial readings (δ, normally in mm) against logarithm of time (t, normally in minutes) for each increment of load.
- Draw a straight line through the points representing the final readings which exhibit a straight-line trend and constant slope.
- Draw a second line tangent to the steepest part of the curve.
- Read the compression and time corresponding to the intersection of the two straight lines that represent 100% primary consolidation and denote them by δ_{100} and t_{100}, respectively.
- Select two early times t_1 and t_2 ($= 4t_1$) such that the compression at t_2 should be greater than one-quarter but less than one-half of the total compression for the load increment.

Figure 5.13 Analysis for c_v using Casagrande's logarithm of time fitting method (adapted from Lambe and Whitman, 1979; Taylor, 1948)

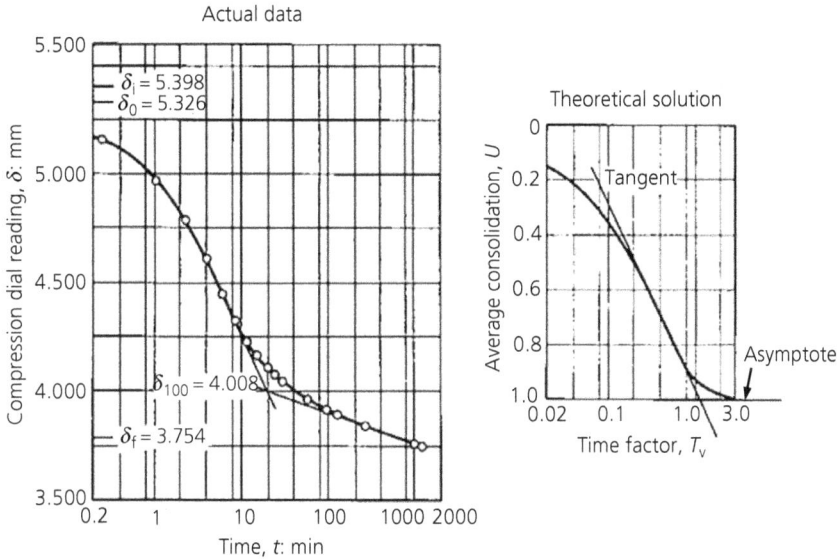

Note: δ_i is the initial dial reading; δ_0 is the corrected zero point; δ_f is the final dial reading.

▨ Obtain the offset between t_1 and t_2, and plot this offset distance above t_1 to obtain the compression ordinate (δ_0) representing 0% primary consolidation.

▨ Take the average of the compressions corresponding to the 0 and 100% consolidations to obtain the compression point δ_{50} ($= (\delta_0 + \delta_{100})/2$) corresponding to 50% primary consolidation.

▨ Read graphically the time (t_{50}) required for 50% consolidation ($T_v = 0.197$).

▨ Finally calculate the coefficient of consolidation (c_v), generally reported in m²/year, for each increment of load using Equation 5.41 as

$$c_v = \frac{0.197 H^2}{t_{50}} \tag{5.51}$$

5.6.2 Taylor's square root of time fitting method

In this method, readings of a compression dial are taken up to a time corresponding to at least 90% of primary compression of the test specimen under the increment of loading being considered. The 90% consolidation time varies, of course, with different specimens and with different thicknesses of specimens, but the compression dial is read frequently for a period of up to 1 h after the initiation of loading. The procedure for determining c_v is as follows (Figure 5.14).

▨ Plot δ against $\sqrt{t}$ for the specific load increment.

▨ Draw a straight line through the points representing the initial readings that exhibit a straight-line trend.

▨ Extrapolate the line back to $t = 0$ to obtain the dial compression ordinate (δ_0) representing 0% primary consolidation.

Analysis for c_v using Taylor's square root of time fitting method (adapted from Lambe and Whitman, 1979; Taylor, 1948)

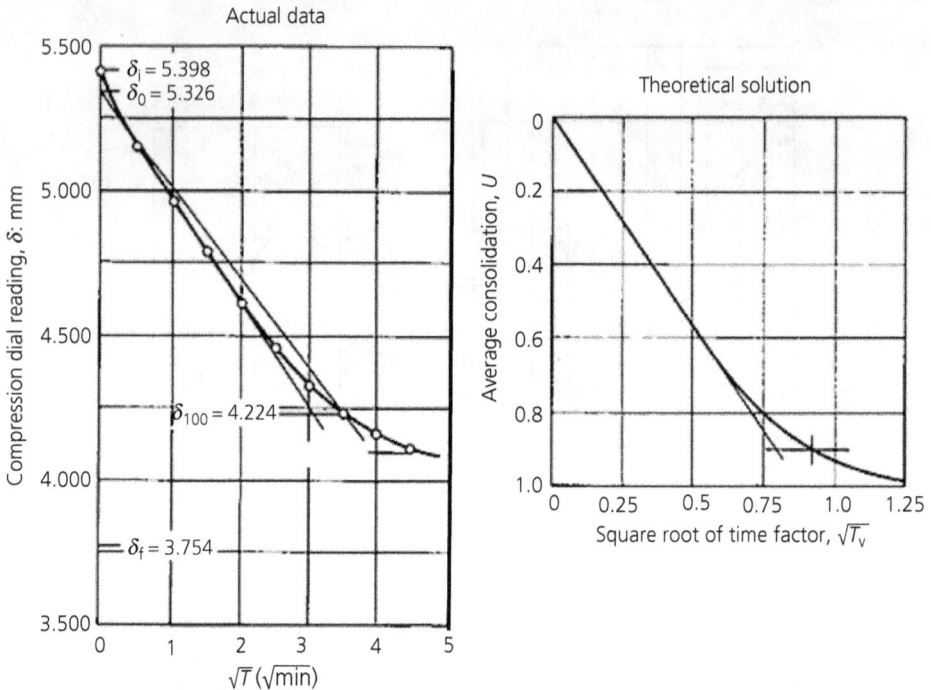

Note: δ_i is the initial dial reading; δ_0 is the corrected zero point; δ_f is the final dial reading.

- Draw a second straight line through the 0% compression ordinate so that the abscissa of this line is 1.15 times the abscissa of the first line through the data.
- Read the compression and time corresponding to the intersection of this second line with the curve that represents 90% primary consolidation ($T_v = 0.848$), and denote them by δ_{90} and t_{90}, respectively.
- Finally calculate c_v for each increment of load using Equation 5.41 as

$$c_v = \frac{0.848H^2}{t_{90}} \tag{5.52}$$

Both the above methods replicate the consolidation process as the test proceeds in many soils and they are, therefore, most widely used in practice despite some limitations. These methods can be used to determine both the initial compression corresponding to 0% consolidation and the compression corresponding to 100% primary consolidation. Casagrande's logarithm of time fitting method can be applied only to typical S-shaped curves but for other curves, the method is unsatisfactory. Taylor's square root of time fitting method requires that the dial gauge readings should be taken at close intervals of time after the specimen has been loaded until 90% consolidation is attained, but this is sometimes difficult. For some soils, the curve does not have an initial linear portion at all, but instead exhibits continuous curvature, or the curve may have a negative value for the initial compression; for all such soils, Taylor's method cannot be applied successfully for reliable values of c_v. If the results of standard methods are compared, it is found that Taylor's method yields values of c_v larger than those obtained from Casagrande's method (Lambe and Whitman, 1979; Leonards, 1962; Shukla et al., 2009). Leroueil (1988) has reported that the ratio of c_v value obtained by Casagrande's method to that obtained by Taylor's method varies between 0.2 and 1. To find corrected initial settlement (δ_0), Taylor's method is expected to give more realistic values. To find δ_{100}, the two methods are both theoretically valid if no secondary consolidation effects occur.

Note that the total compression or settlement in the loading increment of a laboratory test (Figures 5.13 and 5.14) has the following three parts (Taylor, 1948)

- initial compression or settlement, extending from δ_i to δ_0
- primary compression or settlement, extending from δ_0 to δ_{100}
- secondary compression or settlement, extending from δ_{100} to δ_f.

5.7. Secondary consolidation

The previous sections have discussed the reduction in volume of a fully saturated soil mass by the process of primary consolidation, which theoretically ends when the excess pore water pressure, u_e, induced by the applied vertical stress increment, $\Delta\sigma_v$, becomes zero and is completely transferred to the soil skeleton as an increase in effective vertical stress increment, $\Delta\sigma'_v$, with a decrease in water content. The decrease in volume may continue further at a gradually decreasing rate at the constant, $\Delta\sigma'_v$, by the gradual plastic readjustment of soil fabric or structure owing to its viscous nature. This process of volume reduction that causes an additional increase in strength with time is called secondary consolidation, also known as secondary compression, delayed compression or creep. The mechanism of secondary consolidation involves sliding at interparticle contacts, expulsion of water from microfabric elements, and rearrangement of adsorbed water molecules and cations into different positions (Mitchell and Soga, 2005). At this stage, the void ratio of a soil mass is a function of both the effective vertical consolidation stress and the time, and it also depends on the rate of loading. As secondary consolidation refers to the volume change at constant effective stress, Terzaghi's consolidation theory cannot be applied to determine the rate of secondary consolidation.

In the plot of the void ratio (e) against the logarithm of time (t) – that is, the $e - \log_{10} t$ plot – based on the confined compression test data, the $e - \log_{10} t$ relationship is linear for most soils over the time ranges of interest following the primary consolidation, as illustrated in Figure 5.15, with a decrease in e along the vertical axis. The coefficient of secondary compression, also known as the secondary compression index, C_α, that defines the rate of secondary consolidation and gives the amount of secondary compression is expressed as (Terzaghi et al., 1996)

$$C_\alpha = -\frac{de}{d(\log_{10} t)} \tag{5.53a}$$

or

$$C_\alpha = -\frac{\Delta e}{\Delta(\log_{10} t)} \tag{5.53b}$$

To determine C_α, Δe is usually taken over one log cycle or decade of time, t. Note that in Equation 5.53, $t > t_f$, where t_f is assumed time at the end of primary consolidation, as indicated in

Figure 5.15 A typical $e - \log_{10} t$ relationship (Note: t_f is the assumed time at the end of primary consolidation, as explained by Mitchell and Soga, 2005)

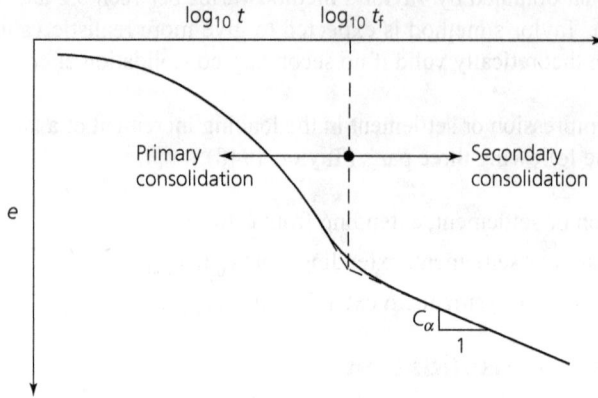

Figure 5.15 A typical $e - \log_{10} t$ relationship (Note: t_f is the assumed time at the end of primary consolidation, as explained by Mitchell and Soga, 2005)

Figure 5.15. The value of C_α/C_c varies from 0.02 ± 0.01 for granular soils, including rockfill, to 0.06 ± 0.01 for peats (Terzaghi et al., 1996). The value of C_α for normally consolidated soils is generally greater than that for overconsolidated soils. For most inorganic soils, the rate of ground surface settlement due to secondary consolidation is significantly lower than that from primary consolidation. As a result, the effects of secondary consolidation are typically neglected in settlement analysis. For organic soils and peats, ground surface settlement due to secondary consolidation may exceed that caused by primary consolidation.

Studies indicate that secondary consolidation occurs simultaneously with primary consolidation, becoming more pronounced after approximately 60% of primary consolidation is complete. The separation of the effects of primary and secondary consolidations to predict the ground surface settlement is not a straightforward task. The effect of secondary consolidation occurring during primary consolidation is to reduce c_v values. The conventional Casagrande's and Taylor's methods rely on the later stages of consolidation. Casagrande's method uses $U = 0\%$ and $U = 100\%$, whereas Taylor's method uses $U = 0\%$ and $U = 90\%$ data points of the laboratory consolidation curve. The values of c_v obtained from these standard methods are more influenced by secondary consolidation effects; therefore, a suitable alternate method such as the revised logarithm of time fitting method proposed by Robinson and Allam (1996), which is based on Casagrande's method, and the inflection point method which have several advantages, may be used to obtain more realistic values of c_v (Shukla et al., 2009).

5.8.   Computation of settlement

The total settlement or compression, ρ, of a loaded soil mass generally consists of the following three components.

▨ Immediate (or distortion) settlement (also known as initial consolidation settlement or initial compression), ρ_i. This occurs due to a relatively sudden reduction in the volume of a soil mass under an applied load, primarily caused by the expulsion and compression of gas or air present in the voids, if any, prior to primary consolidation. In fully saturated cohesive soils, immediate settlement generally occurs because of distortion of the soil without any volume change under undrained condition. As it is commonly estimated using elastic theory, this settlement is sometimes called the elastic settlement, even though the deformation itself is not truly elastic.

▨ Primary consolidation settlement (also known as primary compression), ρ_c. This results from a decrease in the volume of a soil mass under a sustained load, primarily due to the expulsion of water from the void spaces, accompanied by the transfer of load from the pore water to the soil solids.

▨ Secondary consolidation settlement (also known as secondary compression or creep), ρ_s. This results from a reduction in the volume of a soil mass under a sustained load, primarily due to the adjustment of the internal structure of the soil mass after most of the load has been transferred from the pore water to the soil solids.

Thus, the total settlement is

$$\rho = \rho_i + \rho_c + \rho_s \tag{5.54}$$

In cohesionless soils and partially saturated clays, the immediate settlement, ρ_i, predominates with, perhaps, some secondary compression, ρ_s (Bowles, 1996). In inorganic cohesive soils, the consolidation settlement, ρ_c, predominates. The secondary compression, ρ_s, may predominate in peats and highly organic soils.

5.8.1   Immediate settlement

The settlement at the corner of a rectangular surface area of a homogeneous, isotropic, elastic half-space carrying a uniform vertical stress, q, may be computed from (Terzaghi, 1943)

$$\rho_i = qB\frac{1-\mu^2}{E}I_\rho \tag{5.55}$$

where B is the width (least dimension) of the rectangle; μ is the Poisson's ratio of the material; E is the Young's modulus of elasticity of the material; I_ρ is an influence coefficient or factor for the settlement, which accounts for the shape and rigidity of the foundation that applies the uniform vertical stress, $\Delta\sigma_v$. The values of I_ρ for some situations are given in Table 5.1.

Table 5.1 Influence factor, I_ρ, for uniformly loaded areas on an elastic half-space of infinite depth

Shape		I_ρ for flexible foundation			I_ρ for rigid foundation
		Centre	Corner	Average	
Circle[a]		1.00	0.64	0.85	0.79
Square		1.12	0.56	0.95	0.82
Rectangle	$L/B = 2$	1.53	0.76	1.30	1.12
L = length	$L/B = 5$	2.10	1.05	1.82	1.60
B = width	$L/B = 10$	2.56	1.28	2.24	2.0

[a]For a circle, B is the diameter in Equation 5.55 (after Day, 2005; Department of the Navy, 1982).

Note that for a finite depth of the soil stratum, the values of I_p will be lower than those given in Table 5.1, depending on the Poisson's ratio (Department of the Navy, 1982). For foundations embedded in soil, the settlement value will be lower than that computed using Equation 5.55, hence a depth correction is required (Fox, 1948). The settlement at points other than the corner of a rectangular area, as well as for any shape of a loaded area that can be divided into rectangles, can be determined using the method of superposition, as discussed in Chapter 3 for computing stresses. For complex problems, commercially available computer programs based on finite element analysis can be used to compute the immediate settlement assuming the soil to be elastic or inelastic.

For soils, the value of E can be determined as the initial tangent modulus or secant modulus from the stress–strain curve from a consolidated undrained triaxial test (see Chapter 6) by maintaining the consolidation pressure equal to the effective vertical stress at the depth from where the soil sample is collected. As the value of the modulus of elasticity, E, of a cohesionless soil increases with confining pressure and hence increases with depth, and also varies across the width of the loaded area, the settlement of a loaded cohesionless soil stratum cannot be computed accurately using Equation 5.55. Therefore, the settlement in cohesionless soils is generally computed using either field tests or the semi-empirical approaches that consider the variation of E with depth (Schmertmann et al., 1978).

Example 5.5

A loaded rigid rectangular footing, 1.5 m by 3 m in plan, applies a uniform vertical stress of 200 kPa on a very thick saturated cohesive soil layer. The Young's modulus of elasticity and the Poisson's ratio of the soil are 5.5×10^4 kPa and 0.5, respectively. Determine the immediate settlement of the footing.

Solution

Given $B = 1.5$ m, $L = 3$ m, $E = 5.5 \times 10^4$ kPa and $\mu = 0.5$.

Here, $L / B = 2$; hence, from Table 5.1, the influence coefficient, $I_p = 1.12$.

From Equation 5.55, the immediate settlement

$$\rho_i = qB\frac{1-\mu^2}{E}I_\rho = (200)(1.5)\frac{1-(0.5)^2}{5.5\times10^4}(1.12) = 4.58\times10^{-3} \text{ m} = 4.58 \text{ mm}$$

5.8.2 Primary consolidation settlement

To compute the primary consolidation settlement of a clay stratum of depth or thickness H_0, the following must be determined

- overconsolidation ratio, OCR $\left(= \sigma'_{vmax} / \sigma'_{v0}\right)$ (Equation 5.29), which indicates whether the soil stratum is normally consolidated or overconsolidated/precompressed, where σ'_{vmax} is the maximum past effective vertical stress (or preconsolidation pressure), and σ'_{v0} is the present effective vertical stress

- in situ void ratio, e_0, and compression index, C_c, and/or recompression index, C_r, as applicable
- average initial or present effective vertical stress, σ'_{v0}, before any increase in stress due to the loading
- average increase in effective vertical stress, $\Delta\sigma'_v$, within the soil stratum due to loading, $\Delta\sigma_v$, over the ground surface by placement of a structure or fill ($\Delta\sigma'_v = \Delta\sigma_v$ if the width of the loaded surface area is very large compared to H_0 and its depth below the ground surface is very low so that there is no horizontal strain or deformation within the soil stratum and hence 1D consolidation takes place).

Case I: the soil is normally consolidated (OCR $= 1$, i.e. $\sigma'_{v0} = \sigma'_{v\,max}$)

From Equations 5.27 and 5.28, taking $\Delta H = \rho_c$

$$\rho_c = m_v\, \Delta\sigma'_v\, H_0 = \frac{1}{D}\Delta\sigma'_v\, H_0 \tag{5.56}$$

From Equation 5.22 and Equation 5.24b, taking $\Delta H = \rho_c$ and $\sigma'_v = \sigma'_{v0} + \Delta\sigma'_v$

$$\rho_c = \frac{C_c}{1+e_0} H_0 \log_{10}\left(\frac{\sigma'_{v0} + \Delta\sigma'_v}{\sigma'_{v0}}\right) \tag{5.57}$$

To determine the primary consolidation settlement, Equation 5.56 or Equation 5.57 can be used, depending on the availability of compressibility parameters.

Example 5.6

A 9 m-thick normally consolidated clay stratum is subjected to a stress increment of 60 kPa. If the drainage takes place at both the top and the bottom surfaces of the stratum, and it undergoes 50% consolidation in one year, determine the coefficient of consolidation. If the coefficient of permeability is 1.8×10^{-9} m/s, determine the primary consolidation settlement in one year.

Solution

Given $H_0 = 9$ m, $\Delta\sigma'_v = 60$ kPa, $U = 50\%$, $k = 1.8\times10^{-9}$ m/s and $t = 1$ year.

From Equation 5.49, for 50% consolidation – that is, for an average degree of consolidation $U = 50\%$ – the time factor

$$T_v = \frac{\pi}{4}\left(\frac{U}{100}\right)^2 = \frac{\pi}{4}\left(\frac{50}{100}\right)^2 = 0.196$$

$H = H_0/2 = 9/2 = 4.5$ m, because the clay stratum has two drainage surfaces.

From Equation 5.41, the time factor

$$T_v = \frac{c_v t}{H^2}$$

or

$$c_v = \frac{H^2 T_v}{t} = \frac{(4.5)^2 (0.196)}{1} = 3.969 \text{ m}^2/\text{yr} = 1.26 \times 10^{-7} \text{ m}^2/\text{s}$$

From Equation 5.36, the coefficient of consolidation

$$c_v = \frac{k}{\gamma_w m_v}$$

or the coefficient of volume change

$$m_v = \frac{k}{\gamma_w c_v} = \frac{1.8 \times 10^{-9}}{(9.81)(1.26 \times 10^{-7})} = 1.46 \times 10^{-3} \text{ m}^2/\text{kN}$$

From Equation 5.54, the final or total primary consolidation settlement

$$\rho_c = m_v \Delta \sigma_v' H_0 = (1.46 \times 10^{-3})(60)(9) = 0.788 \text{ m}$$

From Equation 5.47, the primary consolidation settlement in one year

$$\rho = U \rho_c = (0.5)(0.788) = 0.394 \text{ m}$$

Case II: the soil is overconsolidated (OCR > 1) and $(\sigma_{v0}' + \Delta \sigma_v') \leq \sigma_{v\max}'$

Following Equation 5.57 with replacement of compression index, C_c, by recompression index, C_r

$$\rho_c = \frac{C_r}{1+e_0} H_0 \log_{10} \left(\frac{\sigma_{v0}' + \Delta \sigma_v'}{\sigma_{v0}'} \right) \tag{5.58}$$

Case III: the soil is overconsolidated (OCR > 1) and $(\sigma_{v0}' + \Delta \sigma_v') > \sigma_{v\max}'$

For the vertical stress increment from σ_{v0}' to $\sigma_{v\max}'$, following Equation 5.57 with replacement of C_c by C_r

$$\rho_{c1} = \frac{C_r}{1+e_0} H_0 \log_{10} \left(\frac{\sigma_{v\max}'}{\sigma_{v0}'} \right) = \frac{C_r}{1+e_0} H_0 \log_{10} (OCR) \tag{5.59}$$

For the vertical stress increment from $\sigma_{v\max}'$ to $\sigma_v' = \sigma_{v0}' + \Delta \sigma_v'$, following Equation 5.57

$$\rho_{c2} = \frac{C_c}{1+e_0} H_0 \log_{10} \left(\frac{\sigma_{v0}' + \Delta \sigma_v'}{\sigma_{v\max}'} \right) \tag{5.60}$$

The total primary consolidation settlement

$$\rho_c = \rho_{c1} + \rho_{c2} = \frac{C_r}{1+e_0} H_0 \log_{10}(OCR) + \frac{C_c}{1+e_0} H_0 \log_{10}\left(\frac{\sigma'_{v0} + \Delta\sigma'_v}{\sigma'_{v\max}}\right)$$

or

$$\rho_c = \frac{H_0}{1+e_0}\left[C_r \log_{10}(OCR) + C_c \log_{10}\left(\frac{\sigma'_{v0} + \Delta\sigma'_v}{\sigma'_{v\max}}\right)\right] \tag{5.61}$$

Note that ρ_{c1} may be very small compared to ρ_{c2} and can, therefore, be neglected in computing the total primary consolidation settlement.

Example 5.7

For a 3 m-thick overconsolidated clay stratum at a construction site, the following details are available: initial void ratio, $e_0 = 1.25$; preconsolidation pressure, $\sigma'_{v\max} = 90$ kPa, compression index, $C_c = 0.07$, recompression index, $C_r = 0.28$, and average initial effective vertical stress, $\sigma'_{v0} = 60$ kPa. Determine the total primary consolidation settlement of the clay stratum due to a vertical stress increment of 75 kPa by the placement of a fill over a large area.

Solution

From Equation 5.29, the overconsolidation ratio

$$OCR = \frac{\sigma'_{v\max}}{\sigma'_{v0}} = \frac{90}{60} = 1.5$$

From Equation 5.61, the total primary consolidation settlement

$$\rho_c = \frac{H_0}{1+e_0}\left[C_r \log_{10}(OCR) + C_c \log_{10}\left(\frac{\sigma'_{v0} + \Delta\sigma'_v}{\sigma'_{v\max}}\right)\right]$$

$$= \frac{3}{1+1.25}\left[(0.28)\log_{10}(1.5) + (0.07)\log_{10}\left(\frac{60+75}{90}\right)\right]$$

$$= (1.33)(0.049 + 0.012) = 0.081 \text{ m} = 81 \text{ mm}$$

Case IV: the soil is underconsolidated (OCR < 1 i.e. $\sigma'_{v\max} < \sigma'_{v0}$)

In this case, the primary consolidation settlement

$$\rho_c = \frac{C_c}{1+e_0} H_0 \log_{10}\left(\frac{\sigma'_{v0} + \Delta\sigma'_v}{\sigma'_{v\max}}\right) \tag{5.62}$$

5.8.3   Secondary consolidation settlement

From Equation 5.53b and Equation 5.22, taking $\Delta H = \rho_s$ and replacing e_0 by e_f

$$C_\alpha = -\frac{\Delta e}{\Delta(\log_{10} t)} = \frac{\rho_s(1+e_f)}{H_0}\frac{1}{\Delta(\log_{10} t)}$$

or

$$\rho_s = \frac{C_\alpha}{1+e_f}H_0\Delta(\log_{10} t) = \frac{C_\alpha}{1+e_f}H_0\log_{10}\left(\frac{t_f+\Delta t}{t_f}\right) \tag{5.63}$$

where Δt is the time interval over which the secondary consolidation settlement is computed, and is measured from the end of the primary consolidation time, t_p as shown in Figure 5.15.

Chapter summary

1. The property of soil that describes its susceptibility to a decrease in volume when subjected to a load is known as compressibility. This property depends on factors such as soil type or composition, effective stress and stress history.

2. Consolidation is defined as the transient process of dissipation of excess pore water pressure accompanied by a change in volume of saturated soil at a constant total stress. Initially the applied stress is resisted by the water in the voids, gradually transferred to the soil skeleton in the form of effective stress because of expulsion of pore water from the voids of the soil, and finally at the end of the transient flow, completely transferred to the soil skeleton.

3. The type of test to study the stress–strain behaviour of a soil varies with the type of loading observed in actual problems. For saturated clays subjected to undrained loading, Poisson's ratio, $\mu = 0.5$, and volumetric strain, $\varepsilon_v = 0$.

4. The best feature of the confined compression test is that the strain or deformation condition of the soil specimen in the test approximates certain field problems. For 1D consolidation, the compressibility parameters are obtained from the confined compression test.

5. The compression index C_c is used to determine the magnitude of primary consolidation settlement in a normally consolidated soil, while the coefficient of consolidation C_v is used to determine the rate of settlement during primary consolidation.

6. The ratio of maximum past effective vertical stress ($\sigma'_{v\,max}$) to present effective vertical stress (σ'_{v0}) is called the overconsolidation ratio (OCR), which is used to express the degree of overconsolidation of a soil.

7. Consolidation theory refers to the mathematical equation that describes the dissipation of excess pore water pressure and associated deformation or settlement of the soil mass.

8. It is common practice to express the total compression or settlement in terms of the average degree of consolidation.

9. The mechanism of secondary consolidation involves sliding at interparticle contacts, expulsion of water from microfabric elements and rearrangement of adsorbed water molecules and cations into different positions.

10. The total settlement or compression (ρ) of a loaded soil mass generally consists of the following three components: immediate settlement, primary consolidation settlement and secondary consolidation settlement.

Questions for practice

Select the most appropriate answer for each of the multiple choice questions from Q5.1 to Q5.10. Questions Q5.11 to Q5.28 require more detailed answers.

5.1 The shear modulus of elasticity, G, Young's modulus of elasticity, E, and Poisson's ratio, μ, of an isotropic material are related as

(a) $E = \dfrac{G}{2(1-\mu)}$

(b) $E = \dfrac{G}{2(1+\mu)}$

(c) $G = \dfrac{E}{2(1-\mu)}$

(d) $G = \dfrac{E}{2(1+\mu)}$.

5.2 The consolidation process in a saturated soil
   (a) is a function of total stress
   (b) is a function of excess pore water pressure
   (c) is a function of effective stress
   (d) does not depend on the present stress.

5.3 When consolidation of a saturated soil mass takes place, its degree of saturation
   (a) decreases
   (b) increases
   (c) remains constant
   (d) first decreases and then increases.

5.4 Terzaghi's 1D consolidation theory assumes
   (a) e against σ'_v relationship is linear
   (b) e against $\log_{10} \sigma'_v$ relationship is linear
   (c) σ'_v against $\log_{10} e$ relationship is linear
   (d) none of the above.

5.5 The coefficient of consolidation is used for
   (a) computing the amount of primary consolidation settlement for a given load increment
   (b) establishing the duration of primary consolidation settlement
   (c) determining the preconsolidation pressure of soil deposits
   (d) determining the depth to which the saturated soil deposit is stressed when a load is applied on its surface.

5.6 A building with a raft foundation is constructed on the ground surface. There is a 2 m-thick clay stratum sandwiched between two highly pervious strata beneath the raft. The building starts settling with time. If the average coefficient of consolidation of the clay is 2.5×10^{-8} m²/s, approximately how many days will it take for the building to reach half of its total primary consolidation settlement?
   (a) 9
   (b) 19
   (c) 91
   (d) none of the above.

5.7 The time for a clay stratum to achieve 80% consolidation is 10 years. If the stratum was half as thick, 10 times more permeable and 4 times more compressible, the time that would be required to achieve the same degree of consolidation is
(a) 1 year
(b) 5 years
(c) 10 years
(d) 16 years.

5.8 The total settlement of a soil stratum under any given stress increment is
(a) dependent on the length of the drainage path
(b) proportional to the thickness of the soil stratum
(c) proportional to the square of the thickness of the stratum
(d) dependent on factors other than the above.

5.9 If the liquid limit of a normally consolidated saturated soil is 45%, the compression index of the soil at the project site for the virgin compression is expected to be
(a) 0.225
(b) 0.245
(c) 0.315
(d) 0.405.

5.10 Select the incorrect statement.
(a) The greater the width of the foundation, the greater the immediate settlement for the same stress increment.
(b) Consolidation time decreases with increasing permeability.
(c) Consolidation time increases with increasing compressibility.
(d) Consolidation time is dependent on the magnitude of the stress change.

5.11 In the laboratory 1D consolidation test, the void ratio of a clay decreases from 1.64 to 1.40 during virgin compression when the effective vertical stress increases from 100 to 166.4 kPa. The coefficient of permeability of the clay is 1.8×10^{-9} m/s. Determine the following
(a) coefficient of compressibility
(b) compression index
(c) coefficient of volume change
(d) coefficient of consolidation.

5.12 A laboratory 1D consolidation test indicated that a 0.02 m-thick clay specimen took 0.5 days to undergo 90% primary consolidation. How many days will a 2.5 m-thick stratum of identical clay sandwiched between sand strata and subjected to an identical stress increment take to undergo 90% primary consolidation?

5.13 In the laboratory 1D consolidation test, a 25 mm-thick cohesive soil specimen drains at the top only, and 50% consolidation occurs in 12 minutes. Determine the time required for the corresponding 2 m-thick soil stratum that drains at both its top and bottom surfaces to undergo 80% consolidation.

5.14 A stratum of sand and a stratum of clay are each 4 m thick. The compressibility of the sand is one-sixth of the compressibility of the clay and the permeability of the sand is 10 000 times that of the clay. What is the ratio of the consolidation time for clay to that for sand?

5.15 A 4.5 m-thick saturated soil stratum has drainage at both its top and bottom surfaces. The time for 50% consolidation is 1 year 6 months. If the coefficient of permeability, k, is 3.5×10^{-8} m/s, determine the following
(a) coefficient of consolidation, c_v, in m²/s
(b) coefficient of volume change, m_v, in m²/kN.

5.16 A loaded flexible rectangular footing, 1.5 m by 3 m in plan, applies a uniform vertical stress of 200 kPa on a very thick saturated cohesive soil layer. The Young's modulus of elasticity and the Poisson's ratio of the soil are 5.5×10^4 kPa and 0.5, respectively. Determine the immediate settlement of the footing at its centre. Compare this value with the rigid footing case considered in Example 5.5.

5.17 For a 2.5 m-thick overconsolidated clay stratum at a construction site, the following details are available: initial void ratio, $e_0 = 1.18$; preconsolidation pressure, $\sigma'_{vmax} = 100$ kPa; compression index, $C_c = 0.06$; recompression index, $C_r = 0.30$; average initial effective vertical stress, $\sigma'_{v0} = 50$ kPa. Determine the total primary consolidation settlement of the clay stratum due to a vertical stress increment of 70 kPa by the placement of a fill over a large area.

5.18 An 8 m-thick normally consolidated clay stratum is subjected to a stress increment of 50 kPa. If the drainage takes place at both the top and bottom surfaces of the stratum, and it undergoes 70% consolidation in three years, determine the coefficient of consolidation. If the coefficient of permeability is 1.1×10^{-9} m/s, determine the primary consolidation settlement in three years.

5.19 A settlement analysis carried out for a proposed structure indicates that 10 cm of primary consolidation settlement will occur in 5 years and the final primary consolidation settlement will be 50 cm if the consolidating layer has drainage at both its top and bottom surfaces. A detailed subsurface investigation indicates that the drainage takes place only at the top of the stratum. Estimate the settlement at the end of 5 years based on the correct information.

5.20 Define consolidation and explain its importance. How does it differ from compaction?

5.21 Draw typical stress–strain curves for soil and describe the various elastic constants associated with them.

5.22 Describe the procedure for conducting an oedometer test to determine the compressibility of soils. What parameters can be obtained from this test?

5.23 What are the various causes of preconsolidation in soils, and how does it influence settlement behaviour?

5.24 Discuss Terzaghi's theory of consolidation and explain why it continues to be widely used despite its limitations.

5.25 What is the time factor and why is it important?

5.26 Define the parameters used to determine the following
(a) magnitude of primary consolidation settlement
(b) rate of primary consolidation settlement
(c) degree of overconsolidation.

5.27 Explain the different types of settlement (immediate, consolidation and secondary) with examples.

5.28 Under what conditions does secondary consolidation become significant?

REFERENCES

ASTM (2001) ASTM D2435-04: Standard test methods for one-dimensional consolidation properties of soils using incremental loading. ASTM International, West Conshohocken, PA, USA.

Atkinson J (2007) *The Mechanics of Soils and Foundations*, 2nd edn. Taylor and Francis, London, UK.

Bowles JE (1996) *Foundation Analysis and Design*, 5th edn. McGraw-Hill, New York, NY, USA.

BSI (2022) BS 1377-2:2022: Method of test for soils for civil engineering purposes – classification tests and determination of geotechnical properties. BSI, London, UK.

Bureau of Indian Standards (2021) IS 2720 (Part 15): 1965 (Reaffirmed 2021)). Determination of consolidation properties. Bureau of Indian Standards, New Delhi, India.

Casagrande A (1936) The determination of the preconsolidation load and its practical significance. *Proceedings of the 1st International Conference of Soil Mechanics and Foundation Engineering*, Cambridge, MA, USA **3**: 60–64.

Casagrande A and Fadum RE (1940) *Notes on Soil Testing for Engineering Purposes*. Harvard Graduate School of Engineering, Cambridge, MA, USA, Soil Mechanics Series, No. 8.

Day RW (2005) *Foundation Engineering Handbook*. McGraw-Hill, New York, NY, USA.

Department of the Navy (1982) NAVFAC DM-7.1 (1982) *Soil Mechanics Design Manual 7.1*, Department of the Navy, Naval Facilities Engineering Command, Alexandria, VA, USA.

Fox EN (1948) The mathematical solution for the early stages of consolidation. *Proceedings of the Second International Conference of Soil Mechanics and Foundation Engineering* **1**: 129–132.

Lambe TW and Whitman RV (1979) *Soil Mechanics, SI Version*. Wiley, New York, NY, USA.

Leonards GA (1962) Engineering Properties of Soils. In *Foundation Engineering* (Leonards GA (ed.)). McGraw Hill, New York, USA, pp. 164–169.

Leroueil S (1988) Tenth Canadian Geotechnical Colloquium: Recent developments in consolidation of natural clays. *Canadian Geotechnical Journal* **25(1)**: 85–107.

Mitchell JK and Soga K (2005) *Fundamentals of Soil Behaviour*, 3rd edn. Wiley, New York, NY, USA.

Robinson RG and Allam MM (1996) Determination of coefficient of consolidation from early stage of log t plot. *Geotechnical Testing Journal, ASTM* 19(3): 316–320.

Schmertmann, JH, Hartman JP and Brown PR (1978) Improved strain influence factor diagrams. *Journal of Geotechnical Engineering, ASCE* **104(GT8)**: 1131–1135.

Shukla SK, Sivakugan N and Das BM (2009) Methods for determination of the coefficient of consolidation and the time-rate of settlement – An overview. *International Journal of Geotechnical Engineering, USA* **3(1)**: 89–108.

Skempton AW (1944) Notes on the compressibility of clays. *Quarterly Journal of the Geological Society of London* **100**: 119–135.

Standards Australia (2020) AS 1289.6.6.1:2020: Determination of the one-dimensional consolidation properties of a soil – Standard method. Standards Australia, Strathfield, NSW, Australia.

Taylor DW (1948) *Fundamentals of Soil Mechanics*. Wiley, New York, NY, USA.

Terzaghi K (1925) *Erdbaumechanik auf bodenphysikalischer Grundlage*. Franz Deuticke, Vienna, Austria. (In German.)

Terzaghi K (1943) *Theoretical Soil Mechanics*. Wiley, New York, NY, USA.

Terzaghi K and Peck RB (1967) *Soil Mechanics in Engineering Practice*, 2nd edn. Wiley, New York, NY, USA.

Terzaghi K, Peck RB and Mesri G (1996) *Soil Mechanics in Engineering Practice*, 3rd edn. Wiley, New York, NY, USA.

FURTHER READING

Das BM (2022) *Principles of Geotechnical Engineering*, 10th edn. Cengage, Boston, MA, USA.

Holtz RD, Kovacs WD and Sheahan TC (2023) *An Introduction to Geotechnical Engineering*, 3rd edn. Pearson Education, Hoboken, NJ, USA.

emerald PUBLISHING   ice

Sanjay Kumar Shukla
ISBN 978-1-83608-519-5
https://doi.org/10.1108/978-1-83608-516-420252009
Emerald Publishing Limited: All rights reserved

Chapter 6
Shear strength of soil

Learning aims

This chapter explores the core concepts of the following topics

- shear strength and its types
- Mohr–Coulomb failure criterion for soils
- state of stress at a point within the soil mass, and concept of Mohr circle and stress path
- fundamentals of shear tests
- determination of strength parameters by laboratory and field tests.

6.1. Introduction

The maximum internal resistance of a soil per unit area to an applied shear stress, τ, is called its shear strength. It is usually denoted by τ_f because the shear strength appears as the shear stress on the failure plane within the soil mass at failure. Its SI unit is N/m^2 or pascal (Pa). When a dense sand or an overconsolidated clay is sheared in a laboratory shear test or in the field by loading, it fails when the applied shear stress, τ, reaches τ_f with some shear displacement, which is generally called the peak shear strength. If the shear displacement continues after failure, the soil still has some lower shear strength and may attain a constant value, which is called the ultimate shear strength, τ_{ult}, for sands and other cohesionless soils, and often residual shear strength, τ_r, for clays and other cohesive soils. For a loose sand under shear, there is no significant difference between peak and ultimate shear strengths because they are almost equal, but when a normally consolidated clay is sheared, its residual shear strength can be significantly less than its peak shear strength. The ultimate or residual condition of a soil is referred to as its critical state. This chapter discusses three more shear strength terms: drained shear strength, consolidated-undrained shear strength and undrained shear strength. Note that the strength of a soil generally means its shear strength, which can be any one of its types, depending on the stress condition in the field application under consideration. This is because in most geotechnical structures (see Figure 1.3 in Chapter 1), the failure occurs from excessive applied shear stresses. This chapter primarily presents the basic concepts of soil shear strength and its measurement.

6.2. Mohr–Coulomb failure criterion

There are several ways or criteria to define failure in a soil mass and assess its stability. For most geotechnical stability problems, the Mohr–Coulomb failure criterion is commonly used by engineers. This criterion is stated as

$$\tau_f = c + \sigma_f \tan \phi \tag{6.1}$$

where τ_f is the shear strength of the soil; σ_f is the applied total normal stress on the failure plane within the soil mass at failure; c is the cohesion (also known as the cohesion intercept); ϕ is the angle of internal friction (also known as the angle of shearing resistance, or simply friction angle).

The parameters c and ϕ are called the shear strength parameters, or simply strength parameters, of the soil. If the soil is dry, there is no difference between total and effective stresses, but when the soil is saturated, c and ϕ are known as the total stress–strength parameters or undrained strength parameters, and are also denoted by c_u (or s_u) and ϕ_u, respectively. In terms of effective normal stress σ_f' ($= \sigma_f - u$) on the failure plane at failure, where u is the pore water pressure, the Mohr–Coulomb failure criterion is stated as

$$\tau_f = c' + \sigma_f' \tan \phi' \qquad (6.2)$$

where c' is the effective cohesion (also known as the effective cohesion intercept), and ϕ' is the effective angle of internal friction (also known as the effective angle of shearing resistance, or simply effective friction angle). c' and ϕ' are known as the effective stress–strength parameters or the drained strength parameters of the soil. As shear stress can only be resisted by the soil skeleton, the shear strength is a function of effective normal stress, and hence Equation 6.2 should be used.

Note that c' and ϕ' are not inherent properties of the soil material, but are dependent on the stress condition within the soil mass. For most soils, for a given state of stress, the shear strength is a unique function of the effective normal stress, σ_f', on the failure plane at failure. Also note that Equations 6.1 and 6.2 can be applicable to express any type of shear strength (peak, ultimate or residual, drained, consolidated-drained or undrained strength), with corresponding values of strength parameters.

The following points regarding the Mohr–Coulomb failure criterion are noteworthy.

- The Mohr–Coulomb failure criterion is based on the early studies of CA Coulomb in 1776 and Otto Mohr in 1882.
- From Equation 6.2, the shear strength of soil has two components, namely cohesive and frictional. The cohesive component is independent of σ_f', while the frictional component is directly proportional to σ_f'.
- Equation 6.2 gives a linear relationship between τ_f and σ_f', but the soils may exhibit a nonlinear relationship. Thus, Equation 6.2 does not always work well for soils; however, it is often used in geotechnical engineering practice primarily due to its simplicity.
- The angle between the horizontal and the maximum slope that a granular or cohesionless soil assumes through natural processes or by pouring it from any point above the ground is called the angle of repose, which represents the angle of shearing resistance. The angle of repose depends on the type of granular soil and other factors such as the water content.
- The stability analysis of geotechnical structures can be provided by either the total stress analysis or the effective stress analysis. The total stress analysis uses the total stress–strength or undrained strength parameters, which can be determined by either laboratory or field tests. For effective stress analysis, effective stress–strength or drained strength parameters are used, which are generally measured through laboratory shear tests. For effective stress analysis, the measurement of pore water pressure is required both in the laboratory and in the field to determine the effective stress within the soil mass.

Example 6.1

A total normal stress of 100 kPa acts at a point on a plane within a soil mass. The effective stress–strength parameters, namely effective cohesion intercept and effective angle of internal friction, of the soil are 20 kPa and 30°, respectively. If the pore water pressure at the point under consideration is 35 kPa, and the soil tends to fail along the plane, what shear strength of soil in terms of effective stress will be observed?

Solution

Given $\sigma_f = 100$ kPa, $c' = 20$ kPa, $\phi' = 30°$ and $u = 35$ kPa

From Equation 6.2, the shear strength in terms of effective stress

$$\tau_f = c' + \sigma_f' \tan\phi' = c' + (\sigma_f - u)\tan\phi' = 20 + (100 - 35)\tan 30° = 57.5 \text{ kPa}$$

6.3. Representation of state of stress at a point

As described in Chapter 3, for most geotechnical problems, the state of stress can be considered in a 2D plane that contains the effective major and effective minor principal stresses, σ_1' and σ_3', respectively, even though the effective intermediate principal stress, σ_2', may be between σ_1' and σ_3' in some problems. Considering a cylindrical soil element within the geostatically stressed soil mass, in many problems, the 3D state of stress can be completely defined in terms of σ_1' and $\sigma_3' (= \sigma_2')$ as an axisymmetric state of stress. The difference between σ_1' and σ_3' – that is, $(\sigma_1' - \sigma_3')$ – is the principal stress difference, referred to as the deviator stress. Note that $(\sigma_1' - \sigma_3') = (\sigma_1 - u) - (\sigma_3 - u) = (\sigma_1 - \sigma_3)$, where u is the pore water pressure, and σ_1 and σ_3 are total major and total minor principal stresses, respectively.

Figure 6.1(a) shows a 2D soil element *abcd* with a failure plane *ef* under the applied stresses σ_1' and σ_3', with horizontal plane *ab* (or *cd*) and vertical plane *bc* (or *ad*) as the major principal and minor principal planes (no shear stresses on these planes), respectively. Note that σ_1' and σ_3' cause a shear failure of the soil element, but they have not been represented as σ_{1f}' and σ_{3f}' with a subscript f, indicating that these definitions are also valid for any non-failure condition. With known magnitude and direction of σ_1' and σ_3', the effective normal stress, σ', and shear stress, τ, can be computed in any other direction, say on a plane *mn* that is inclined at an angle α to *ab*, by considering the static equilibrium of a part *mnb* of the soil element as

$$\sigma' = \sigma_1' \cos^2\alpha + \sigma_3' \sin^2\alpha = \frac{\sigma_1' + \sigma_3'}{2} + \frac{\sigma_1' - \sigma_3'}{2}\cos 2\alpha \tag{6.3}$$

$$\tau = (\sigma_1' - \sigma_3')\sin\alpha\cos\alpha = \frac{\sigma_1' - \sigma_3'}{2}\sin 2\alpha \tag{6.4}$$

Note that normal stresses are considered positive when compressive. The shear stress is positive when counterclockwise. α is measured counterclockwise from the direction of σ_3'.

Equations 6.3 and 6.4 provide a complete description of the state of stress at a point, and they can be represented by a circle on a plot of shear stress against effective normal stress, as shown in Figure 6.1(b). The stresses on a plane *mn* are given by the coordinates (σ', τ) of the point, *P*, on

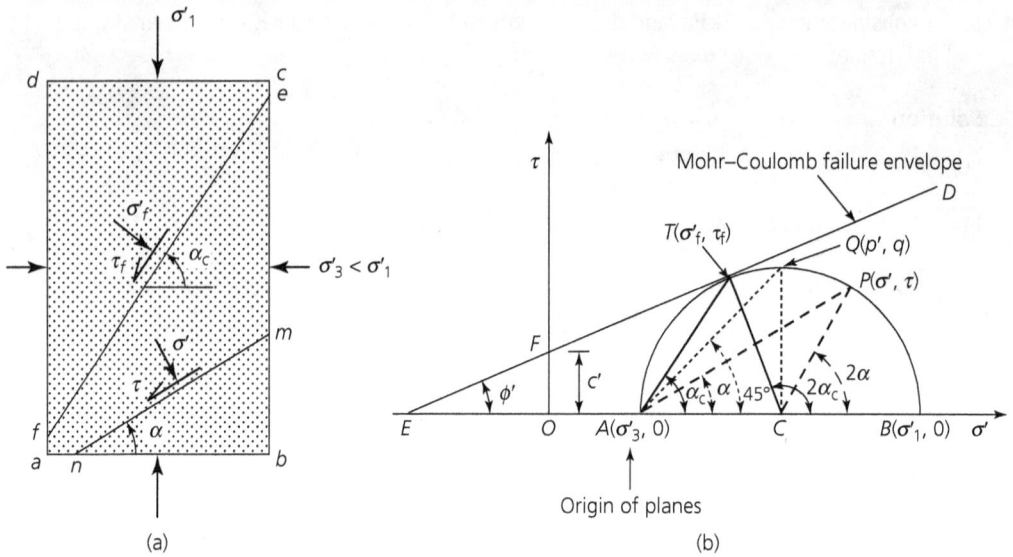

(a)   (b)

the circle. Note that $OA = \sigma'_3$ and $OB = \sigma'_1$ with $AB = (\sigma'_1 - \sigma'_3)$ as the diameter of the circle. This graphical representation of the stresses acting on various planes at a given point – that is, the state of stress at a point within the stressed soil mass – is called the Mohr circle. The concept of the Mohr circle is very useful in soil mechanics. If σ'_1 and σ'_3 are known, by constructing the Mohr circle, stresses σ' and τ acting on any plane can be found without using Equations 6.3 and 6.4. Also, if stresses σ' and τ acting on any two planes are known, the magnitude and direction of σ'_1 and σ'_3 can be found. This task is easily done with the concept of the origin of planes (also known as pole). In Figure 6.1(b), point A is the origin of planes. The property of the origin of planes A is that the line AP of the Mohr circle is parallel to plane mn, and this is valid for any other plane within the soil element with a corresponding point on the Mohr circle.

In Figure 6.1(b), the radius of the Mohr circle is $(\sigma'_1 - \sigma'_3)/2$, which is equal to the maximum shear stress τ_{max} at a point lying on the planes inclined at $\pm 45°$ to the major principal plane ab. Thus,

$$\tau_{max} = \frac{\sigma'_1 - \sigma'_3}{2} = \frac{(\sigma_1 - u) - (\sigma_3 - u)}{2} = \frac{\sigma_1 - \sigma_3}{2} \qquad (6.5)$$

In Figure 6.1(b), Equation 6.2 is also presented graphically as a straight line DE tangent to the Mohr circle at the point T. Line DE is called the Mohr–Coulomb failure envelope, which is practically obtained using the triaxial compression test results, described in Section 6.6, as an envelope of a sequence of Mohr circles representing different stress conditions at failure for the soil. The failure plane ef shown in Figure 6.1(a), will be parallel to line AT in Figure 6.1(b). The failure plane ef will be inclined at a critical angle α_c to the horizontal plane ab.

From the geometry of Figure 6.1(b)

$$\angle TCB = \angle CTE + \angle TEC$$

or

$$2\alpha_c = 90° + \phi'$$

or

$$\alpha_c = 45° + \frac{\phi'}{2} \tag{6.6}$$

In Figure 6.1(b)

$$\sin \phi' = \frac{\overline{TC}}{\overline{EC}} = \frac{\overline{TC}}{\overline{EO} + \overline{OC}} = \frac{\dfrac{\sigma_1' - \sigma_3'}{2}}{c' \cot \phi' + \dfrac{\sigma_1' + \sigma_3'}{2}} \tag{6.7}$$

or

$$\sigma_1' = \sigma_3' \left(\frac{1 + \sin \phi'}{1 - \sin \phi'} \right) + 2c' \left(\frac{\cos \phi'}{1 - \sin \phi'} \right) \tag{6.8}$$

or

$$\sigma_3' = \sigma_1' \left(\frac{1 - \sin \phi'}{1 + \sin \phi'} \right) - 2c' \left(\frac{\cos \phi'}{1 + \sin \phi'} \right) \tag{6.9}$$

Equation 6.8 can be expressed as

$$\sigma_1' = \sigma_3' \tan^2 \left(45° + \frac{\phi'}{2} \right) + 2c' \tan \left(45° + \frac{\phi'}{2} \right) = \sigma_3' \tan^2 \alpha_c + 2c' \tan \alpha_c \tag{6.10}$$

Equation 6.9 can be expressed as

$$\sigma_3' = \sigma_1' \tan^2 \left(45° - \frac{\phi'}{2} \right) - 2c' \tan \left(45° - \frac{\phi'}{2} \right) = \sigma_1' \cot^2 \alpha_c - 2c' \cot \alpha_c \tag{6.11}$$

Note that if a Mohr circle for a given state of stress lies entirely below the Mohr–Coulomb failure envelope for a soil, then the soil will be stable for that state of stress. If the Mohr circle just touches the Mohr–Coulomb failure envelope as considered in Figure 6.1(b), then the shear strength has appeared fully on some plane (plane *ef*) within the soil. No stress condition exists in soil for which the Mohr circle intersects the Mohr–Coulomb failure envelope because such a state of stress results in unlimited strains or deformations – that is, a failure.

If many states of stress at a point within a soil mass caused by loading and/or unloading are represented by Mohr circles on a single diagram, it becomes quite confusing and difficult to clearly see the details. In this situation, it is better to plot each state of stress as a stress point and connect all such points by a line or curve, called the stress path. The coordinates of the stress point are given by the coordinates of point *Q* shown in Figure 6.1(b) as (Lambe and Whitman, 1979)

$$p' = \frac{\sigma_1' + \sigma_3'}{2} = \frac{\sigma_1 + \sigma_3}{2} - u \qquad (6.12)$$

and

$$q = \frac{\sigma_1' - \sigma_3'}{2} = \frac{\sigma_1 - \sigma_3}{2} \qquad (6.13)$$

As τ and q do not depend on the pore water pressure, u, there is no need to denote them as τ' and q' as some authors do. Note that p' and q as defined in Equations 6.12 and 6.13 cause a shear failure of the soil element, but they have not been represented as p_f' and q_f with a subscript f, indicating that these definitions are also valid for any non-failure condition.

Equation 6.7 can be expressed as

$$q_f = a' + p_f' \tan \psi \qquad (6.14)$$

where

$$c' = \frac{a'}{\cos \phi'} \qquad (6.15a)$$

and

$$\sin \phi' = \tan \psi \qquad (6.15b)$$

Equation 6.14 is graphically presented in Figure 6.2. The curve obtained is called the K_f line, which is similar to the Mohr–Coulomb failure envelope. Thus, Equation 6.14 presents an alternative way of expressing the failure criterion and determining the strength parameters by plotting the values of p_f' and q_f from the triaxial compression tests, described in Section 6.6, in place of constructing several Mohr circles for different states of failure using the same test data and then drawing a tangent line to all of them to get the Mohr–Coulomb failure envelope. Figure 6.2 also shows a stress path AB for a constant σ_3' and increasing σ_1' – that is, for $\Delta\sigma_3' = 0$ and $\Delta\sigma_1' > 0$. Note that the soil fails at $B(p_f', q_f)$. Further details about the stress path and its applications can be found in Lambe and Whitman (1979).

For a soil with $c' = 0$, from Equation 6

$$\sin \phi' = \frac{\dfrac{\sigma_1' - \sigma_3'}{2}}{\dfrac{\sigma_1' + \sigma_3'}{2}} = \frac{q_f}{p_f'} = \frac{1 - K}{1 + K} \qquad (6.16)$$

where $K = \sigma_3'/\sigma_1'$, known as the coefficient of lateral stress, as defined in Section 3.6 of Chapter 3. Equation 6.16 gives

$$K = \frac{1 - \sin \phi'}{1 + \sin \phi'} = \tan^2\left(45° - \frac{\phi'}{2}\right) \qquad (6.17)$$

Figure 6.2 A typical p'–q diagram with an example of the effective stress path (ESP)

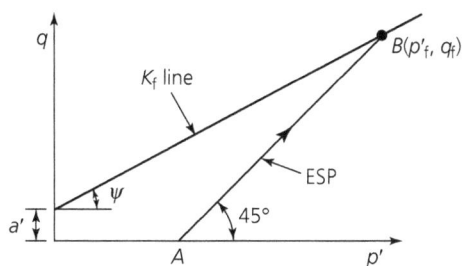

Note that Equation 6.17 applies to $\sigma_3' < \sigma_1'$, which occurs within the soil backfill element behind a retaining wall tending to move away from the soil backfill. This is the case of active earth pressure condition, described in more detail in Chapter 8 along with other lateral earth pressure conditions. In this case, the coefficient of lateral stress, K, is denoted by K_a, called the coefficient of active earth pressure.

The following points regarding the representation of state of stress at a point are noteworthy.

- The basic concepts of the Mohr circle presented in this section are applicable to any set of σ_1' and σ_3', whether they cause failure of the soil element or not. The concepts are also applicable for the total principal stresses σ_1 and σ_3.
- A stress path need not be a straight line, and may consist of a series of straight lines and/or curves joined together.

6.4. Fundamentals of shear tests

The shear strength of soil materials, such as natural soils, soils mixed with fibres, and waste materials (e.g. coal ashes and mine tailings), is determined in their undisturbed, remoulded (natural structure altered by manipulation while maintaining the water content) or compacted conditions, depending on their applications, through shear tests conducted in the laboratory or field. Common laboratory tests are direct shear test, triaxial compression test and unconfined compression test, as described below. The basic principles of these tests are illustrated in Figures 6.3(a)–(c). The vane shear test is another laboratory test, but it is generally conducted as an in situ test, as shown in Figure 6.3(d). An arrangement similar to the direct shear test can be made in the field to determine the strength properties of an in situ soil consisting of large particles such as coarse gravels, cobbles and boulders. Field penetration tests, such as the standard penetration test and the static/dynamic cone penetration test indirectly provide the strength parameters, generally undrained parameters, through crude correlations with penetration values obtained from these tests.

In a laboratory shear test, the soil specimen is tested by closely simulating the drainage and stress condition observed in the field situations. In general, there are two stages of the laboratory shear test, especially for the direct shear and triaxial compression tests, as described below.

First stage. A normal stress, σ, or an all-around confining stress $\sigma_c (= \sigma_3)$ is applied to the soil specimen. A vertical normal stress to the soil specimen is applied in the direct shear test, while an all-around confining stress is applied to the soil specimen in the triaxial compression test. No

Figure 6.3 Shear tests: (a) direct shear test; (b) triaxial compression test; (c) unconfined compression test; (d) vane shear test

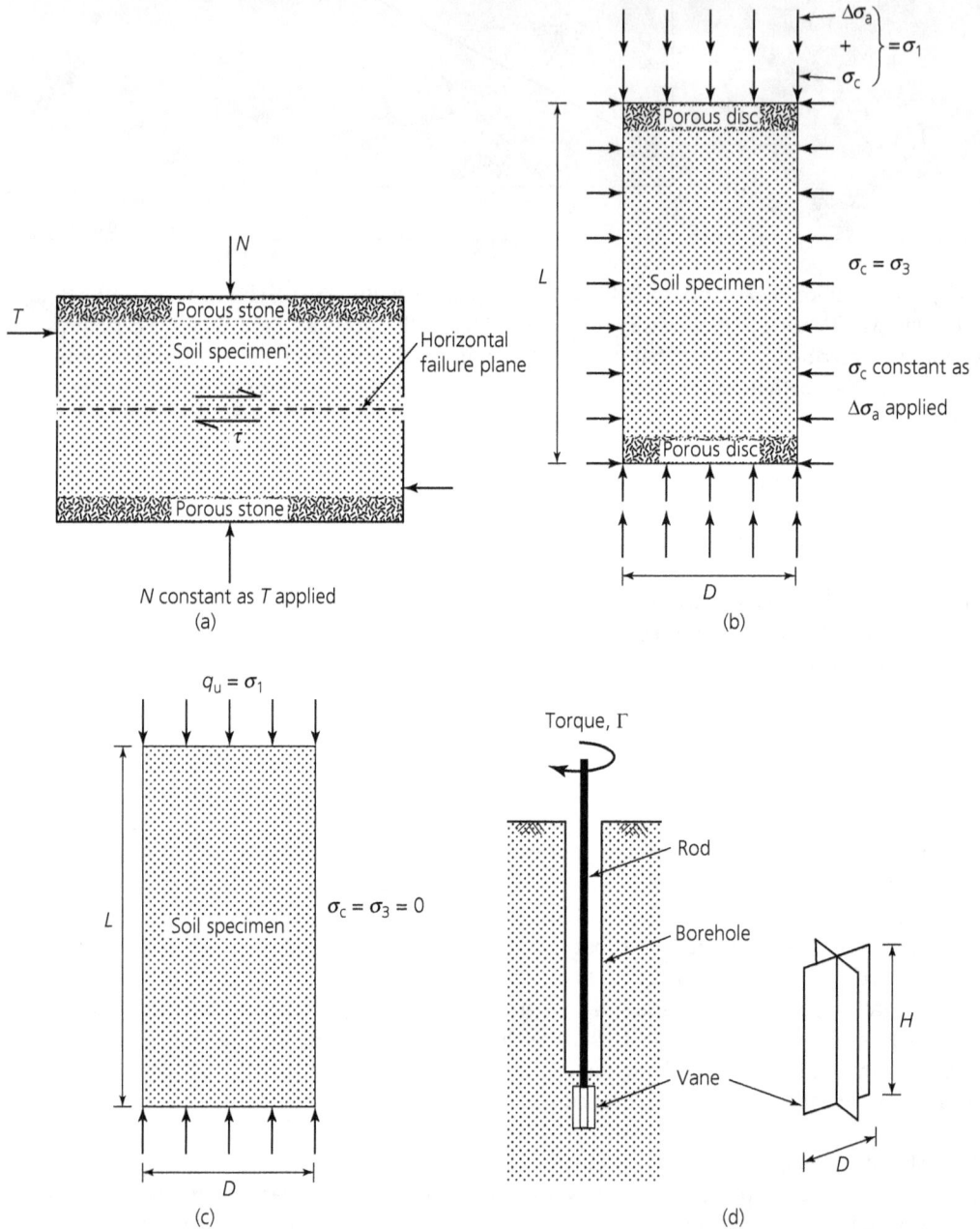

normal or confining stress is applied in an unconfined compression test, which is generally conducted on fully saturated cohesive soils.

Second stage (also known as the shear stage). A a shear stress or an additional axial stress – that is, the deviator stress $\Delta\sigma_a(=\sigma_1-\sigma_3)$ – is applied to the soil specimen at a constant normal, σ, or confining stress, $\sigma_c(=\sigma_3)$. A shear stress is applied to the soil specimen at a constant normal stress in the direct shear test, while in the triaxial compression test, the deviator stress is applied at a constant confining stress. Only axial stress $\sigma_1(=q)$ is applied to the soil specimen in the unconfined compression test.

Depending on the drainage conditions in the two stages of the test, there are three types of test.

■ Consolidated-drained (CD) test or simply drained (D) test. The drainage of water from the soil specimen is allowed during both stages. During the second stage, the specimen is sheared at a very slow rate to ensure that the specimen remains drained, with induced pore water pressure equal to zero during the entire second stage of the test. This test requires several hours; thus, the test is known as the slow (S) test. The strength parameters obtained are used to analyse the stability of an embankment constructed very slowly in layers over a soft clay deposit, an earth dam with steady-state seepage, a strip footing on a clay deposit after a long time from the end of construction, or an excavated or natural slope in clay (Ladd, 1971; Winterkorn and Fang, 1991).

■ Consolidated-undrained (CU) test. The drainage of water from the soil specimen is allowed during the first stage only; thus, the water content of the test specimen remains practically unchanged during the second stage of the test. If the pore water pressures developed during the test are measured, this test is denoted as the C̄U test. This test is known as the consolidated quick (Q_c or R, simply because R falls between Q and S in the alphabet) test. This test at in situ confining stress overestimates the strength parameters, because the disturbance leads to smaller water content on reconsolidation. The strength parameters obtained are used to analyse the stability of an embankment raised subsequent to consolidation under its original height, rapid drawdown behind an earth dam, or rapid construction of an embankment on a natural slope in clay (Ladd, 1971; Winterkorn and Fang, 1991).

■ Unconsolidated-undrained (UU) test. The drainage of water from the soil specimen is not allowed during any stage of the test; thus, the water content of the test specimen remains practically unchanged during the entire test. If the pore water pressures developed during the test are measured, this test is denoted as the ŪU test, but usually the pore water pressure is not measured in this test. The test can be completed within a few minutes (e.g. 10 to 20 min); thus, the test is known as the quick (Q) test. This test may be viewed as an opposite extreme from the CD test. The test at in situ confining stress is the most representative of the laboratory tests because of compensating errors. The strength parameters obtained are used to analyse the stability of an embankment constructed rapidly over a soft clay deposit, a large earth dam constructed rapidly with no change in water content of the clay core, or a loaded footing placed rapidly on a clay deposit, as well as the short-term slope stability (Ladd, 1971; Winterkorn and Fang, 1991).

Note that any effort to have the unconsolidated-drained (UD) test is meaningless. The unconfined compression test is a special case of the UU test with confining stress $\sigma_c=\sigma_3=0$. In the unconfined compression test, when the specimen is compressed (and hence sheared) rapidly by increasing the vertical axial stress, the drainage of water from the voids of soil rarely takes place (undrained situation) if the soil has a very low permeability. It is the best general-purpose test, although it underestimates the strength parameters because the disturbance decreases the effective stress. The unconfined compression test is usually conducted for short-term stability analysis and field control of soil materials. The vane shear test is usually considered to give the best results of undrained strength parameters, but its application is limited to only clayey soils.

The direct shear test is a simple test and is suitable for determining the consolidated-drained properties of soil materials because the test can be completed relatively rapidly. In the direct shear test, the drainage paths through the test specimen are short, thereby allowing excess pore water pressure to be dissipated more rapidly than with the triaxial compression test (ASTM, 2012). In the laboratory, the strength behaviour of a soil under undrained condition is usually studied by means of the triaxial compression test in which drainage can easily be prevented by the valve provided in the test equipment. The triaxial compression test is more complex than the direct shear test. The complete details of the triaxial compression tests are described by Bishop and Henkel (1962).

In a shear test, the shear stress can be applied either by increasing the shear displacement or strain at a constant rate, called the strain-controlled test, or by increasing the shear force or stress at a constant rate, called the stress-controlled test. Most shear tests are conducted as strain-controlled because the stress–strain behaviour of soil can be observed beyond its peak shear strength so that the ultimate or residual strength can be obtained. In a stress-controlled test, the behaviour can be studied only up to the peak shear strength, although it closely simulates field conditions.

It is important to note that cyclic shear strength tests are available to evaluate the dynamic behaviour of soil. These tests are essential for assessing soil liquefaction potential, and the stability of slopes, foundations and other structures subjected to seismic and vibratory loads, as they provide critical insights into soil resistance to cyclic loading and its deformation characteristics under repeated stress conditions. Further details can be found in relevant test standards and books on soil dynamics and geotechnical earthquake engineering. Chapter 9 discusses the core principles and concepts of soil liquefaction as a special topic.

6.5. Direct shear test

The direct shear test is the oldest method for assessing shear strength. It is used to evaluate the shear strength of all types of soil materials in undisturbed, remoulded or compacted conditions. In the test, a soil specimen placed between porous stones under an applied normal load in a box, called the shear box, either square or circular in plan and split across its middle, is stressed to failure by causing relative lateral displacement, called the shear displacement, δ, between the two parts of the box at a controlled strain rate between 0.0025 and 1.0 mm/min, as shown in Figure 6.3(a). Figure 6.4 shows the test set-up for the direct shear test.

The minimum diameter of a circular soil specimen or the minimum width of a square soil specimen is generally 50 mm. The minimum initial specimen thickness or depth is generally taken as 12.5 mm. Note that porous stones are not necessary for tests on dry soil. The specimen is allowed to consolidate under the applied normal load, N, and then the shear failure is reached slowly under drained conditions so that the induced pore water pressures are dissipated in order to obtain the consolidated-drained properties of the soil. Generally, three or more specimens, each under a different normal load, are tested.

Figure 6.4 Test set-up for the direct shear test

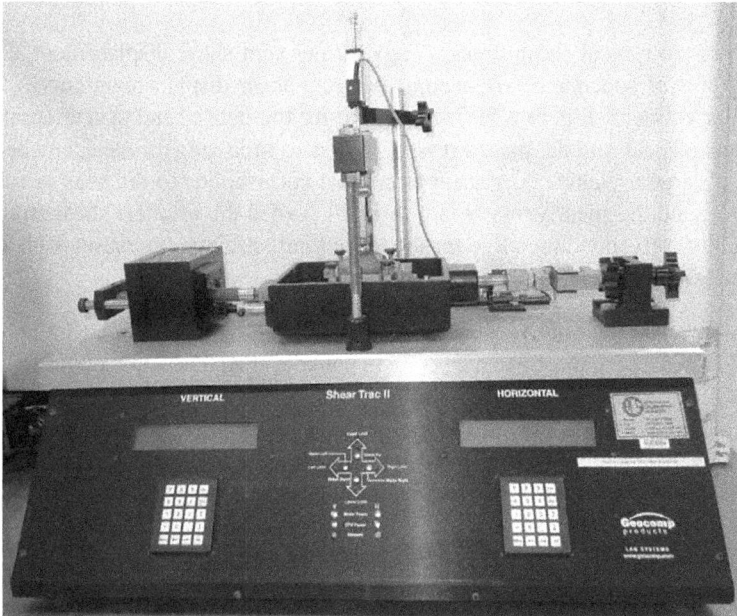

Figure 6.4 Test set-up for the direct shear test

The failure is often taken to correspond to the maximum shear stress attained, or the shear stress at 15 to 20% shear displacement, whichever occurs first. Note that the per cent shear displacement – that is, the per cent relative lateral displacement, δ_p – for a given shear force is calculated by dividing the incremental displacement, $\Delta\delta$, by the relative lateral displacement, δ. The magnitude of the shear force, T, is recorded as a function of shear displacement. Usually, the change in thickness, ΔH, of the soil specimen is also recorded. The normal stress and the shear stress are generally calculated by diving the normal load, N, and the shear force, T, respectively, by the initial cross-sectional area, A, of the specimen. The complete details of the test can be found in the test standards (ASTM D 3080-04 (ASTM, 2012), IS 2720 (Part 13) (Bureau of Indian Standards, 2021) and AS 1289.6.2.2:2020 (Standards Australia, 2020)).

The direct shear test has an advantage in determining the shear strength of a weak joint or bedding plane within the soil specimen, as well as the strength of the interface between dissimilar materials – for example, soil–concrete, soil–steel and soil–geosynthetic interfaces. This test has several limitations, as outlined below.

- During the test, there is a rotation of principal stresses, which may not be the reality in field situations.
- The failure is forced to occur on or near a horizontal plane at the middle of the specimen, which may not be the weakest plane within the soil specimen.
- There is a non-uniform distribution of shear stresses and displacements within the specimen.
- An appropriate height of specimen cannot be defined.

▨ The stress path cannot be drawn exactly, although there is a K_0 loading as the normal load is applied because only the stresses on a horizontal plane are known; thus, the entire state of stress is unknown (Lambe and Whitman, 1979).

Figure 6.5 shows the typical shear stress, τ, against per cent shear displacement, δ_p, curves and change in thickness of specimen, ΔH, against per cent shear displacement curves for loose and dense dry sand specimens. Figure 6.5(a) shows that for the dense sand specimen, the τ–δ_p curve shows a pronounced peak and the shear stress, τ, decreases following this peak and becomes essentially constant at large δ_p values. The shear stress that corresponds to the peak is called the peak shear strength, τ_f, and the shear stress at large strain is called the ultimate shear strength, τ_{ult}. On the other hand, the corresponding curve for the loose sand specimen increases with an increase in δ_p and becomes essentially constant after a certain δ_p value.

Figure 6.5(b) shows that the thickness of the dense sand specimen increases with an increase in δ_p except at very low δ_p value, indicating an increase in volume – that is, dilation of the dense sand specimen – whereas the thickness of the loose sand specimen decreases with an increase in δ_p, indicating a decrease in volume – that is contraction or compression of the loose sand specimen. At large shear displacement, both the dense and loose sands attain the ultimate strength, τ_{ult}, and also practically the same void ratio, called the critical void ratio, as introduced by Casagrande (1948). In brief, when a

Figure 6.5 Typical results from the direct shear test on loose and dense dry sands: (a) shear stress, τ, against per cent shear displacement, δ_p, curves; (b) thickness of specimen, ΔH, against per cent shear displacement, δ_p, curves; and (c) plot of shear stress, τ, against effective normal stress, σ', for peak and ultimate strength conditions

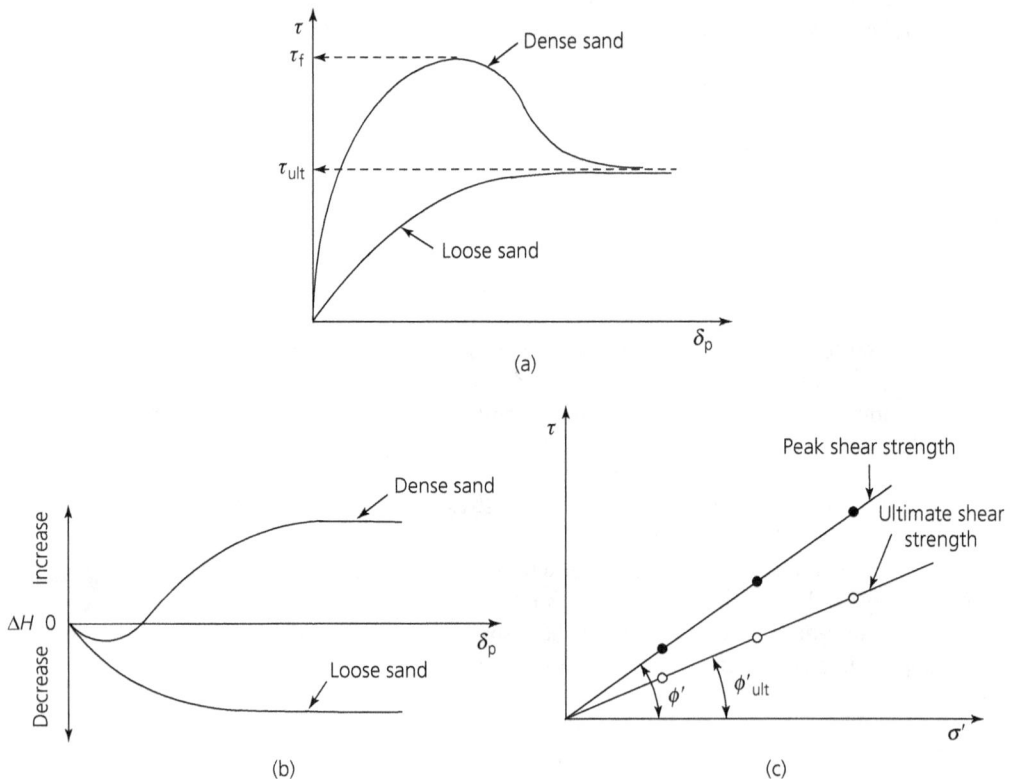

dense sand specimen is sheared, it tends to become loose, and when a loose sand is sheared it tends to become dense, and finally both reach the same void ratio. This behaviour can be explained by assuming they are spherical particles in their loosest and densest arrangements, as explained in Chapter 2, and subjecting them to shear displacements. Note that if a sand is at an in situ void ratio greater than its critical void ratio, it is highly susceptible to liquefaction and flow slides.

Figure 6.5(c) shows the typical plots of shear stress, τ, against effective normal stress, σ', for peak and ultimate strength conditions. These plots provide the values of drained shear strength parameters ϕ' (effective peak friction angle) and ϕ'_{ult} (effective ultimate friction angle). Note that the direct shear test is a plane strain test, and the shear strength values obtained are generally higher than those from the triaxial compression test.

Also note that in shear tests, loose sand behaves similarly to normally consolidated clay, whereas dense sand behaves like overconsolidated clay. Hence, the curves for normally consolidated and overconsolidated clays under consolidated-drained (CD) conditions are similar to those shown in Figures 6.5(a) and (b) for loose and dense sands, respectively, but the residual strength, τ_r, is often used in place of the ultimate strength, τ_{ult}, term. In the plots of shear stress, τ, against effective normal stress, σ', for peak and residual strength conditions for clays, the lines may have cohesion intercepts on the τ axis, especially for overconsolidated clays. The residual shear strength parameters are used for the stability analysis of slopes in overconsolidated clays.

Example 6.2

A series of consolidated-drained direct shear tests is conducted on a compacted sand in a 60 mm-square shear box, and the following results are obtained at failure
Determine the shear strength parameters.

normal load, N (N):	100	200	300
peak shear force, T (N):	85	169	254
ultimate shear force, T_{ult} (N):	61	126	193

Solution

Cross-sectional area of the soil specimen, $A = 60 \times 60$ mm^2 = 3600 mm^2.

From the given data, the vertical normal stress (N / A), peak shear stress at failure (peak shear strength) (T / A) and ultimate shear stress at failure (ultimate shear strength) (T_{ult}/A) are calculated as given below

normal stress, σ (kPa):	27.8	55.6	83.3
peak shear strength, τ_f (kPa):	23.6	46.9	70.6
ultimate shear stress, τ_{ult} (kPa):	16.9	35.0	53.6

Figure E6.2 shows the plot of shear stress, τ, against effective normal stress, σ', for peak and ultimate strength conditions. From the plot, the values of shear strength parameters are obtained as effective peak friction angle, $\phi' = 40°$ and effective ultimate friction angle, $\phi'_{ult} = 32.5°$. Note that effective cohesion is zero.

Figure E6.2

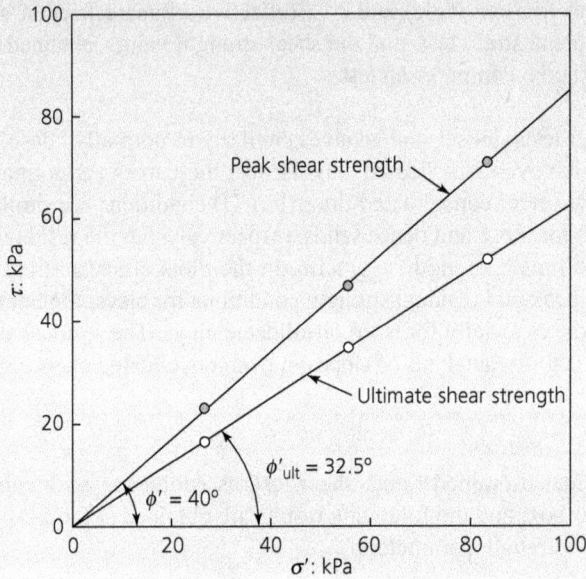

6.6.   Triaxial compression test

The triaxial compression test is the most common and versatile method for studying the stress–strain behaviour and strength properties of soil. In this test, a cylindrical specimen (minimum diameter of 38 mm with a ratio of length or height, L, to diameter, D, between 2 and 2.5) of soil with porous discs at the top and the bottom and encased in an impervious (rubber) membrane is first subjected to an all-around confining stress $\sigma_c (= \sigma_3)$ from the liquid (generally water) in a triaxial chamber of the triaxial test equipment in which the specimen is mounted (Figure 6.6). Then, the specimen is axially loaded by a piston at a constant rate of axial deformation with an increment of axial stress, $\Delta\sigma_a$, keeping σ_c constant, to failure (Figure 6.3(b)). The total axial stress $(\sigma_c + \Delta\sigma_a)$ and confining stress, σ_c, are the total major and total minor principal stresses, σ_1 and σ_3, respectively. Thus, the increment of axial stress, $\Delta\sigma_a = \sigma_1 - \sigma_3$, which is the principal stress difference, also called the deviator stress. Generally, three specimens are tested at different confining stresses. The data obtained from the triaxial compression tests are used to determine the stress–strain behaviour and strength properties of soils, such as the Young's modulus of elasticity and the strength parameters.

Note that a full saturation of the specimen before the stress application is obtained by applying a back pressure to the specimen pore water. A check for saturation is required before initiating the first stage of the test. This can be done by measuring the pore pressure parameter, B, which is defined as

Figure 6.6 Test set-up for the triaxial compression test

$$B = \frac{\Delta u}{\Delta \sigma_3} \qquad (6.18)$$

where Δu is the change in the specimen pore water pressure because of a change in the chamber pressure (confining stress), $\Delta \sigma_3$, when the specimen drainage valves are closed. The parameter B varies from 0 for a dry soil to 1 for a fully saturated soil, and is a nonlinear function of degree of saturation.

In the triaxial compression tests, the failure is free to take place on the weakest plane, or sometimes the soil specimen bulge, and the stress paths to the K_f line can be controlled effectively as required for the field conditions. Although the failure is generally defined as the maximum deviator stress $(\sigma_1 - \sigma_3)_{max}$ – that is, the deviator stress at failure $(\sigma_1 - \sigma_3)_f$ – it can also be defined as the shear stress $\tau = (\sigma_1 - \sigma_3)/2$ at the prescribed strain.

A triaxial compression test can be classified into the following three types: CD (or D) test, CU (or $\overline{CU}$) test, and UU (or $\overline{UU}$) test, providing the drained strength parameters (c' and ϕ'), the consolidated-undrained strength parameters [c' and ϕ', or c_u (or s_u) and ϕ_u], and the undrained strength parameters [c_u (or s_u) and $\phi_u = 0$], respectively. The isotropic consolidation is allowed during the first stage of the test in CD and CU tests only. The CD test results (measured axial deformation and volume change with applied σ_1' and σ_3') are used to assess the stress–strain and strength properties of a soil that has been completely consolidated in the field situation under existing normal stresses. In $\overline{CU}$ tests, the axial deformation of the specimen and the induced pore water pressure for the applied σ_1 and σ_3 are measured. These two tests allow for the computation of axial strain, ε_a, and the total and effective principal stresses. The volume change, ΔV, in the CD tests, and induced pore water pressure, Δu, in the CU tests (also in the UU tests) are measured with suitable devices. For a saturated soil specimen, ΔV is simply the volume of water that enters or leaves it. The direction of the volume change – that is, compression ($\Delta V < 0$) or dilation ($\Delta V > 0$) – of the soil specimen in the CD test depends on the density or relative density, the confining pressure and the stress history. The value of ΔV is equal to the void ratio change during the shear stage of the test. Complete details of the CU and UU tests can be found in relevant test standards, such as ASTM D4767-11(2020) (ASTM, 2020) and ASTM D2850-23 (ASTM, 2024), respectively.

During the shear stage of the CD tests, loose sand and normally consolidated clay compress or consolidate (decrease in volume), whereas dense sand and overconsolidated clay dilate or expand (increase in volume). During the shear stage of the CU and UU tests, loose sand and normally consolidated clay tend to compress or consolidate (decrease in volume); hence, due to undrained condition, a positive pore water pressure $\Delta u > 0$ is induced, resulting in a decrease in the effective stress, whereas dense sand and overconsolidated clay tend to dilate (increase in volume); hence, due to undrained condition, a negative pore water pressure $\Delta u < 0$ is induced, resulting in an increase in the effective stress.

Figure 6.7 shows the typical results for the CD triaxial compression tests on normally consolidated and overconsolidated clays. The stress–strain (Figure 6.7(a)) and the volume change (Figure 6.7(b)) results are based on the tests for the same effective confining stress, σ_3'. Note that overconsolidated clay has a higher strength, higher modulus and lower strain at failure than those for normally consolidated clay. During the shear stage of the test, overconsolidated clay expands or dilates, while normally consolidated clay contracts or compresses. Similar behaviour exhibited by loose and dense sands can be observed, corresponding to normally consolidated and overconsolidated clays, respectively. Figures 6.7(c) and 6.7(d) show how the effective stress–strength parameters c' and ϕ' of soils can be determined by drawing a common tangent to all Mohr circles.

Example 6.3

In a consolidated-drained triaxial test conducted on a saturated sand, failure occurs when the deviator stress is 387 kPa at a confining stress of 100 kPa. What is the effective angle of shearing resistance of sand and the inclination of failure plane to the major principal plane?

Solution

Given $\Delta\sigma_a = 387$ kPa and $\sigma_3 = \sigma_3' = 100$ kPa

As $\Delta\sigma_a = \sigma_1 - \sigma_3 = \sigma_1' - \sigma_3'$

$\sigma_1' = \sigma_3' + \Delta\sigma_a = 100 + 387 = 487$ kPa

For the sand under consolidated-drained condition, effective cohesion, $c' = 0$; therefore, from Equation 6.10

$$\sigma_1' = \sigma_3' \tan^2\left(45° + \frac{\phi'}{2}\right)$$

or

$$487 = (100) \tan^2 \left(45° + \frac{\phi'}{2} \right)$$

or

$$45° + \frac{\phi'}{2} = 65.6°$$

$$\phi' = 41.2°$$

From Equation 6.6, the inclination of the failure plane to the major principal plane

$$\alpha_c = 45° + \frac{\phi'}{2} = 45° + \frac{41.2°}{2} = 65.6°$$

Values of ϕ' and α_c can be obtained by constructing the Mohr circle and drawing the failure envelope passing through the origin, as illustrated in Figure E6.3.

Figure E6.3

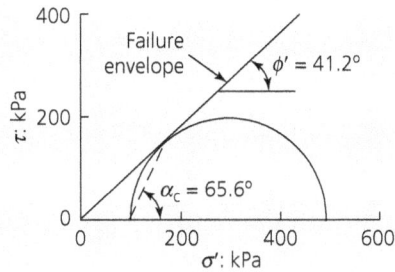

Figure 6.8 shows the typical results for the $\overline{CU}$ triaxial compression tests on normally consolidated and overconsolidated clays. Both the effective stress and the total stress failure envelopes can be obtained from the test results. It may be noticed that $\phi' > \phi$, and often $\phi' \approx 2\phi$ for normally consolidated clays. As an overconsolidated clays expands during shear, the induced pore water pressure, Δu, decreases and may become negative – that is, $-\Delta u_f$ – at failure, resulting in a shift of the effective stress circle at failure to the right (Holtz et al., 2023). There is also a possibility that the effective friction angle, ϕ', may be lower than the total friction angle, ϕ, for overconsolidated clays. In general, the CD tests give a greater value of ϕ', while the UU tests provide the minimum value of ϕ'.

The UU test is generally conducted on a fully saturated soil ($S = 100\%$) to determine its undrained shear strength, c (also denoted by c_u or s_u) in terms of total stress. The failure envelope for such soils is usually a horizontal straight line over the entire range of confining stresses applied to the soil specimens; thus, $\phi = 0$, theoretically resulting in $\alpha_c = 45°$ from Equation 6.6, which is also applicable for total stress conditions with replacement of ϕ' by ϕ, although in reality effective stress will

Typical results from the CD triaxial tests on normally consolidated and overconsolidated clays: (a) plot of deviator stress, $\Delta\sigma_a$, against axial strain, ε_a; (b) plot of volume change, ΔV, during the deviator stress application against ε_a; (c) Mohr–Coulomb failure envelope for a normally consolidated clay; (d) Mohr–Coulomb failure envelope for an overconsolidated clay

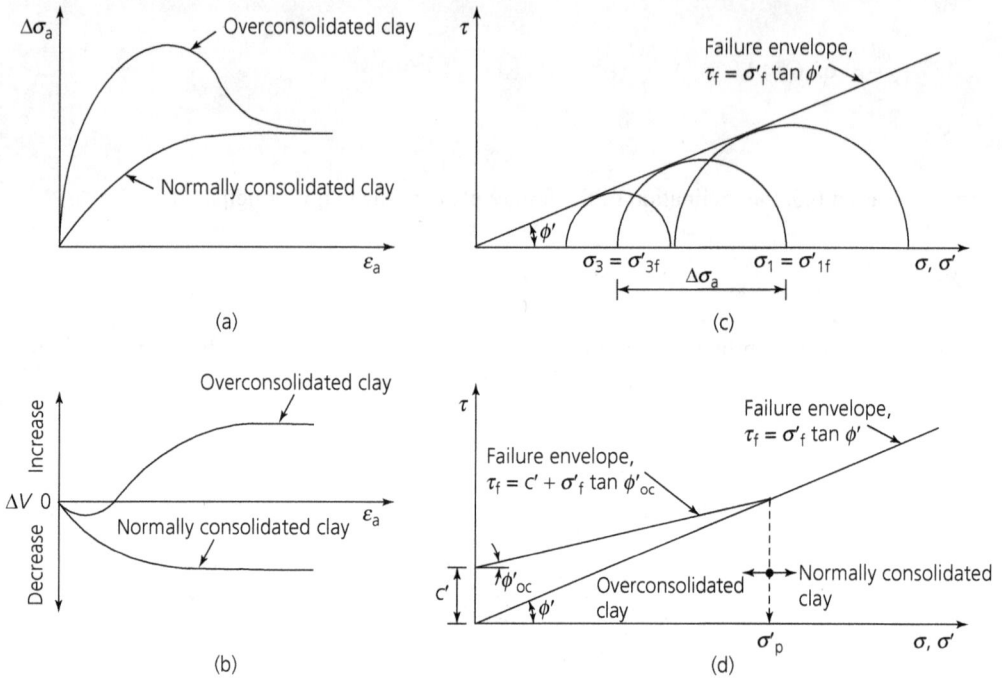

(a)

(b)

(c)

(d)

control α_c. If the soil is partially saturated, the failure envelope is usually curved initially until the soil becomes fully saturated due to the applied confining pressure alone. The stress–strain curves are typically similar to those from the CD and CU tests. Figure 6.9 shows typical results from the UU tests. Note that all the total stress Mohr circles at failure have the same diameter equal to the deviator stress, $\Delta\sigma_{af}$. Due to the difficulty in measuring the pore water pressure in the UU test, it is generally not measured. If the pore water pressure is determined, there will be only one effective stress Mohr circle at failure with the same effective minor principal stress for all the total stress Mohr circles; thus it may not be possible to determine the effective stress failure envelope to determine effective stress–strength parameters c' and ϕ' from the UU tests.

Note that during the shear stage – that is, undrained loading in $\overline{\text{CU}}$ tests – the induced pore water pressure, Δu, can be related to the deviator stress, $\Delta\sigma_a$, as

$$A = \frac{\Delta u}{\Delta\sigma_a} = \frac{\Delta u}{\sigma_1 - \sigma_3} \tag{6.19}$$

where A is a nondimensional parameter, which depends on soil type and its stress history, type of stress and magnitude of shear strain. At failure, the A parameter is

$$A_f = \frac{\Delta u_f}{\Delta\sigma_{af}} = \frac{\Delta u_f}{(\sigma_1 - \sigma_3)_f} \tag{6.20}$$

Figure 6.8 Typical results from the $\overline{CU}$ triaxial tests on normally consolidated and overconsolidated clays: (a) plot of deviator stress, $\Delta\sigma_a$, against axial strain, ε_a; (b) plot of induced pore water pressure, Δu, during the deviator stress application against ε_a; (c) Mohr–Coulomb failure envelopes for a normally consolidated clay; (d) Mohr–Coulomb failure envelopes for an overconsolidated clay

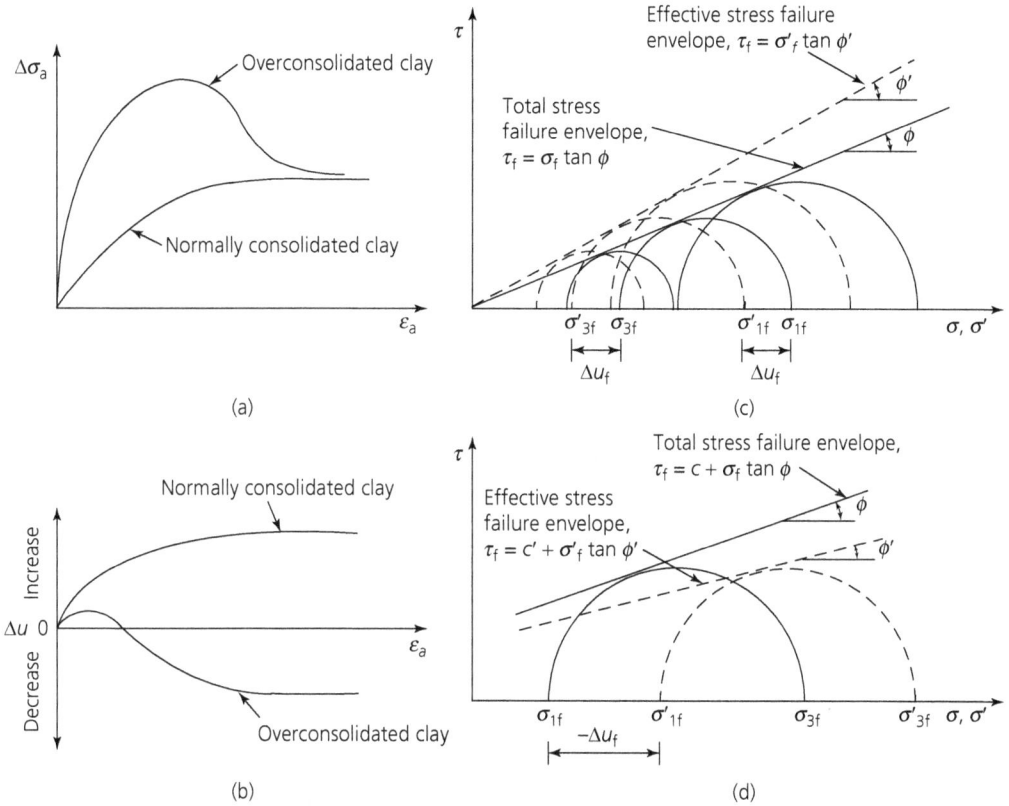

(a)

(b)

(c)

(d)

Figure 6.9 Total stress failure envelope from the UU triaxial compression tests for a fully saturated clay

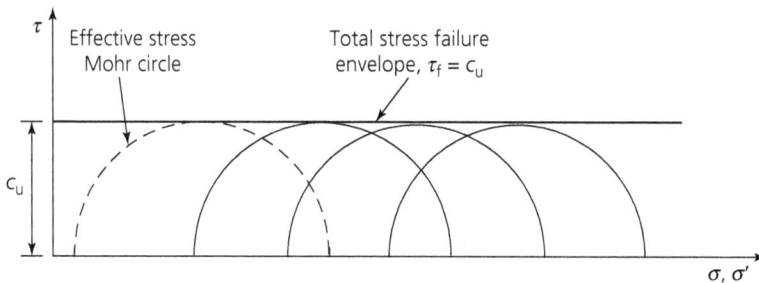

As Δu_f can be negative for highly overconsolidated clays and dense sands, A is also negative for them. Typically, A_f ranges from 1 to -0.5, and may be greater than 1 for sand or clay having a loose structure which collapses on load application. In general, soft loose soils have high values of A and the higher the shear strain, the greater the value of A (Lambe and Whitman, 1979).

During the $\overline{UU}$ tests on saturated clays, the pore water pressure is induced during both the consolidation and the shear stages. From Equations 6.18 and 6.19, the total pore water pressure induced by application of both the confining stress, σ_3, and the deviator stress, $\Delta\sigma_a$, is

$$\Delta u = (\Delta u)_{\text{first stage}} + (\Delta u)_{\text{second stage}} = \sigma_3 + A\Delta\sigma_a = \sigma_3 + A(\sigma_1 - \sigma_3) \tag{6.21}$$

The general expression relating induced pore water pressure to the changes in total stresses during undrained loading is (Skempton, 1954)

$$\Delta u = B\left[\Delta\sigma_3 + A(\Delta\sigma_1 - \Delta\sigma_3)\right] = B\Delta\sigma_3 + \overline{A}(\Delta\sigma_1 - \Delta\sigma_3) \tag{6.22}$$

where $\overline{A} = BA$. Equation 6.22 is known as Skempton's equation, and B and $\overline{A}$ are called Skempton's pore pressure parameters or coefficients. This equation is very useful in predicting the pore water pressures induced by undrained loading for the stability analysis in field problems, such as embankment construction over a soft clay deposit. Note that for a saturated soil, $\overline{A} = A$ because $B = 1$.

For cohesionless, granular or frictional soils, such as sands and gravels, and for normally consolidated clays and silts, $c' = 0$. For overconsolidated clays, $c' > 0$. The drained friction angle, ϕ', ranges from $26°$ to $48°$ for sands and silts (Das, 2022). For cohesionless soils, ϕ' has been correlated with their relative density, total unit weight and standard penetration test (SPT) value, N, called the standard penetration resistance or the blow count (Table 6.1).

Note that the standard penetration test is an in situ test, originally developed in 1927 in the USA. This test is aimed at driving a split-barrel sampler (internal diameter $= 35$ mm, outer diameter $= 50$ mm) to obtain a representative soil sample and a measure of the resistance of the soil to penetration of the sampler driven by repeatedly dropping a 63.5 kg hammer over a distance of 760 mm. The number of blows required to achieve three subsequent 150 mm penetrations is recorded, but the number of blows taken to penetrate the final 300 mm is termed the SPT value, N, at that depth. The test is conducted in a borehole in the soil deposit at the desired depth. The measured N value is corrected

Table 6.1 Correlation of relative density, total unit weight, SPT value and drained friction angle for cohesionless soils

Relative density (descriptive term)	Relative density, D_r: %	Total unit weight, γ: kN/m³	SPT value, N: blows/300 mm	Drained friction angle, ϕ': °
Very loose	0–15	11–13	0–4	<30
Loose	15–35	13–16	4–10	30–35
Medium dense	35–65	16–19	10–30	35–40
Dense	65–85	19–21	30–50	40–45
Very dense	85–100	>21	>50	>45

suitably to account for effects of dilatancy, overburden stress and some other factors on the meas-ured value (Bowles, 1996; Shukla 2025).

6.7.   Unconfined compression test

The unconfined compression test involves compressing a cylindrical soil specimen, generally as used in the triaxial compression test, by applying vertical normal stress without any lateral sup-port or confinement. This test is essentially a special case of the UU triaxial compression test, in which the confining pressure, $\sigma_c = \sigma_3 = 0$, and the major principal stress, $\sigma_1 = q_u$, called the unconfined or uniaxial compressive strength (UCS) or simply compressive strength. The value of q_u is measured as the load per unit area at which the soil specimen fails, or the load per unit area at 15% axial strain, whichever occurs first. The time required for the failure of the specimen during the test generally does not exceed 15 min. The details of the test are outlined in relevant standards, such as ASTM D2166-06 (ASTM, 2010a). This test is most suitable for fully saturated cohesive soils having low permeability, such as non-fissured clays. Figure 6.10 shows the Mohr circle with total stress failure envelope, which is horizontal; thus, $\phi = \phi_u = 0$ and, theoretically, the inclination of the failure plane to the horizontal, $\alpha_c = 45°$. From Figure 6.10, and also from Equation 6.10 for the total stress case

$$\tau_f = c_u = \frac{q_u}{2} \tag{6.23}$$

If the unconfined compression test is conducted on both the undisturbed and remoulded specimens of the same cohesive soil at its natural water content, the effect of remoulding on the consistency can be measured in terms of sensitivity, defined as

$$S_t = \frac{q_{u(\text{undisturbed})}}{q_{u(\text{remoulded})}} = \frac{c_{u(\text{undisturbed})}}{c_{u(\text{remoulded})}} \tag{6.24}$$

Based on the sensitivity value, the soil deposits can be classified into six categories (Table 6.2). The sensitivity of most clays ranges from 2 to 4, but marine clays may have a very high sensitivity value and hence they are generally classified as quick. They often lose their strength when dis-turbed or subjected to loading, leading to structural collapse.

Note that during the UU and unconfined compression test, to determine the deviator stress or axial stress for any measured applied axial load, Q, at a given axial strain, ε_a, the following relationship can be used.

Figure 6.10 Typical results from the unconfined compression test

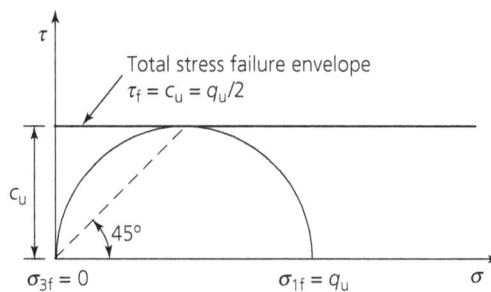

189

Table 6.2 Classification of clays based on sensitivity

Sensitivity, S_t	Classification
1	Insensitive
1–2	Low sensitive
2–4	Medium sensitive
4–8	Highly sensitive
8–16	Extra sensitive
>16	Quick

$$\Delta\sigma_a = \sigma_1 - \sigma_3 = \frac{Q}{A} \tag{6.25a}$$

with

$$A = \frac{A_0}{1 - \varepsilon_a} \tag{6.25b}$$

where A is the average cross-sectional area of the specimen at axial strain, ε_a, and A_0 is the initial average cross-sectional area of the specimen. Equation (6.25) is also applied for the CU test, but A_0 should be taken as the average cross-sectional area of the specimen at the end of the first stage – that is, the consolidation stage of the test. Note that Equation (6.25) is based on the assumption that there is no change in volume of the saturated soil during undrained loading.

Example 6.4

An unconfined compression test is conducted on a cylindrical specimen of saturated clay. The diameter and length of the specimen are 38 mm and 76 mm, respectively. At failure, the axial load is 156 N and the corresponding axial deformation is 10.3 mm. Determine the unconfined compressive strength and the shear strength of the clay.

Solution

Given diameter, $D = 38$ mm, length, $L_0 = 76$ mm, axial load at failure, $Q = 156$ N, and axial deformation at failure, $\Delta L = 10.3$ mm.

The initial cross-sectional area

$$A_0 = \frac{\pi}{4}D^2 = \frac{\pi}{4}(38)^2 = 1134.1 \text{ mm}^2$$

The axial strain at failure

$$\varepsilon_a = \frac{\Delta L}{L_0} = \frac{10.3}{76} = 0.136$$

From Equation 6.25b, the cross-sectional area at failure

$$A = \frac{A_0}{1-\varepsilon_a} = \frac{1134.1}{1-0.136} = 1312.6 \text{ mm}^2$$

Using Equation 6.25a, the unconfined compressive strength of clay

$$q_u = \sigma_{1f} = \frac{Q}{A} = \frac{156}{1312.6} = 0.1188 \text{ N/mm}^2 \text{ or } 118.8 \text{ kPa}$$

From Equation 6.23, the shear strength of clay

$$\tau_f = c_u = \frac{q_u}{2} = \frac{118.8}{2} = 59.4 \text{ kPa}$$

Note: the error can be estimated by using A_0 in place of A in calculating q_u and c_u.

6.8. Vane shear test

The field vane shear test, illustrated schematically in Figure 6.3(d), is widely employed for determining the in situ undrained shear strength, c_u, of saturated clays and silts, particularly those of soft to medium consistency with c_u values typically below 200 kPa. The test is most suitable for sensitive saturated cohesive soil deposits which are highly susceptible to sampling disturbance and is also suitable for other geomaterials, such as mine tailings and coal ashes, when the undrained shear strength is required for designs. The vane used in the test consists of two rectangular thin (less than 3 mm) blades that are perpendicular to each other. Typically, the ratio of height or length, H, to breadth, D, of the blade is 2:1, with D in the range of 35–100 mm for field vanes. Miniature vanes are used in laboratories to determine undrained shear strength of clay samples in sampling tubes (ASTM D4648-05 (ASTM, 2010b)). The details of the field vane shear test can be found in relevant test standards, such as ASTM D2573-08 (ASTM, 2015).

The vane is pushed without a hole from the ground surface or inserted from the bottom of a borehole at the desired depth where the test is to be carried out. The test can be conducted at depths even greater than 50 m. The vane is rotated at the rate of 0.1° per second by applying a torque at the surface through a torque applicator that also measures the torque. This rotation will initiate shearing of the clay along a cylindrical surface of height, H, and diameter, D, surrounding the vanes. On the assumption that the shear stress distributions are uniform on the vertical as well as horizontal failure surfaces, and they are equal to c_u, the expression for the torque, Γ, required to turn the vane at failure can be derived as

$$\Gamma = \left(\pi D H c_u\right) \times \frac{D}{2} + 2\int_0^{D/2} (2\pi r dr)c_u \times r = c_u \pi D^2 \left(\frac{H}{2} + \frac{D}{6}\right) \tag{6.26}$$

where r is the radius of the elemental horizontal shear surfaces varying from 0 to $D/2$.

For a rectangular vane with $H/D = 2$, Equation 6.26 reduces to

$$c_u = \frac{6\Gamma}{7\pi D^3} \tag{6.27}$$

The expression in Equation 6.27 is used to compute c_u from the applied maximum or peak torque, Γ(ASTM, 2015; Terzaghi *et al.*, 1996). The test can be continued by rotating the vane rapidly after shearing the clay to determine the remoulded shear strength, $c_{u(remoulded)}$, from the corresponding value of Γ using Equation 6.27. The value of $c_{u(remoulded)}$ can be used with $c_{u(undisturbed)}$ to calculate the sensitivity of the soil deposit using Equation 6.24. The back analysis of several failed embankments, foundations and excavations in clays have shown that the vane shear test overestimates the undrained shear strength for design, and therefore a suitable correction may be required.

Chapter summary

1. The strength of a soil in terms of effective stress is given as $\tau_f = c' + \sigma'_f \tan \phi'$.
2. The representation of the state of stress at a point within a soil mass by the Mohr circle or the $p' - q$ diagram is very useful in soil mechanics.
3. The triaxial compression test is widely regarded as the most versatile and effective method for studying the stress–strain behaviour and strength properties of a soil. This test is classified into three types: CD, CU and UU tests.
4. The stress–strain behaviour of loose sand is similar to that of normally consolidated clay, whereas dense sand behaves similarly to overconsolidated clay.
5. A clay subjected to undrained loading with a constant volume behaves as a purely cohesive material, which has an angle of shearing resistance equal to zero.
6. The stress–strain behaviour of soil depends on several factors, such as soil type, magnitude and type of initial consolidation, stress history, disturbance and type of loading.
7. Skempton's pore pressure parameters B and A are useful in expressing the portion of a stress increment carried by the pore water.
8. The undrained shear strength of sensitive saturated clays and silts can be determined by the field vane shear test.
9. For most problems, the value of ϕ' based on the peak of the stress–strain curve is used to represent the strength of a soil, thus avoiding large strains or displacements in the soil mass.

Questions for practice
Select the most appropriate answer for each of the multiple choice questions from Q6.1 to Q6.10. Questions Q6.11 to Q6.26 require more detailed answers.

6.1 Figure Q6.1 shows a stress path in the $p' - q$ diagram. Which of the following conditions satisfy?

(a) $\Delta\sigma'_1 = 0; \Delta\sigma'_3 < 0$

(b) $\Delta\sigma'_1 > 0; \Delta\sigma'_3 = 0$

(c) $\Delta\sigma'_1 = -\Delta\sigma'_3$

(d) none of the above.

Figure Q6.1

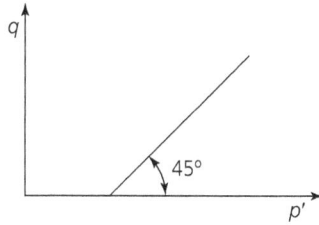

6.2 If the average angle of shearing resistance for a homogenous dry sand deposit is 30°, the ratio of major principal stress to minor principal stress at any point within the soil mass at failure will be
(a) 1/3
(b) 1/2
(c) 2
(d) 3.

6.3 The unconfined compressive strength of a soft clay falls in the range of
(a) < 50 kPa
(b) 50–100 kPa
(c) 100–200 kPa
(d) > 200 kPa.

6.4 A direct shear test is conducted on a dry sand soil specimen under a vertical normal stress of 300 kPa. The specimen fails at a shear stress of 250 kPa. The angle of shearing resistance of the sand is
(a) 11.3°
(b) 30.8°
(c) 39.8°
(d) 50.2°.

6.5 A saturated clay specimen is tested in a triaxial compression test apparatus under consolidated-drained conditions at a confining pressure of 150 kPa. What will be the pore water pressure at a deviator stress of 50 kPa?
(a) 0 kPa
(b) 50 kPa
(c) 100 kPa
(d) 150 kPa.

6.6 Skempton's pore pressure parameter B for a partially saturated soil is
(a) 0
(b) 0.5
(c) 1.0
(d) between 0 and 1.

6.7 If the shear stress is zero on any two planes within a soil mass, then the angle between these planes will be
(a) 0°
(b) 45°
(c) 90°
(d) 135°.

6.8 The ratio of unconfined compressive strength of an undisturbed specimen of soil to that of a remoulded specimen at the same water content is known as
(a) activity
(b) sensitivity
(c) permittivity
(d) plasticity.

6.9 In a consolidated-drained triaxial compression test on an overconsolidated clay, the volume of the soil specimen during the shear stage
(a) remains unaltered
(b) decreases
(c) increases
(d) first increases and then decreases.

6.10 Select the incorrect statement.
(a) The appropriate field test for determining the undrained shear strength of a soft clay is the vane shear test.
(b) In a direct shear test, the plane of shear failure is predetermined.
(c) The unconfined compression test is a special case of unconsolidated-undrained triaxial compression test.
(d) If the Mohr circle for a given state of stress lies entirely below the Mohr–Coulomb envelope for a soil, then it will be unstable for that state of stress.

6.11 A total normal stress of 150 kPa acts at a point on a plane within a soil mass. The effective stress–strength parameters, namely effective cohesion intercept and effective angle of internal friction, of the soil are 15 kPa and 32°, respectively. If the pore water pressure at the point under consideration is 60 kPa, and the soil tends to fail along the plane, what shear strength of soil in terms of effective stress will be observed?

6.12 In a consolidated-drained triaxial test conducted on a saturated sand, failure occurs when the deviator stress is 220 kPa at a confining stress of 60 kPa. What is the effective angle of shearing resistance of sand, and the inclination of failure plane to the major principal plane? What would be the deviator stress and the effective major principal stress at failure for another identical specimen of sand if it is tested under a confining stress of 120 kPa?

6.13 Two identical soil specimens were tested in a triaxial compression test apparatus under consolidated drained conditions. The first specimen failed at a deviator stress of 350 kPa when the confining stress was 100 kPa. The second specimen failed at a deviator stress of 650 kPa when the confining stress was 200 kPa. Determine the shear strength parameters. If the same soil is tested in a direct shear apparatus under consolidated-drained conditions with a vertical normal stress of 500 kPa, what shear strength will be observed?

6.14 For a silty clay, the cohesion intercept and the angle of internal friction obtained from the undrained direct shear test are 600 kPa and 15°, respectively. If a specimen of this soil is subjected to an unconsolidated-undrained triaxial compression test with a confining pressure of 300 kPa, determine the total axial stress at which the failure would be expected.

6.15 An unconfined compression test is conducted on a cylindrical specimen of saturated clay. The diameter and length of the specimen are 38 mm and 76 mm, respectively. At failure, the axial load is 46.5 N, and the corresponding axial deformation is 11.7 mm. Determine the unconfined compressive strength and the shear strength of the clay.

6.16 An unconfined compression test conducted on a clay provided a strength of 100 kPa. If the failure plane is inclined at 51° to the horizontal, what are the values of shear strength parameters?

6.17 In a field vane shear test, a 150 mm-long and 75 mm-wide vane is pressed into a soft clay at the bottom of a borehole. If the clay fails at a torque of 50 Nm, determine the shear strength of the clay.

6.18 Define the shear strength of soil. Is it possible to tabulate the shear strength values for different soils?

6.19 What are the different types of triaxial compression tests? Explain their significance and applications in specific contexts.

6.20 Construct Mohr circles for the consolidation and shear stages of the triaxial test. Additionally, construct stress paths corresponding to these stages.

6.21 Can you construct the typical Mohr circles and failure envelope for a partially saturated clay under the unconsolidated-undrained loading condition?

6.22 Explain the importance of the critical void ratio of a soil.

6.23 Working with Skempton's pore pressure parameters B and A for saturated soils, which one requires more attention and analysis to obtain the realistic value of induced pore water pressure caused by changes in total stresses? Justify your answer.

6.24 In a consolidated-drained triaxial test on a normally consolidated clay, under what conditions is the effective stress path vertical?

6.25 Is there any difference between the ultimate strength behaviour of sand and clay? Explain briefly.

6.26 Can the field vane shear test be used to estimate the sensitivity of marine clays in field projects? Justify your answer.

REFERENCES

ASTM (2010a) ASTM D2166-06: Standard test method for unconfined compressive strength of cohesive soil. ASTM International, West Conshohocken, PA, USA.

ASTM (2010b) ASTM D4648-05: Standard test method for laboratory miniature vane shear test for saturated fine-grained clayey soil. ASTM International, West Conshohocken, PA, USA.

ASTM (2012) ASTM D3080-04: Standard test method for direct shear test of soils under consolidated drained conditions. ASTM International, West Conshohocken, PA, USA.

ASTM (2015) ASTM D2573-08: Standard test method for field vane shear test in cohesive soil. ASTM International, West Conshohocken, PA, USA.

ASTM (2020) ASTM D4767-11(2020): Standard test method for consolidated undrained triaxial compression test for cohesive soils. ASTM International, West Conshohocken, PA, USA.

ASTM (2024) ASTM D2850-23: Standard test method for unconsolidated-undrained triaxial compression test on cohesive soils. ASTM International, West Conshohocken, PA, USA.

Bishop AW and Henkel DJ (1962) *The Measurement of Soil Properties in the Triaxial Test*, 2nd edn. Edward Arnold, London, UK.

Bowles JE (1996) *Foundation Analysis and Design*, 5th edn. McGraw-Hill, New York, NY, USA.

Bureau of Indian Standards (2021) IS 2720 (Part 13): 1986: Direct shear test. Bureau of Indian Standards, New Delhi, India.

Casagrande A (1948) Classification and identification of soils. *Transactions*, ASCE **113**: 901–930.

Das BM (2022) *Principles of Geotechnical Engineering*, 10th edn. Cengage, Boston, MA, USA.

Holtz RD, Kovacs WD and Sheahan TC (2023) *An Introduction to Geotechnical Engineering*, 3rd edn. Pearson Education, Hoboken, NJ, USA.

Ladd CC (1971) *Strength Parameters and Stress-Strain Behaviour of Saturated Clays*. Research Report R71-23, Soils Publication 278, Department of Civil Engineering, MIT, Cambridge, MA, USA.

Lambe TW and Whitman RV (1979) *Soil Mechanics, SI Version*. Wiley, New York, NY, USA.

Shukla SK (2025) *ICE Core Concepts: Geotechnical Engineering*, 2nd edn. Emerald/ICE Publishing, London, UK.

Skempton AW (1954) The pore-pressure coefficients A and B. *Geotechnique* **4(4)**: 143–147.

Standards Australia (2020) AS 1289.6.2.2:2020: Determination of the shear strength of a soil – direct shear test using a shear box. Standards Australia, Strathfield, NSW, Australia.

Terzaghi K, Peck RB and Mesri G (1996) *Soil Mechanics in Engineering Practice*, 3rd edn. Wiley, New York, NY, USA.

Winterkorn HF and Fang H-Y (1991) Soil technology and engineering properties of soils. In *Foundation Engineering Handbook*, 2nd edn (Fang H-Y (ed.)). Van Nostrand Reinhold, New York, NY, USA, Chapter 3, pp. 88–143.

FURTHER READING

Atkinson J (2007) *The Mechanics of Soils and Foundations*, 2nd edn. CRC Press, Taylor and Francis, London, UK.

emerald
PUBLISHING

ice

Sanjay Kumar Shukla
ISBN 978-1-83608-519-5
https://doi.org/10.1108/978-1-83608-516-420252010

Chapter 7
Compaction of soil

Learning aims

This chapter explores the core concepts of the following topics

- laboratory compaction details
- compaction parameters and their importance
- factors affecting compaction
- process of compaction
- effect of compaction on engineering behaviour of soil
- field compaction of soil and quality control.

7.1.   Introduction

In construction projects such as roads, airfield taxiways and runways, railway tracks, earth dams, embankments, landfill and pond liners, and structural fills, the soil is often unsuitable as a foundation or construction material. Therefore, it is frequently necessary to improve the soil to achieve the required engineering properties for the project. One of the common requirements for most field projects is a dense or hard soil with enhanced characteristics such as high strength and stiffness, low permeability, minimal compressibility and volume change, reduced liquefaction potential, and long-term durability. A loose or soft soil can be improved to dense or hard soil through densification methods, commonly referred to as soil or ground improvement techniques, including the following

- compaction (densification with mechanical equipment, usually rollers and rammers)
- preloading (densification by placing a temporary load)
- dewatering (densification by removal of pore water and/or reduction of pore water pressure)
- heavy tamping (densification by impact of a large mass dropped from some height)
- vibration (densification by a vibrating probe or pile)
- compaction grouting (densification by injection of a zero-slump mortar into the soil or ground)
- explosion (densification by detonating explosives on the ground surface or in boreholes)
- saturation (densification by saturating the soil, mainly cohesionless soil)
- water jetting (densification by applying water jets on soil, such as rock fills).

Compaction refers to the reduction in volume of an unsaturated soil mass due to the expulsion of air from its voids or pores, caused by the application of an external compressive load or stress. It is almost instantaneous and, therefore, a time-independent process of densification. As a result of compaction, with reduction in total volume of the soil mass, the void volume of the soil mass reduces while the volume of solids and the water content of the soil mass remain essentially unchanged. Some of the soil and rock particles may get crushed during the compaction process, resulting in a modification of the gradation of the soil. As discussed in Chapter 5, note that compaction differs from consolidation, which is a time-dependent process of volume reduction, primarily in saturated soil masses, caused by the expulsion of water from their voids or pores. Consequently, the water content of the soil mass decreases over time during the consolidation process.

This chapter presents the basic features of the compaction process and its field application, focusing on its soil mechanics aspects. The details of other ground improvement techniques can be found in Hausmann (1990) and Moseley (1993).

7.2. Laboratory compaction

In field compaction, the compaction by mechanical equipment employs the static, dynamic or kneading process, or a combination thereof. With the intention of simulating the field compaction process, several types of laboratory compaction tests have been developed in the past. The dynamic (or impact), kneading, static (oedometric) and vibratory compaction tests are some examples. The dynamic compaction test that consists of placing the soil in a cylindrical mould and then dropping a steel rammer a specified number of times is the earliest and most common type of compaction test, which is described here in detail.

7.2.1 Basic test details

For a given compactive effort (mechanical energy input) by a mechanical equipment such as a roller (Figure 7.1) in the field compaction of soil, the maximum possible densification can be achieved only when the soil has a particular amount of water content, w_{opt}, known as the optimum water content (OWC) or the optimum moisture content (OMC). The maximum densification of a soil is usually described in terms of the maximum dry unit weight, γ_{dmax} (or maximum dry density ρ_{dmax}). The aim of any laboratory compaction is to obtain w_{opt} and γ_{dmax} (or ρ_{dmax}), called the compaction parameters, by simulating the field compaction procedures. Note that the compactive effort is a measure of the

Figure 7.1 Compaction of soil in field by a roller

mechanical energy input to the soil mass per unit its volume; thus, its SI unit is joules per cubic metre (J/m^3); however, it is commonly reported in kilojoules per cubic metre (kJ/m^3).

In the compaction test, a steel rammer of mass, M_r (or weight, W_r), is dropped freely by a manual or mechanical operation from a height, H_d, on loose soil placed in a cylindrical mould of internal volume, V. The soil at a selected moulding water content is compacted in the mould in three or five layers of about equal thickness, with each layer compacted by a specific number of blows, N_b, of the rammer, depending on the specific values of V, M_r and H_d, as recommended in the relevant standards adopted by the country or state where the compaction is carried out. Figure 7.2 shows the equipment for the compaction test, while Table 7.1 provides the details of the equipment and the compaction test procedures adopted in some countries. Note that the extension collar shown in Figure 7.2(b) is attached to the top of the compaction mould while compacting the final soil layer.

The laboratory test to determine the compaction parameters was first developed by RR Proctor in the USA and the test procedure was published in 1933 (Proctor, 1933; Taylor, 1948). The compaction equipment details and test procedure for this test have been adopted as the light compaction test using a cylindrical mould with a diameter of 4″ (101.6 mm) and a volume of 1/30 cubic feet (944 cc) in ASTM D698-12-2021 (ASTM, 2021a), and with some modifications in other standards. This originally developed compaction test is commonly known as the standard Proctor compaction test. With the introduction of heavier earth-moving and compaction machineries, especially for large dams, the standard Proctor compaction test has been modified to achieve a higher dry unit weight (or dry density) by increasing the mass or weight and drop height of the rammer as well as the number of layers of soil to be compacted, as given in Table 7.1. The compaction test that uses a

Figure 7.2 Equipment for the laboratory compaction test: (a) cylindrical mould with base plate; (b) extension collar; (c) rammer

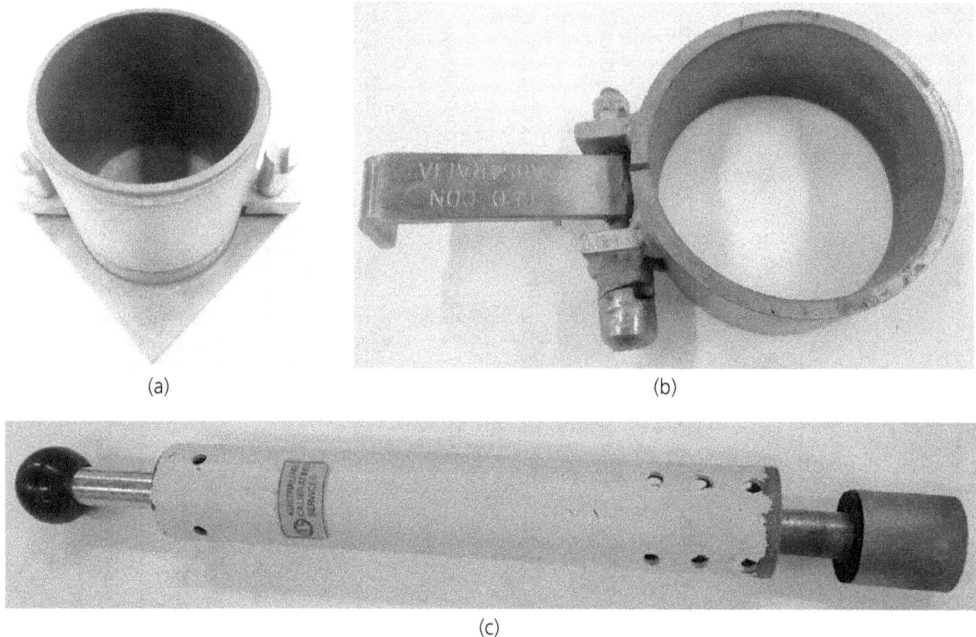

(a)

(b)

(c)

Table 7.1 Details of compaction moulds and rammers, and compaction test procedures adopted in some countries

Standard	Type of compaction test	Details of compaction mould			Details of rammer		Compaction procedure	
		Inside diameter, D_m: mm	Height, Hm: mm	Internal volume, V: cm3 or cc	Mass, Mr: kg	Drop height, Hd: mm	Number of layers, Nl	Number of blows per layer, Nb
Standards Australia (2017a)	AS 1289.5.1.1-2017	Light compaction	105	115.5	1000	2.7	300	3	25
			152	132.5	2400	2.7	300	3	60
Standards Australia (2017b)	AS 1289.5.2.1-2017	Heavy compaction	105	115.5	1000	4.9	450	5	25
			152	132.5	2400	4.9	450	3	100
ASTM (2021a)	ASTM D698-12-2021	Light compaction	101.6	116.4	944	2.495	304.8	3	25
			152.4	116.4	2124	2.495	304.8	3	56
ASTM (2021b)	ASTM D1557-12-2021	Heavy compaction	101.6	116.4	944	4.536	457.2	5	25
			152.4	116.4	2124	4.536	457.2	5	56
BSI (2022)	BS 1377-2-2022	Light compaction	105	115.5	1000	2.5	300	3	27
			152	127	2305	2.5	300	3	62
		Heavy compaction	105	115.5	1000	4.5	450	5	27
			152	127	2305	4.5	450	5	62
Bureau of Indian Standards (2021)	IS 2720 (Part 7)-1980	Light compaction	100	127.3	1000	2.6	310	3	25
			150	127.3	2250	2.6	310	3	55
Bureau of Indian Standards (2020)	IS 2720 (Part 8)-1983	Heavy compaction	100	127.3	1000	4.9	450	5	25
			150	127.3	2250	4.9	450	5	55

Note: the striking face of the rammer is planar and circular with a diameter of about 50mm. Drop height is measured from the surface of the soil specimen in the compaction mould.

high compactive effort is known as the modified Proctor compaction test. Currently, there are two common types of laboratory compaction test

- standard compaction test (also known as the standard Proctor compaction test or light compaction test)
- modified compaction test (also known as the modified Proctor compaction test or heavy compaction test).

The mechanical energy input, E_{in}, to the soil in terms of the work done in operating the rammer while conducting the compaction test in accordance with any test standard given in Table 7.1 can be computed from

$$E_{in} = \frac{W_r H_d N_b N_l}{V} \tag{7.1}$$

by substituting the values of rammer weight, W_r, rammer drop height, H_d, number of blows, N_b, number of layers, N_l, and internal volume of the mould, V.

Note that the use of a large-size compaction mould with a diameter of 150–152.4 mm allows the compaction of a soil specimen with coarser particles, generally up to 37.5 mm in diameter. In a mould with a diameter of 100–105 mm, it is generally recommended to compact soil containing particles smaller than 19 mm in diameter. If the soil sample contains a significant proportion of oversized particles, such as 30% or more of the total sample, it is not possible to determine its compaction parameters in the geotechnical laboratory using standard compaction moulds. This is because no existing standard provides a laboratory compaction procedure for such soils. This problem can be solved either by conducting a field test adopting the sand or water replacement method in a test pit in the compacted fill (ASTM, 2013, 2016; Shukla and Sivakugan, 2011), using the replacement technique in which the oversized fraction is replaced with a finer fraction, or using the analytical expressions with some laboratory measurements (Shukla et al., 2009; Standards Australia, 2007, Hausmann, 1990). The correction for the oversized fraction, as recommended by Shukla et al. (2009), is detailed in Section 7.3.

Example 7.1

Compute the mechanical energy input (compactive effort) in kilojoules per cubic metre (kJ/m³) to the soil in terms of the work done by operating a rammer during the light compaction test in accordance with AS 1289.5.1.1-2017 (Standards Australia (2017a), ASTM D698-12-2021 (ASTM (2021a), BS 1377-2-2022 (BSI, 2022) and IS 2720 (Part 7)-1980 (Bureau of Indian Standards, 2021) for the compaction mould of small size. Compare the values and record your observations.

Solution

In accordance with AS 1289.5.1.1-2017 (Standards Australia, 2017a)

$W_r = M_r g = (2.7)(9.81) = 26.49$ N, $H_d = 300$ mm $= 0.3$ m, $N_b = 25$, $N_l = 3$ and
$V = 1000$ cc $= 0.001$ m³

From Equation 7.1

$$E_{in} = \frac{W_r H_d N_b N_l}{V} = \frac{(26.49)(0.3)(25)(3)}{0.001} = 596025 \text{ J/m}^3 \approx 596 \text{ kJ / m}^3$$

In accordance with ASTM D698-12-2021 (ASTM, 2021a)

$W_r = M_r g = (2.495)(9.81) = 24.476$ N, $H_d = 304.8$ mm $= 0.3048$ m, $N_b = 25$, $N_l = 3$ and $V = 944$ cc $= 9.44 \times 10^{-4}$ m^3

From Equation 7.1

$$E_{in} = \frac{W_r H_d N_b N_l}{V} = \frac{(24.476)(0.3048)(25)(3)}{9.44 \times 10^{-4}} = 592713.3 \text{ J/m}^3 \approx 593 \text{ kJ / m}^3$$

In accordance with BS 1377-2-2022 (BSI, 2022)

$W_r = M_r g = (2.5)(9.81) = 24.52$ N, $H_d = 300$ mm $= 0.3$ m, $N_b = 27$, $N_l = 3$ and $V = 1000$ cc $= 0.001$ m^3

From Equation 7.1

$$E_{in} = \frac{W_r H_d N_b N_l}{V} = \frac{(24.52)(0.3)(27)(3)}{0.001} = 595836 \text{ J/m}^3 \approx 596 \text{ kJ / m}^3$$

In accordance with IS 2720 (Part 7)-1980 (Bureau of Indian Standards, 2021)

$W_r = M_r g = (2.6)(9.81) = 25.51$ N, $H_d = 310$ mm $= 0.310$ m, $N_b = 25$, $N_l = 3$ and $V = 1000$ cc $= 0.001$ m^3

From Equation 7.1

$$E_{in} = \frac{W_r H_d N_b N_l}{V} = \frac{(25.51)(0.310)(25)(3)}{0.001} = 593108 \text{J/m}^3 \approx 593 \text{ kJ / m}^3$$

The energy input values to the soil are almost the same, although there is some variation in the compaction equipment and test procedure. This difference between the standards helps analyse and use the compaction test data reported for similar soils in slightly different ways.

Note: similar computations for light compaction can be made using the large-size compaction mould. The calculation steps can also be repeated for heavy compaction.

7.2.2 Compaction curve and parameters

A sufficient number of compaction tests (at least four) are conducted on soil with different moulding water content to establish the relationship between dry unit weight, γ_d (or dry density, ρ_d), and water content, w, for the soil. The last test is identified when an increase in water content results in a decrease in the total unit weight of the compacted soil in the mould. To compute the value of γ_d for each test, the weight of the compacted wet or moist soil, W, in the mould and its water content, w, are determined. The computations are conducted using Equations 1.7 and 1.22 as follows.

Total unit weight of the compacted soil

$$\gamma = \frac{W}{V} \tag{7.2}$$

where V is the internal volume of the mould.

Dry unit weight of the compacted soil

$$\gamma_d = \frac{\gamma}{1+w} \tag{7.3}$$

A plot of γ_d (or ρ_d) against w values are joined by a smooth curve, called the compaction curve, which represents a curvilinear relationship between γ_d (or ρ_d) and w. Figure 7.3 shows a typical compaction curve for cohesive soil. The dry unit weight, γ_d (or dry density, ρ_d), and the water content corresponding to the peak of the compaction curve are the maximum dry unit weight, γ_{dmax} (or maximum dry density, ρ_{dmax}), and the optimum water content, w_{opt}, respectively, which are the compaction test parameters. Note that each data point, called the compaction point, on the compaction curve represents a single compaction test, and the compaction points may not lie exactly on the compaction curve. The following can be observed from Figure 7.3.

Figure 7.3 Typical compaction curve for cohesive soil with ZAV curve

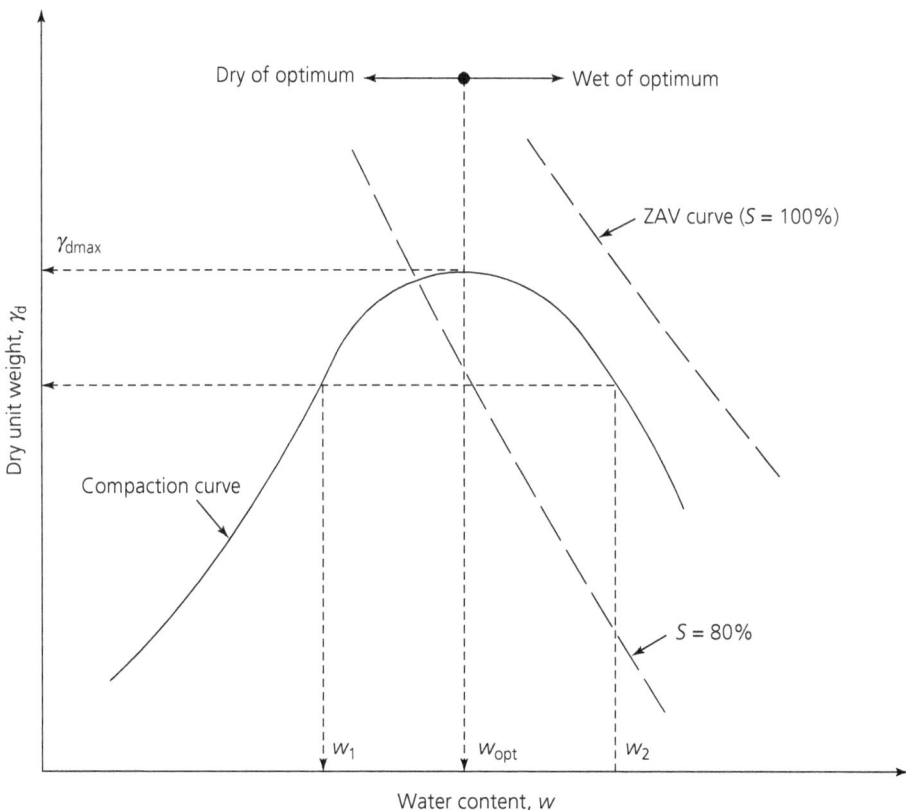

- As the water content of the soil is increased, the dry unit weight of the soil increases to a peak and then decreases.
- Corresponding to a point on the compaction curve to the left of the peak, the soil is called dry of optimum, while corresponding to a point to the right of the peak, the soil is called wet of optimum. Thus, with respect to the optimum water content, a given compacted soil is dry of optimum, near optimum, at optimum or wet of optimum.
- Any dry unit weight other than γ_{dmax} can be achieved by compacting the soil at two different water contents w_1 and $w_2 (w_1 < w_2)$ with $w_1 < w_{opt}$ and $w_s > w_{opt}$.
- The compaction curve and the values of γ_{dmax} and w_{opt} are unique for a soil only for the given compactive effort and the type of compaction used. Thus, the compaction process deals with five variables: dry unit weight, water content, soil type, compactive effort and type of compaction.

It is almost essential that a compaction curve should be accompanied on the same scale by the saturation curve, often called the zero-air-voids (ZAV) curve, which represents the dry unit weights corresponding to a degree of saturation $S = 100\%$ at given water contents and for the known value of specific gravity, G, of soil solids. The points of the ZAV curve are calculated from Equation 1.21a with Equation 1.17 (see Chapter 1) as

$$\gamma_d = \frac{G\gamma_w}{1+e} = \frac{G\gamma_w}{1+\dfrac{wG}{S}} \tag{7.4}$$

where γ_w is the unit weight of water.

Note that for the known values of w and G, γ_d can be maximum when $S = 100\%$ – that is, the void ratio is minimum, given by $e_{min} = wG$. As $S = 100\%$, and the achievement of zero air voids cannot be achieved because of practical difficulties, the ZAV curve for a soil is a theoretical curve. The ZAV curve supports plotting the compaction curve and helps identify erroneous test results. The compaction curve lies always to the left of the ZAV curve, and practically it will never intersect the ZAV curve. Curves similar to the ZAV curve but for $S < 100\%$, say $S = 80\%$, can be drawn along with the compaction curve and the ZAV curve, as shown in Figure 7.3. These curves assist in estimating the degree of saturation of the compacted soil specimen at the desired dry unit weight.

Example 7.2

What is the theoretical maximum dry unit weight of a soil with solid particles having a specific gravity of 2.66?

Solution

Given $G = 2.66$.

From Equation 7.4, the dry density, γ_d, will be maximum when the water content $w \to 0$ with the degree of saturation $S = 100\%$. Thus, the theoretical maximum dry unit weight is

$$\gamma_{dmax(th)} = \frac{G\gamma_w}{1+\dfrac{wG}{S}} = \frac{(2.66)(9.81)}{1+\dfrac{(0)(2.66)}{(1)}} = 26.09\text{kN/m}^3$$

Note that $e_{min} = wG = (0)(2.66) = 0$ when the soil is at the theoretical maximum dry unit weight. If a document reports γ_{dmax} of soil greater than or even very close to $\gamma_{dmax(th)}$, the reported value is incorrect. Thus, this example illustrates how the accuracy of the maximum dry unit weight values obtained through the laboratory compaction test can be verified in a simple way and erroneous test results can be avoided.

Sometimes, the compaction curve may be accompanied by curves representing different percentage air voids, n_a, in place of degree of saturation, S, along with the ZAV curve. Percentage air voids is defined as the ratio of air void volume, V_a, to the total volume, V, of the soil mass. Thus,

$$n_a = \frac{V_a}{V} \tag{7.5}$$

Percentage air voids is normally expressed as a percentage.

Referring to Figure 1.4(b), $V = V_a + V_w + V_s$; therefore,

$$n_a = \frac{V_a}{V} = \frac{V - V_w - V_s}{V} = 1 - \frac{V_w}{V} - \frac{V_s}{V} = 1 - \frac{W_w}{\gamma_w V} - \frac{W_s}{\gamma_s V} =$$

$$1 - \frac{wW_s}{\gamma_w V} - \frac{W_s}{G\gamma_w V} = 1 - \frac{w\gamma_d}{\gamma_w} - \frac{\gamma_d}{G\gamma_w}$$

or

$$\gamma_d = \frac{(1 - n_a)G\gamma_w}{1 + wG} \tag{7.6}$$

Equation 7.6 is used to plot the curves representing different values of n_a (e.g. $n_a = 0\%$, 5%, 10% etc.). Note that $n_a = 0\%$ refers to $S = 100\%$; hence, Equation 7.4 with $S = 100\%$ and Equation 7.6 with $n_a = 0\%$ give the same ZAV curve. However, it can be seen by computation that the 20% air void line is not identical to the 80% saturation line because $n_a = 20\%$ does not refer to $S = 80\%$. Sometimes the term air content, a_c ($= V_a/V_v$), is used, usually represented as a decimal, which is related to the degree of saturation as $a_c = V_a/V_v = (V_v - V_w)/V_v = 1 - S$. A value of $a_c = 0.2$ always refers to $S = 80\%$. It is important to understand the difference between n_a and a_c and their relationship with $n_a = 20\%$ in order to be able to utilise them suitably. Also note that n, n_a and a_c are related by $n_a = na_c$.

The following points regarding laboratory compaction are noteworthy.

- In the kneading compaction test, the soil is placed in a mould and then pushed a specified number of times with a tamper at a specified stress. In the static compaction test that is most useful for research purposes, the soil is subjected to a static stress of a given magnitude by simply pressing in a mould. For these types of compaction test, it is not possible to compute the compactive effort in a simple way, as shown in Example 7.1 for the dynamic compaction test.
- The blows are applied at a uniform rate of about 25 blows per minute to ensure complete uniform coverage of the specimen surface in the compaction mould. Figure 7.4 shows the commonly used sequence of blows using the hand rammer in a mould with a diameter of

Sequence of blows using the hand rammer in a compaction mould with a diameter of 100–105 mm (Based on ASTM D698-12:2021 (ASTM, 2021a) and ASTM D1557-12:2021 (ASTM, 2021b))

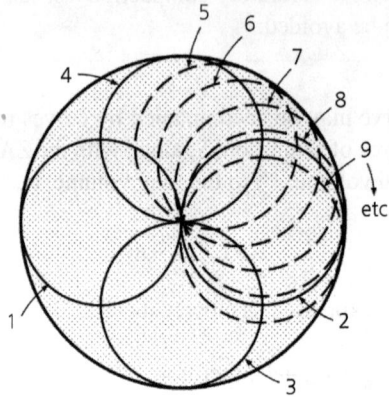

100–105 mm. In a mould of large size (diameter of 150–152.4 mm), a similar blow pattern with more initial blows (e.g. 1st to 9th blows like 1st to 4th blows, but 5th blow in the centre of the mould) can be adopted along with blows around the mould. For a mechanical rammer, the blow pattern consists of systematic blows around the mould, as indicated by the 5th, 6th and subsequent blows.

- The optimum water content and the maximum dry unit weight are generally reported to the nearest 0.1% and 0.02 kN/m³, respectively.
- Most soils, in particular cohesive soils, show a distinct peak that corresponds to γ_{dmax} and w_{opt} in the compaction curve, probably in the range of $S = 80$ to 90% (Hausmann, 1990).
- The optimum water content of soils ranges from 5 to 45% with a typical range from 10 to 20%. For some soils, the specific ranges are as follows: well-graded sands (SW) 9–16%; silty sands (SM) 11–16%; clayey sands (SC) 11–19%; silts (ML) 12–24%; lean clays (CL) 12–24%; elastic silts (MH) 24–40%; fat clays (CH) 19–36%; and organic clays/silts (OH) 21–45%.
- The maximum dry unit of soils ranges from 13 to 24 kN/m³ with a typical range from 16 to 20 kN/m³. For some soils, the specific ranges are as follows: well-graded sands (SW) 18–21 kN/m³; silty sands (SM) 18–20 kN/m³; clayey sands (SC) 17–20 kN/m³; silts (ML) 15–19 kN/m³; lean clays (CL) 15–19 kN/m³; elastic silts (MH) 11–15 kN/m³; fat clays (CH) 12–17 kN/m³; and organic clays/silts (OH) 10–16 kN/m³.
- Waste materials such as fly ashes are used as soil materials. Fly ash consisting of mainly silt-sized particles is a by-product of the combustion of coal obtained during power generation in thermal power plants. The typical value of specific gravity of fly ash particles is about 2.14, which is lower than the typical value of 2.65 of most soil solid particles. The maximum dry unit weight and the optimum water content obtained from the standard compaction test are typically about 11.5 kN/m³ and 33.6%, respectively (Shukla, 1992; Shukla and Gokhale, 1995).
- Soil compacted dry of optimum has generally a flocculated structure, while soil compacted wet of optimum has a dispersed structure (Figure 1.12(b)) (Lambe, 1958).
- Laboratory compaction tests are conducted primarily because they are cheaper and faster to perform than field compaction tests.

Example 7.3

The maximum dry unit weight and the optimum water content of a cohesive soil obtained from the standard compaction test are 17.52 kN/m³ and 16.1%, respectively. If the specific gravity of soil solids is 2.65, what is the degree of saturation and percentage air voids of the soil at its maximum dry unit weight? What is the maximum dry unit weight to which it can be further compacted?

Solution

Given $\gamma_d = \gamma_{d\,max} = 17.52 \text{ kN/m}^3$, $w = w_{opt} = 16.1\%$ and $G = 2.65$.

From Equation 7.4

$$\gamma_d = \frac{G\gamma_w}{1 + \dfrac{wG}{S}}$$

or

$$S = \frac{wG}{\dfrac{G\gamma_w}{\gamma_d} - 1} = \frac{(0.161)(2.65)}{\dfrac{(2.65)(9.81)}{17.52} - 1} = 0.882 \text{ or } 88.2\%$$

From Equation 7.6

$$\gamma_d = \frac{(1 - n_a)G\gamma_w}{1 + wG}$$

or

$$n_a = 1 - \frac{(1 + wG)\gamma_d}{G\gamma_w} = 1 - \frac{(1 + 0.161 \times 2.65)(17.52)}{(2.65)(9.81)} = 0.038 \text{ or } 3.8\%$$

For the maximum possible dry unit weight, the degree of saturation $S = 100\%$ and the percentage air voids $n_a = 0\%$. Thus, from Equation 7.4 or Equation 7.6, the soil can be further compacted to achieve the following maximum dry unit weight.

$$\gamma_d = \frac{G\gamma_w}{1 + wG} = \frac{(2.65)(9.81)}{1 + (0.161)(2.65)} = 18.22 \text{ kN/m}^3$$

Note: achieving $S = 100\%$ (or $n_a = 0\%$) through the compaction process is practically difficult.

7.3.   Correction of laboratory compaction parameters

The previous sections explained that the compaction test is generally performed to obtain the values of compaction test parameters, optimum water content and maximum dry unit weight that are required to achieve maximum densification of the soil in the field with a given compactive effort. In most compaction test procedures, depending on the size of the compaction mould, a fraction of the soil sample having particle size larger than a specific value, say d_0, is discarded – for example, in the standard compaction test, soil particles coarser than 19 mm are generally discarded before compacting soil in a compaction mould of a smaller size (diameter 100–105 mm). Similarly, in a larger size (diameter 150–152.4 mm) mould, soil particles coarser than 37.5 mm are generally discarded. If the fraction removed is significant, the laboratory optimum water content and the maximum dry unit weight determined for the remaining soil are not directly comparable with the field values. The removal of a small fraction of particles (up to 5% by weight) retained on a 19 mm or 37.5 mm sieve, as applicable, will affect the maximum dry unit weight only marginally, comparable to the experimental error in measuring compaction parameters.

However, the exclusion of a large proportion of coarser particles from the laboratory test specimen may have a significant effect on the compaction test parameters compared with those obtainable with field soil specimens as a whole. There is currently no realistic and generally accepted procedure for calculating the field values of compaction test parameters from the laboratory values, especially when the oversized materials consist of a significant part of the soil to be compacted in the field. The inaccurate estimation of field compaction parameters for granular materials used in subbase and base courses has likely contributed to pavement settlement failures, resulting in deep ruts observed in some past projects. Figure 7.5 illustrates a typical failure with deep ruts in a long section of newly constructed bituminous pavement.

The field maximum dry unit weight and the field water content can be calculated from the laboratory values using the following equations (Hausmann, 1990; Standards Australia, 2007).

$$\frac{1}{\gamma_{d\,max(F)}} = \frac{(1-p)}{\gamma_{d\,max}} + \frac{p}{G_c \gamma_w} \tag{7.7}$$

and

$$w_{opt(F)} = (1-p)w_{opt} \tag{7.8}$$

Figure 7.5 Typical settlement failure with deep ruts in a newly constructed bituminous pavement (Shukla *et al.*, 2012)

where $\gamma_{dmax(F)}$ is the field maximum dry unit weight; γ_{dmax} is the laboratory maximum dry unit weight; p is the percentage of coarser fraction (larger than d_0) discarded from the soil; G_c is the specific gravity of discarded coarser soil solid particles; γ_w is the unit weight of water; $w_{opt(F)}$ is the field optimum water content; w_{opt} is the laboratory optimum water content.

Equations 7.7 and 7.8 are based on the assumptions that the coarse fraction (larger than d_0) is dry and there is no change in the volume of pore air after removal of the coarse fraction. These assumptions cannot always be appropriate for their field applications. An assumption of zero moisture in the coarse fraction may lead to overestimating the field dry unit weight (Hausmann, 1990). Relatively more rational analytical expressions have been presented by Shukla *et al.* (2009), which are presented here.

Figure 7.6 shows the phase diagrams for field- and laboratory-compacted soil specimens, with their particle size distributions given in Figure 7.7. In Figure 7.6, in addition to the weights and volumes of the three phases, unit weights are also shown with the phase labels. When the coarser fraction (larger than size d_0, e.g. 19 mm or 37.5 mm) is removed, it also takes away some water associated with its water content. In addition, there is also the possibility of some change in the air void volume when the soil is compacted without this coarse fraction. All these are reflected in Figure 7.6, where G_f is the specific gravity of the fine soil solid particles (smaller than d_0) in the field or laboratory soil specimen; V_a is the volume of the air in voids of the field soil specimen; V_F is the total volume of field soil specimen; V is the total volume of the laboratory soil specimen; w_c is the water content of the coarse soil particles in the field soil specimen; W_s is the weight of the soil particles in the field specimen; W_{wc} is the weight of water with coarser soil particles in the field soil specimen; W_{wf} is the weight of water with finer soil particles in the field or laboratory soil specimen; α is the ratio of volume of air in the voids of the laboratory soil specimen to that in the field soil specimen; $G_{c\gamma w}$ is

Figure 7.6 Phase diagrams: (a) field-compacted soil specimen; (b) laboratory-compacted soil specimen (after Shukla et al., 2009)

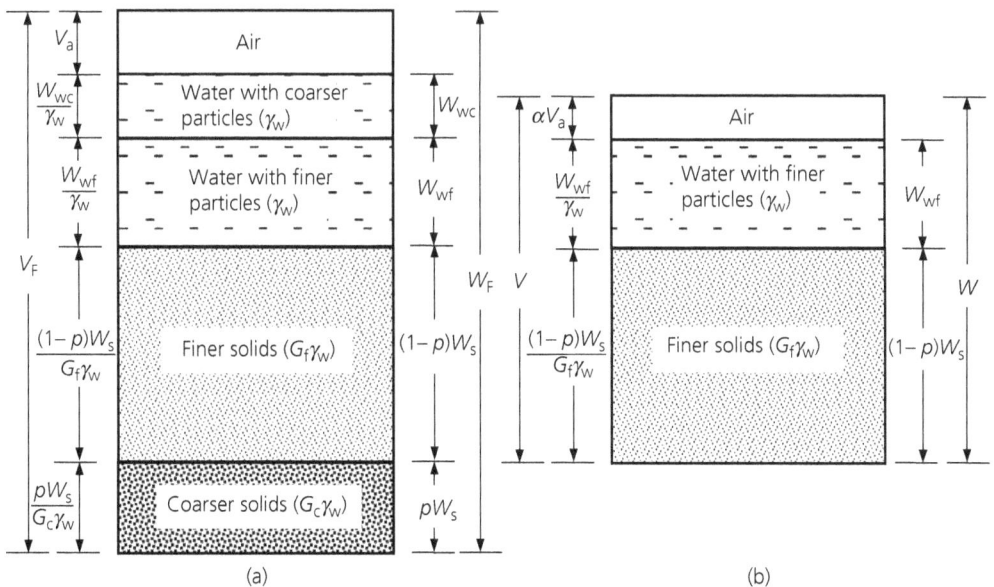

Figure 7.7 Particle size distribution of field- and laboratory-compacted soil specimens (after Shukla *et al.*, 2009)

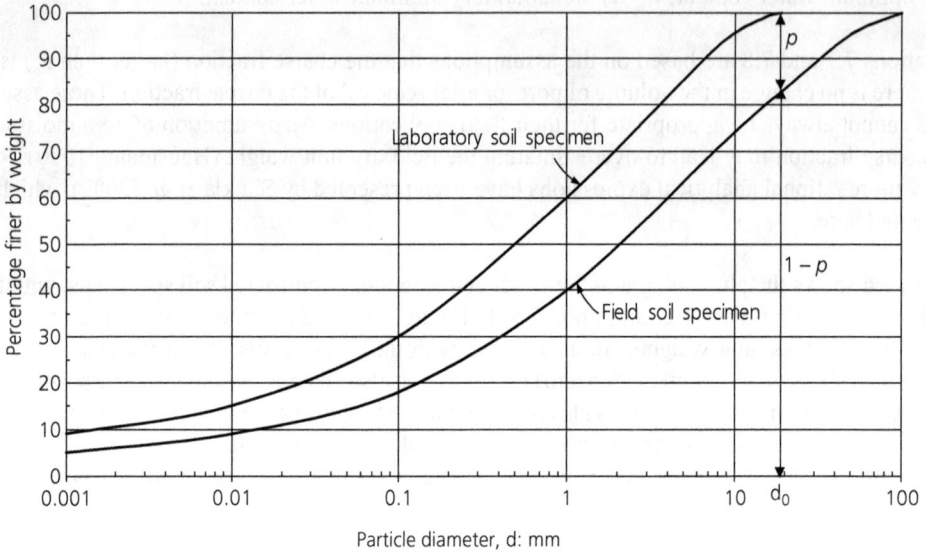

the unit weight of the coarser fraction of soil particles in the field soil specimen; $G_{f\gamma w}$ is the unit weight of the finer fraction of soil particles in the field or laboratory soil specimen.

From Figure 7.6(b), the laboratory maximum dry unit weight and optimum water content can be obtained as

$$\gamma_{d\,max} = \frac{(1-p)W_s}{V} \qquad (7.9)$$

and

$$w_{opt} = \frac{W_{wf}}{(1-p)W_s} \qquad (7.10)$$

From Figure 7.6(a), the corresponding field maximum dry unit weight can be obtained as

$$\gamma_{d\,max(F)} = \frac{W_s}{V_F}$$

where

$$V_F = V + (1-\alpha)V_a + \frac{W_{wc}}{\gamma_w} + \frac{pW_s}{G_c\gamma_w}$$

with

$$V = \frac{(1-p)W_s}{\gamma_{d\,max}} \text{ from Equation 7.9 or}$$

$$\gamma_{d\,max(F)} = \frac{1}{\dfrac{1-p}{\gamma_{d\,max}} + \dfrac{(1-\alpha)V_a}{W_s} + \dfrac{pw_c}{\gamma_w} + \dfrac{p}{G_c\gamma_w}} \qquad (7.11)$$

where

$$w_c = \frac{W_{wc}}{pW_s} \qquad (7.12)$$

From Figure 7.6(b)

$$\frac{\alpha V_a}{W_s} = \frac{V}{W_s} - \frac{W_{wf}}{\gamma_w W_s} - \frac{1-p}{G_f\gamma_w}$$

or using Equations 7.8 and 7.9

$$\frac{V_a}{W_s} = \frac{1}{\alpha}\left[\frac{1-p}{\gamma_{dL}} - \frac{(1-p)w_L}{\gamma_w} - \frac{1-p}{G_f\gamma_w}\right] \qquad (7.13)$$

From Equations 7.11 and 7.13

$$\frac{1}{\gamma_{d\,max(F)}} = \frac{1-p}{\alpha\gamma_{d\,max}} + \frac{p}{G_c\gamma_w} + \frac{pw_c - \left(\dfrac{1-\alpha}{\alpha}\right)(1-p)w_{opt}}{\gamma_w} - \left(\frac{1-\alpha}{\alpha}\right)\left(\frac{1-p}{G_f\gamma_w}\right) \qquad (7.14)$$

Assuming

$$\frac{1-\alpha}{\alpha} = \beta \qquad (7.15)$$

Equation 7.14 can be expressed as

$$\frac{1}{\gamma_{d\,max(F)}} = \frac{(1-p)(1+\beta)}{\gamma_{d\,max}} + \frac{p}{G_c\gamma_w} + \frac{pw_c - (1-p)\beta\,w_{opt}}{\gamma_w} - \frac{(1-p)\beta}{G_f\gamma_w} \qquad (7.16)$$

From Figure 5.6(a), the field optimum water content is

$$w_{opt(F)} = \frac{W_{wf} + W_{wc}}{W_s} = \frac{W_{wf}}{W_s} + \frac{W_{wc}}{W_s}$$

or using Equations 7.10 and 7.12

$$w_{opt(F)} = (1-p)w_{opt} + pw_c \qquad (7.17)$$

Equations 7.16 and 7.17 provide improved expressions for calculating the field maximum dry unit weight and the field optimum water content, respectively, based on the test values obtained from the laboratory compaction test on the laboratory soil specimen, which does not contain soil particles larger than the maximum size limit of the compaction mould.

If the removal of the coarse fraction from the field soil specimen does not alter the volume of the air present in voids of the remaining soil for the laboratory test, then $\alpha = 1$. For this case, Equation 7.16 reduces to

$$\frac{1}{\gamma_{d\max(F)}} = \frac{(1-p)}{\gamma_{d\max}} + \frac{p}{G_c\gamma_w} + \frac{pw_c}{\gamma_w} \tag{7.18}$$

and Equation 7.17 remains unaltered.

If the removal of the coarse fraction from the field soil specimen does not alter the volume of the air present in the voids and the removed coarse particles are dry, then $\alpha = 1$ and $w_c = 0$. For this case, Equations 7.16 and 7.17 reduce to Equations 7.7 and 7.8, respectively, as presented by Hausmann (1990).

The expressions in Equations 7.16 and 7.17 proposed by Shukla *et al.* (2009) are quite suitable for field applications. Note that these proposed expressions require the values of the parameters α and w_c in addition to the laboratory values of the compaction parameters ($\gamma_{d\max}$ and w_{opt}) for calculating the field maximum dry unit weight ($\gamma_{d\max(F)}$) and the field optimum water content ($w_{opt(F)}$). The water content (w_c) of the coarse fraction removed from the soil specimen for the test can be determined in the laboratory as a routine test, but the appropriate value of α should be considered with caution.

Example 7.4

A soil sample consists of 20% particles coarser than 19 mm. A standard compaction test is carried out in accordance with ASTM D698-12-2021 (ASTM, 2021a) using a 101.6 mm mould. The coarse fraction of the soil (soil particles coarser than 19 mm) is removed from the test specimen in view of the particle size limit for the test mould. The maximum dry unit weight and the optimum water content are obtained as 15.73 kN/m³ and 19.1%, respectively. For the coarser particles, $G_c = 2.66$ and $w_c = 2.8\%$. Determine the field values of maximum dry unit weight and optimum water content, if the volume of the air in voids of the test specimen remains the same after removal of the coarse fraction. Also determine the values ignoring the water content of the coarse fraction.

Solution

Given $p = 20\%$, $\gamma_{d\max} = 15.73$ kN/m³, $w_{opt} = 19.1\%$, $G_c = 2.66$, $w_c = 2.8\%$ and $\alpha = 1$.

From Equation 7.18, the field maximum dry unit weight

$$\frac{1}{\gamma_{d\max(F)}} = \frac{(1-0.2)}{15.73} + \frac{0.2}{(2.66)(9.81)} + \frac{(0.2)(0.028)}{9.81} = 5.08\times10^{-2} +$$

$$7.66\times10^{-3} + 5.71\times10^{-4} = 5.91\times10^{-2}$$

or

$$\gamma_{d\,max(F)} = 16.92 \text{ kN} / \text{m}^3$$

From Equation 7.17, the field optimum water content

$$w_{opt(F)} = (1-0.2)(19.1) + (0.2)(0.028) = 15.3\%$$

With $\alpha = 1$ and $w_c = 0\%$, from Equation 7.18, the field maximum dry unit weight

$$\frac{1}{\gamma_{d\,max(F)}} = \frac{(1-0.2)}{15.73} + \frac{0.2}{(2.66)(9.81)} = 5.08 \times 10^{-2} + 7.66 \times 10^{-3} = 5.85 \times 10^{-2}$$

or

$$\gamma_{d\,max(F)} = 17.10 \text{ kN} / \text{m}^3$$

With $\alpha = 1$ and $w_c = 0\%$, from Equation 7.17, the field optimum water content

$$w_{opt(F)} = (1-0.2)(19.1) = 15.3\%$$

7.4.   Factors affecting compaction
The primary factors that affect compaction are water content, compactive effort, soil type and method of compaction.

7.4.1   Water content
Figure 7.3 shows that for a given compactive effort, the maximum densification of a cohesive soil is achieved only when it has the optimum water content. A cohesionless soil, such as a sand mass, can be densified to the maximum dry unit weight when it is either dry or fully saturated. In unsaturated conditions, the surface tension forces cause an apparent cohesion in cohesionless soils and densify loose, dry, cohesionless soils to some extent on wetting.

7.4.2   Compactive effort
Figure 7.8 shows the typical compaction curves drawn with data from standard and modified compaction tests conducted on the same soil. Since the compactive effort for the modified compaction test is greater than that for the standard compaction test, it may be stated that, as the compactive effort is increased, the maximum dry unit weight of the soil increases – that is, $\gamma_{dmax(mod)} > \gamma_{dmax(std)}$ – as expected, but its optimum water content decreases – that is, $w_{opt(mod)} < w_{opt(std)}$. By joining the peaks of the curves, a line can be formed, called the line of optimums, which is almost parallel to the ZAV curve, indicating that the degree of saturation at the optimum water content remains almost the same for all the compactive efforts. As indicated at two different water contents, w_1 and $w_2 (> w_1)$ that $(\Delta\gamma_d)_{w_1} > (\Delta\gamma_d)_{w_2}$, as the water content increases, the influence of the compactive effort on the dry unit weight tends to decrease.

Figure 7.8 Compaction curves for a cohesive soil compacted with different compactive efforts

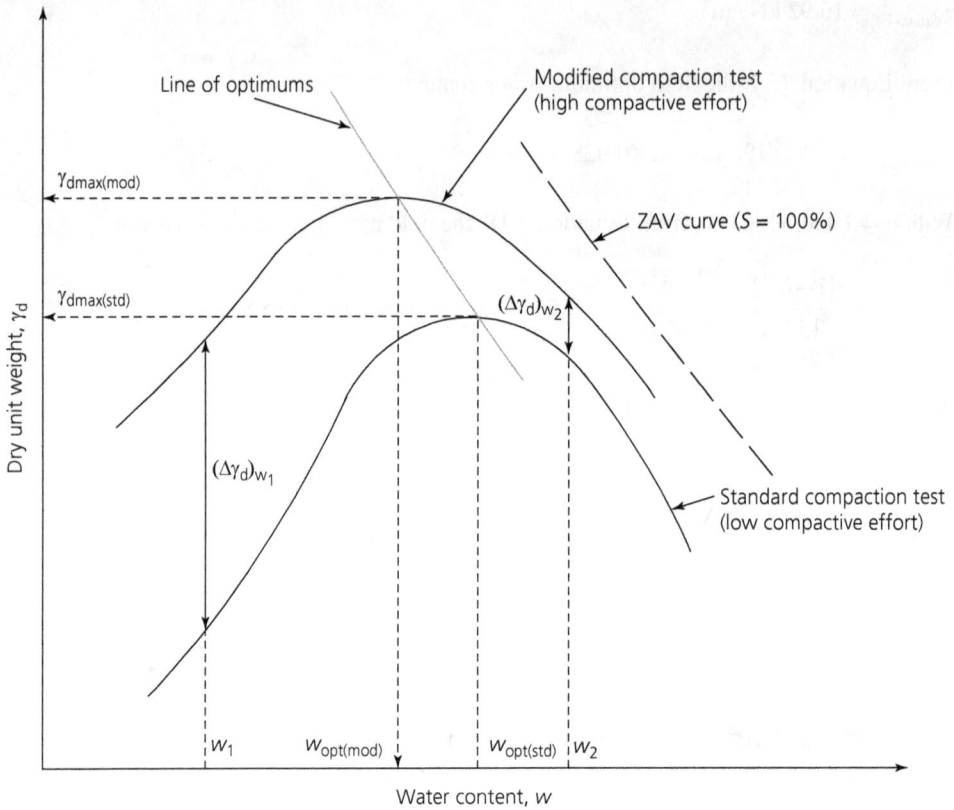

7.4.3 Type of soil

Figure 7.9 shows typical compaction curves for cohesionless and cohesive soils. The compaction curve for the cohesive soil is similar to that shown in Figures 7.3 and 7.8. Cohesionless soil, such as clean sand with less than 5% fines, does not respond to changes in water content and compactive effort in the same way as cohesive soil. The compaction curve for sand is concave upwards at low water contents with a maximum dry unit weight at degree of saturation, $S = 0\%$ – that is, when the sand is dry – and close to degree of saturation, $S = 100\%$ – that is, when the sand is fully saturated – as indicated in the figure. The low unit weight obtained at low water content is due to capillary forces resisting rearrangement of the sand particles. This phenomenon is known as bulking (Lambe and Whitman, 1979). The densification of compacted cohesionless soil is typically measured in terms of relative density, as described in Chapter 2.

7.4.4 Method of compaction

Since field compaction methods involve rolling, kneading, vibration, or a combination of these with the application of a static load, it is challenging to select a laboratory test that accurately replicates a specific field compaction procedure. The optimum water content obtained from the laboratory compaction test is generally somewhat lower than the field optimum water content. It is said that the laboratory kneading compaction test may closely simulate the soil structure in the field

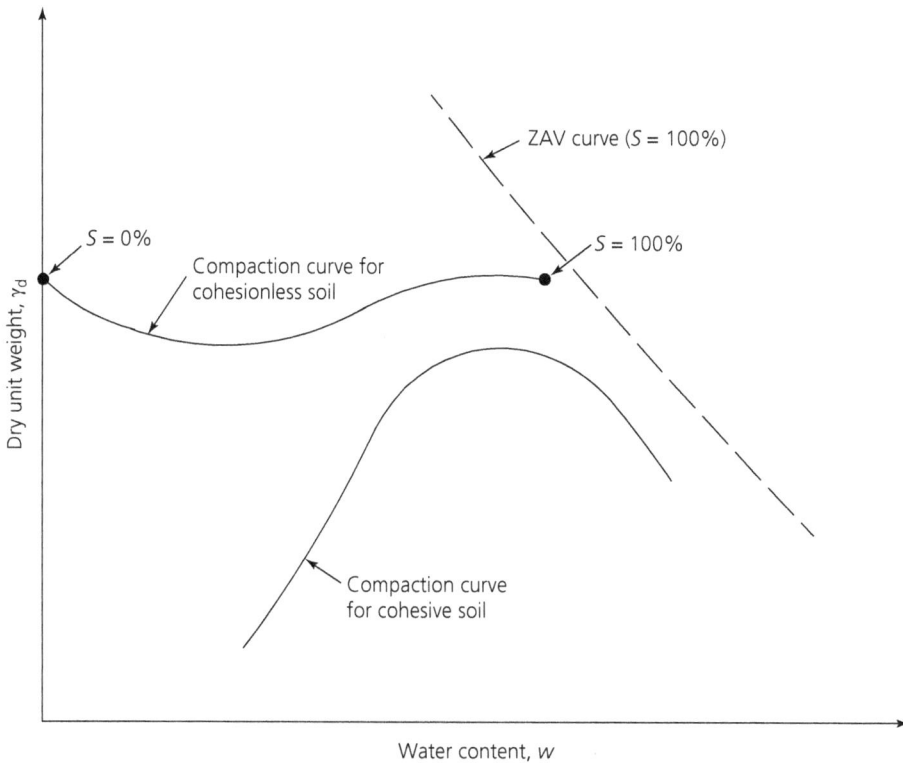

Figure 7.9 Typical compaction curves for cohesionless and cohesive soils

condition, especially when the soil is compacted using a sheepsfoot roller. However, the laboratory dynamic compaction test, as described in detail in this chapter, is routinely conducted by engineers to determine the field compaction parameters.

7.5.   Process of compaction

The compaction curve giving the dry unit weight against water content relationship for a soil is obtained from either the laboratory or the field compaction test. Researchers have tried to propose theories to describe the process of compaction so that the dry unit weight against water content relationship can be explained satisfactorily.

Proposing what was probably the first theory, Proctor (1933) believed that the water in soil acts as a lubricant to reduce the friction between the soil particles as well as the capillary effects. There-fore, as the water content is increased, the compactive effort easily causes densification, but for a very high water content greater than the water requirement to fill the volume of voids in soil, the compactive effort causes a reduction in the dry unit weight with an increase in the volume of the soil. However, this theory is no longer accepted.

According to Hogentogler (1936), the compaction process involves the following: hydration, lubri-cation, swelling and saturation, occurring as the water content is increased. Hydration followed by lubrication takes place on the dry side of optimum, while swelling followed by saturation occurs on the wet side of optimum. This theory assumes that all air in the voids is displaced by water during compaction, resulting in a saturated soil; however, this is not accurate.

The first modern theory was given by Hilf (1956) based on the concept of pore water and pore air pressures. A compactive effort is less effective on the wet side of optimum due to trapping of air and building of pore air pressure. Thus, this theory suggests that compacted soil remains unsaturated. Olson (1963) also discussed pore water and pore air pressures along with the principle of effective stress in explaining the compaction process.

The compaction process in clays was explained by Lambe (1958) in terms of soil structure. At lower water contents, attractive forces between clay particles predominate, creating a flocculated structure (Figure 1.12(a)) with a relatively random orientation of the plate-like clay particles. Increasing the water content tends to increase the interparticle repulsions, thereby permitting a more orderly or parallel arrangement of the soil particles. Section 7.2.2 explained that a soil can be compacted to a particular dry unit weight at two different water contents, one lying on the dry side of optimum, and the other on the wet side. Soil compacted dry of optimum generally has a flocculated structure (Figure 1.12(a)), while soil compacted wet of optimum has a dispersed structure (Figure 1.12(b)). An increase in compactive effort at a given moulding water content tends to disperse the soil, especially when compacted dry of optimum. This may be due to its role in aligning the particles into a more parallel arrangement (Lambe and Whitman, 1979). Lambe and Whitman's theory is highly useful in explaining the effects of compaction on strength and volume change behaviour of cohesive soils when compacted.

7.6.   Effect of compaction on the engineering behaviour of soils

At any construction site, soil is compacted to enhance its engineering behaviour of interest, tailored to the specific requirements of the structure or project. The compaction process may aim to achieve one or more of the following objectives

- increase in the strength and stiffness of soil or ground
- decrease in the compressibility or volume change potential of soil or ground
- decrease or increase in the permeability of soil or ground
- reduction in the liquefaction potential of soil
- increase in the durability of soil or ground against undesirable environmental impacts.

Table 7.2 provides a description of the engineering behaviour or characteristics of fine-grained or cohesive soils when compacted dry of optimum and wet of optimum.

As mentioned in the previous section, the concept of soil structure helps to explain the effects of compaction. A soil on the dry side of optimum is highly permeable because it has a flocculated structure, which consists of connected voids forming continuous channels for the easy flow of water or any other fluid. In a similar way, other characteristics of a compacted soil as listed in Table 7.2 can be explained.

In general, a soil compacted dry of optimum has a flocculated structure, while a soil compacted wet of optimum has a dispersed structure. This means that the soil structure depends on the water content. The soil structure of compacted soil also depends on the type of compaction. When clays are compacted dry of optimum, the soil structure is essentially independent of the type of compaction

Table 7.2 Engineering behaviour of compacted fine-grained soil (modified from Lambe and Whitman, 1979)

Engineering parameter or property		Dry of optimum	Wet of optimum
Soil structure		Flocculated (random) More sensitive to change	Dispersed (parallel or oriented) Less sensitive to change
Water content		Low	High
Coefficient of permeability		High and isotropic High reduction in permeability by permeation	Low and anisotropic (high along particle orientation, low across particle orientation) Low reduction in permeability by permeation
Pore water pressure on loading		Low	High
Compressibility		Low at low stress application High at high stress application Consolidates rapidly	High at low stress application Low at high stress application Consolidates slowly
Shrinkage on drying		Low	High
Swelling with access to water		High	Low
Shear strength as moulded	Undrained	High	Low
	Drained	Somewhat higher	Somewhat lower
Shear strength after saturation	Undrained	Somewhat higher if swelling prevented	Somewhat higher if swelling permitted
	Drained	About the same or slightly higher	About the same or slightly lower
Stress–strain behaviour		Brittle High stress–strain modulus	Plastic or ductile Low stress–strain modulus

Note: this comparison of the engineering behaviour of cohesive soil considers the same dry unit weight at two different water contents, one dry of optimum and the other wet of optimum.

(Seed and Chan, 1959), but when the soil is compacted wet of optimum, the type of compaction has a significant effect on the soil structure and thus on the resulting engineering properties of the soil. If there is a little intimate interaction between the compaction equipment and the soil, as observed during the compaction with a static smooth-wheel roller or an impact roller, the compacted soil may have even a random orientation of particles – that is, a structure closer to a flocculated structure – called the flocculated-dispersed structure.

Note that in the field condition, sometimes highly saturated clays may have higher strength when compacted with light compaction equipment to a lower unit weight than when compacted with heavy compaction equipment to a higher unit weight. A loss in strength with increasing unit weight is a phenomenon sometimes referred to as overcompaction or overstress (Hausmann, 1990). This is probably due to the development of high pore water pressure by compaction at higher unit weights. It may also be explained in terms of the development of a dispersed soil structure, which reduces the strength of a clay under these conditions.

217

7.7. Field compaction and quality control

The compaction parameters, namely the maximum dry unit weight, γ_{dmax}, and the optimum water content, w_{opt}, are highly useful for developing construction specifications aimed at improving soil through compaction. The soil can be compacted either dry of optimum or wet of optimum to achieve a specific dry unit weight. However, the soil is generally compacted dry of optimum except when it is used in landfill liners, dam cores, and places where the volume change caused by a change in water content is not desirable. Table 7.3 provides guidelines for selecting the appropriate moulding or placement water content in compaction-related construction projects.

In the field, soil is generally compacted in lifts or layers using rollers. The lift thickness depends on the type of soil and the compaction equipment, and may vary from a few centimetres to perhaps 0.6 m. The compaction of an existing or in-place soil deposit is usually limited to the top 0.3 m prior to placement of a fill, although a cohesionless soil such as a sand can be densified with rollers for a depth of 1 to 2 m (Lambe and Whitman, 1979).

Table 7.4 lists the types of rollers and their applicability to most and least suitable soils, as described by Hausmann (1990) but with some modifications. Smooth-wheel rollers can assist in bridging uneven surfaces of a clay after its compaction by a sheepsfoot roller or a pneumatic-tyred roller. A smooth-wheel roller has 100% coverage under the wheel with ground contact pressure up to about 500 kPa. Sheepsfoot rollers exert very high pressures (about 1.5–7.0 MPa) on the soil due to low (10 to 15%) coverage under the wheel through the prismatic attachments, called the feet, to the cylindrical steel drum loaded with ballast. Through the kneading action of the feet, a sheepsfoot roller first compacts lower layers and then gradually compacts the upper layers as the strength of the lower layers increases. Pneumatic-tyred rollers having about 80 to 85% coverage under the wheel compact the soil by the static weight and the kneading action of the tyres. A grid roller consists of drums covered with a heavy steel grid, which breaks and rearranges gravel and cobble-size particles by applying a high pressure (about 1.5–6 MPa) with about 50% coverage. Vibratory rollers use rotating weights for vibration action, and the frequency of vibration of heavy vibratory rollers generally ranges from 18 to 25 Hz. An impact roller consists of a non-cylindrical mass, which is towed along the soil surface to be compacted, and has about 40–50% coverage. It applies a very high pressure with some kneading action too in the range of 1.5 to 8 MPa.

In confined areas, such as soil backfill in trenches, around conduits, and behind retaining walls, where access to rollers is difficult, it is common practice to use vibrating tampers, rammers or plate compactors. Photographs of compaction machineries can easily be found by an internet search.

Table 7.3 Selection of moulding or placement water content

Construction project or application area	Water content	Reason
Homogeneous earth dams or embankments	Dry of optimum	To prevent build-up of high pore water pressure
Core of earth dam or landfill liners	Wet of optimum	To reduce coefficient of permeability and to prevent cracking of the core
Subgrade for pavements	Wet of optimum	To limit volume change in the subgrade
Fills	Dry of optimum	To facilitate easy working conditions

Table 7.4 Types of rollers and their applicability

Type of roller	Applicability
Smooth-wheel roller, static or vibrating	Most suitable for well-graded sand–gravel mixtures, crushed rock, and bituminous mixes; least suitable for uniformly graded cohesionless soils
Sheepsfoot roller, static or vibrating	Most suitable for fine-grained/cohesive soils with more than 20% fines if static or vibrating mode is operated, and for sand–gravel mixtures if vibrating mode is used; least suitable for clean, coarse-grained soils, and soils with cobbles
Pneumatic- or rubber-tyred roller	Most suitable for coarse-grained soils with some fines; least suitable for uniformly graded cohesionless soils, and rocks
Impact roller	Most suitable for a wide range of moist and saturated soils, even as natural ground and fills; least suitable for dry cohesionless soils
Grid roller	Most suitable for weathered rock, and well-graded coarse-grained soils; least suitable for clays, silty clays, and uniformly graded materials

The weight of compaction equipment ranges from very low, say 1 kN, to very high, say 1800 kN. The rollers commonly seen in day-to-day construction work typically weigh between 40 kN and 200 kN. A single vibrating plate usually has a weight of not less than 0.9 kN, and the eccentric weight rotates at a frequency not less than 25 Hz. The weight of tampers may range from about 0.13 to 1 kN, and their foot diameters may vary from 0.1 to 0.25 m. The area of one foot of a sheepsfoot roller is about 0.003 to 0.1 m², and the projection of the feet may be at least 0.23 m, especially when the sheepsfoot roller is used in earth dam construction. The diameter of the steel drum of a typical roller is usually about 1.5 m and the length (or width) varies from about 1.8 to 2.7 m (Terzaghi *et al.*, 1996). The inflation pressure of pneumatic-tyred rollers generally ranges from 500 kPa to 1 MPa.

The speed of rollers of all types is usually limited to about 5–6 km/h. A minimum of four to six passes are usually required to compact the soil economically; however, the number of passes can go up to 15–20. At speeds greater than about 5–6 km/h, more passes may be required to achieve the desired compaction, as sufficient deformation requires the load to remain on the soil for an adequate amount of time. Note that a high number of passes may cause increased crushing of particles, and this may lead to undesirable stratification of the soil. Therefore, it is generally recommended to conduct field trials to determine the optimum number of passes that are both technically and economically satisfactory.

The compactive effort of a roller depends on several factors, such as gross weight, wheel or drum width, wheel load, ratio between frame mass and drum mass, tyre width and size, inflation pressure, frequency and amplitude, and roller speed, depending on the type of roller. The compaction capacity, also called the production rate, which assists in the selection of the most economical compaction equipment, is defined as (Hausmann, 1990)

$$P = \frac{BevH}{N} \times 1000 \tag{7.19}$$

where P is the production rate (m³/h); B is the drum width (m); e is the efficiency that may vary from 0.75 to 0.85, allowing for an overlap between adjacent passes and the time required to change direction, stop and start; v is the rolling speed (km/h); H is the lift or layer thickness (m); N is the number of passes.

Example 7.5

Calculate the production rate in m³/h for a roller with the following characteristics

drum width $=2.15$ m
efficiency $=75\%$
rolling speed $=6$ km/h
lift thickness $=0.3$ m
number of passes $=8$.

Solution

Given $B = 2.15$ m, $e = 0.75$, $v = 6$ km/h, $H = 0.3$ m and $N = 8$.

From Equation 7.19, the production rate

$$P = \frac{BevH}{N} \times 1000 = \frac{(2.15)(0.75)(6)(0.3)}{8} \times 1000 = 362.8 \approx 363 \text{ m}^3 / \text{h}$$

Ideally, soil compaction in the field should be specified in terms of actual performance parameters, such as strength, stiffness, permeability and/or compressibility. However, field compaction is generally specified in terms of the minimum dry unit weight to be achieved at the end of the compaction process. A typical specification may state that the soil shall be compacted above a certain minimum dry unit weight, say 98% of the maximum dry unit weight obtained from the modified compaction test. A permissible range of placement water content (e.g. −2 to +2% of the optimum water content) is also indicated, along with the minimum dry unit weight to be achieved in order to have the desired soil structure and properties. The field compaction specification can also be expressed in terms of relative compaction (also referred to as per cent compaction or dry density ratio), R_c, defined as

$$R_c = \frac{\gamma_{d(F)}}{\gamma_{d\,max}} \times 100\,(\%) \tag{7.20}$$

where $\gamma_{d(F)}$ is the dry unit weight of the soil achieved by field compaction and $\gamma_{d\,max}$ is the maximum dry unit weight of soil obtained from the standard or modified compaction test, as applicable.

The value of relative compaction typically ranges from 90 to 98%, depending on the importance of the construction project.

The compaction specification in terms of R_c is an end-product specification, which may also include the permissible range of water content, lift thickness and the size of the soil particles to be compacted. Method specification is another specification type, which specifies the type and weight of the compaction equipment, the number of passes, the lift thickness and the maximum allowable size of the soil particles (Holtz *et al.*, 2011). The engineers responsible for preparing the specifications must have prior knowledge of the borrow soils or materials, as they will be accountable for the quality of compaction.

For the control of field compaction – that is, to check whether the compaction specifications have been met in the field – economical in situ or field tests, such as the determination of total unit weight, and laboratory or field determination of the water content are conducted so that the field dry unit weight can be computed using Equation 7.3 and compared with the specified values. Typically, one set of in situ tests per specified volume (e.g. 3000 m^3) of placed fill may be sufficient.

As many cohesionless soils are fairly insensitive to water content during compaction, and can also be best compacted either dry or fully saturated, their compaction specifications are stated appropriately in terms of the engineering property of interest as per the field project requirement. As a compaction control check, any suitable in situ test may be conducted to determine the engineering property of interest. In order to make the control of field compaction economical, the relative density, D_r, of the compacted soil is often determined as described in Chapter 2 because the engineering properties of cohesionless soils show a better correlation with D_r values than their absolute density or unit weight values. The compactibility of a cohesionless soil can also be described by the factor, F, defined as (Terzaghi, 1925)

$$F = \frac{e_{max} - e_{min}}{e_{min}} \tag{7.21}$$

where e_{min} is the minimum void ratio (i.e. void ratio in the densest possible state of soil) and e_{max} is the maximum void ratio (i.e. void ratio in the loosest possible state of soil).

Typically, F ranges from 0.5 to 2. Poorly graded cohesionless soils have low values of F, say 0.5 to 1, which indicates that it is difficult to compact them. Since well-graded cohesionless soils with some fines can easily be compacted to a higher unit weight, they have high values of F, say 1.5 to 2.

For cohesionless soils, a correlation between R_c and D_r is also available, as given below (Lee and Singh, 1971)

$$R_c = 80 + 0.2D_r \tag{7.22}$$

If D_r is known for a compacted cohesionless soil, R_c can be computed using Equation 7.22.

Generally, the field compaction process may involve any one or all of the following steps.

- Select a borrow pit near the construction site, load the borrow soil, haul it to the site, and dump it.
- Spread the dumped soil at the site in the appropriate lift thickness.
- If the water content of the placed soil is higher than its specified value, allow it to dry, but if lower, increase the water content by adding water.
- Mix the placed soil properly to make the water content uniform.
- Compact the soil using compaction equipment to achieve the specified properties and/or procedures.

Example 7.6

A highway embankment is constructed at a total unit weight of 20.53 kN/m³ with a water content of 17.6%. For construction, the soil is collected from a borrow pit near the highway alignment. The total unit weight and water content of soil in its natural condition in the pit are 16.54 kN/m³ and 13.2%, respectively. Assuming no water is lost while transporting the soil from the pit to the embankment location, determine the volume and weight of soil to be excavated from the borrow pit. Additionally, calculate the weight of water to be added to the borrow soil for every cubic metre of compacted soil.

Solution

Given, for the compacted soil in the embankment, $\gamma_{emb} = 20.53$ kN/m³ and $w_{emb} = 17.6\%$, and, for the soil in its natural condition in the pit, $\gamma_{pit} = 16.54$ kN/m³ and $w_{pit} = 13.2\%$. And, also, the volume of compacted soil for the analysis, $V = 1$ m³.

Dry unit weight of compacted soil

$$\gamma_{d(emb)} = \frac{\gamma_{emb}}{1 + w_{emb}} = \frac{20.53}{1 + 0.176} = 17.46 \text{ kN/m}^3$$

Weight of dry soil in 1 m³ of compacted soil

$$W_s = \gamma_{d(emb)} V_{emb} = (17.46)(1) = 17.46 \text{ kN}$$

Dry unit weight of the soil in its natural condition in the borrow pit

$$\gamma_{d(pit)} = \frac{\gamma_{pit}}{1 + w_{pit}} = \frac{16.54}{1 + 0.132} = 14.61 \text{ kN/m}^3$$

For $W_s = 17.46$ kN in the embankment, the volume of soil to be excavated from the borrow pit

$$V_{pit} = \frac{W_s}{\gamma_{d(pit)}} = \frac{17.46}{14.61} = 1.20 \text{ m}^3$$

Weight of soil to be excavated from the borrow pit to obtain one cubic metre of compacted soil

$$W_{pit} = W_s(1 + w_{pit}) = (17.46)(1 + 0.132) = 19.76 \text{ kN}$$

Weight of one cubic metre of compacted soil in the embankment

$$W_{emb} = W_s(1 + w_{emb}) = (17.46)(1 + 0.176) = 20.53 \text{ kN}$$

Note that W_{emb} is also known directly from the total unit weight of the compacted soil. Weight of water to be added to the borrow soil for every cubic metre of compacted soil

$$W_w = W_{emb} - W_{pit} = 20.53 - 19.76 = 0.77 \text{ kN}$$

Chapter summary

1. Compaction is the process of volume reduction of an unsaturated soil mass due to expulsion of air from its voids caused by the external compressive stress application.
2. The dynamic compaction test, which consists of placing soil in a cylindrical mould and then dropping a steel rammer a specified number of times, is the earliest and most common type of laboratory compaction test. The purpose of the laboratory compaction test on soil is to determine its compaction parameters: maximum dry unit weight, γ_{dmax}, and optimum water content, w_{opt}.
3. A plot of dry unit weight, γ_d, against water content, w, values are joined by a smooth curve, called the compaction curve.
4. The optimum water content of soils ranges from 5 to 45% with a typical range of 10 to 20%. The maximum dry unit of soils ranges from 13 to 24 kN/m³ with a typical range of 16 to 20 kN/m³.
5. Equations 7.16 and 7.17 proposed by Shukla *et al.* (2009) provide expressions for calculating the field maximum dry unit weight and the field optimum water content based on the test values obtained from the laboratory compaction test on the soil specimen, which excludes soil particles larger than the maximum size limit of the compaction mould.
6. In general, a soil compacted dry of optimum has a flocculated structure, while a soil compacted wet of optimum has a dispersed structure .
7. Soil is generally compacted dry of optimum except when it is used in landfill liners, dam cores, and places where the volume change caused by a change in water content is not desirable.
8. In cohesive soils, high dry unit weights can be obtained with most types of roller; however, vibratory rollers are least effective and sheepsfoot rollers are usually the most effective. In cohesionless soils, high dry unit weights can be obtained with vibratory rollers.
9. The speed of rollers of all types is usually limited to 5–6 km/h. A minimum of four to six passes are usually required to compact the soil economically.
10. The field compaction specification is usually expressed in terms of the relative compaction along with the permissible range of water content, lift thickness and size of the soil particles to be compacted.

Questions for practice

Select the most appropriate answer for each of the multiple choice questions from Q7.1 to Q7.10. Questions Q7.11 to Q7.22 require more detailed answers.

7.1 The water content of a soil mass reduces with time during its volume reduction by
 (a) the compaction process
 (b) the consolidation process
 (c) both (a) and (b)
 (d) none of the above.

7.2 According to ASTM D698-12-2021, the internal volume of a compaction mould used for laboratory compaction is
 (a) 944 cc
 (b) 2124 cc
 (c) both (a) and (b)
 (d) 1000 cc.

7.3 In a laboratory compaction test, as the compaction effort is increased, the optimum water content of a soil

(a) decreases
(b) increases
(c) increases first and then decreases
(d) remains unaltered.

7.4 The ZAV curve is nonlinear because

(a) the compaction curve is nonlinear
(b) the water content changes during the compaction process
(c) the dry unit weight at 100% saturation is a nonlinear function of the void ratio
(d) the soil is compacted with a lightweight rammer.

7.5 The theoretical maximum dry unit weight of a soil having solid particles with a specific gravity of 2.65 is
(a) $26.0 \, \text{kN/m}^3$
(b) $26.0 \, \text{N/m}^3$
(c) $26.5 \, \text{kN/m}^3$
(d) $26.5 \, \text{N/m}^3$.

7.6 If the degree of saturation of a compacted soil is 85%, which of the following will be 15%?
(a) porosity
(b) void ratio
(c) air content
(d) percentage air voids.

7.7 Compaction using a vibratory roller is the most effective method for
(a) wet silty sand
(b) lean clay
(c) silty clay
(d) well-graded dry sand.

7.8 Cohesive soil compacted dry of optimum has
(a) a dispersed structure
(b) a flocculated structure
(c) both (a) and (b)
(d) a granular structure.

7.9 The speed of rollers of all types is usually limited to about
(a) $6 \, \text{m/s}$
(b) $60 \, \text{m/s}$
(c) $100 \, \text{m/min}$
(d) $60 \, \text{km/h}$.

7.10 Select the incorrect statement.
    (a) During compaction, soil always remains unsaturated.
    (b) The clay core of an earth dam is generally compacted wet of optimum.
    (c) Clays are best compacted with sheepsfoot rollers.
    (d) The measure of soil compaction is its total unit weight.

7.11 Compute the mechanical energy input (compactive effort) in kilojoules per cubic metre (kJ/m^3) to the soil in terms of the work done by operating a rammer during the light compaction test in accordance with AS 1289.5.1.1-2017 (Standards Australia, 2017a), ASTM D698-12 (ASTM, 2021a), BS 1377-2-2022 (BSI, 2022) and IS 2720 (Part 7)-1980 (Bureau of Indian Standards, 2021) for a large-size compaction mould. Compare the values and record your observations. Are the values the same as those computed for the small-size compaction mould in Example 7.1?

7.12 Compute the mechanical energy input (compactive effort) in kilojoules per cubic metre (kJ/m^3) to the soil in terms of the work done by operating a rammer during the heavy compaction test in accordance with AS 1289.5.2.1-2017 (Standards Australia, 2017b), ASTM D1557-12) (ASTM, 2021b), BS 1377-2-2022 (BSI, 2022) and IS 2720 (Part 8)-1983 (Bureau of Indian Standards, 2020) for a small-size compaction mould. Compare the values and record your observations. Are the values the same as those computed for the light compaction in Example 7.1?

7.13 An embankment is constructed by compacting the soil at a total unit weight of $21.35\,kN/m^3$ and a water content of 11.8%. The specific gravity of soil solids is 2.65. The water table is below the foundation level of the embankment. For the compacted soil of the embankment, determine the following
    (a) dry unit weight
    (b) void ratio
    (c) degree of saturation
    (d) percentage air voids
    (e) air content.

7.14 A soil deposit has a void ratio of 0.91. If the void ratio is reduced to 0.65 by compaction, determine the percentage volume loss due to compaction.

7.15 A soil sample consists of 25% particles coarser than 37.5 mm. A standard compaction test is carried out in accordance with ASTM D698-12 (ASTM, 2021a) using a 152.4 mm mould. The coarse fraction of the soil (soil particles coarser than 37.5 mm) is removed from the test specimen in view of the particle size limit for the test mould. The maximum dry unit weight and the optimum water content are obtained as $16.18\,kN/m^3$ and 18.0%, respectively. For the coarser particles, $G_c = 2.65$ and $w_c = 1.9\%$. Determine the field values of maximum dry unit weight and optimum water content if the volume of the air in the voids of the test specimen remains the same after removal of the coarse fraction. Also determine the values ignoring the water content of the coarse fraction.

7.16 Calculate the production rate in m³/h for a roller with the following characteristics

drum width $= 2.0$ m
efficiency $= 75\%$
rolling speed $= 5$ km/h
lift thickness $= 0.25$ m
number of passes $= 6$.

7.17 What similarities exist among different test standards for compaction tests?

7.18 Explain the importance of the ZAV curve. As the project engineer, would you accept the compaction curve reported in the soil report without it? Justify your answer.

7.19 Conduct a compaction test in the laboratory on the local soil at your location and plot the data to obtain the compaction curve along with the ZAV curve. Report the compaction parameters.

7.20 How do the permeability characteristics of soil compacted dry of optimum differ from those of soil compacted wet of optimum?

7.21 For the subgrades of highways, would you prefer to compact the soil dry of optimum or wet of optimum? Justify your answer.

7.22 What is the practical approach to ensuring quality control in field soil compaction?

REFERENCES

ASTM (2013) ASTM D5030-04: Standard test method for density of soil and rock in place by the water replacement method in a test pit. ASTM International, West Conshohocken, PA, USA.

ASTM (2016) ASTM D4914-08: Standard test methods for density and unit weight of soil and rock in place by the sand replacement method in a test pit. ASTM International, West Conshohocken, PA, USA

ASTM (2021a) ASTM D698-12: Standard test methods for laboratory compaction characteristics of soil using standard effort (12,400 ft-lbf/ft3 (600 kN-m/m3)). ASTM International, West Conshohocken, PA, USA.

ASTM (2021b) ASTM D1557-12: Standard test methods for laboratory compaction characteristics of soil using modified effort (56,000 ft-lbf/ft3) (2700 kN-m/m3)). ASTM International, West Conshohocken, PA, USA.

BSI (2022) BS 1377-2:2022: Method of test for soils for civil engineering purposes – classification tests and determination of geotechnical properties. BSI, London, UK.

Bureau of Indian Standards (2020) IS 2720 (Part 8)-1983: Determination of water content – dry density relation using heavy compaction (Second Revision). Bureau of Indian Standards, New Delhi, India.

Bureau of Indian Standards (2021) IS 2720 (Part 7)-1980: Determination of water content – dry density relation using light compaction (Second Revision). Bureau of Indian Standards, New Delhi, India.

Hausmann MR (1990) Engineering Principles of Ground Modification. McGraw-Hill, New York, NY, USA.

Hilf JW (1956) *An Investigation of Pore-Water Pressure in Compacted Cohesive Soils.* Technical Memo. 654, US Deptartment of the Interior, Bureau of Reclamation, Denver, CO, USA.

Hogentogler CA Jr (1936) Essentials of soil compaction. *Proceedings of the Highway Research Board* **16**: 209–216.

Holtz RD, Kovacs WD and Sheahan TC (2011) *An Introduction to Geotechnical Engineering*, 2nd edn.Pearson, Upper Saddle River, NJ, USA.

Lambe TW (1958) The engineering behaviour of compacted clay. *Journal of the Soil Mechanics and Foundation Engineering Division, ASCE* **84(SM2)**: 1654-1 to 1654-34.

Lambe TW and Whitman RV (1979) *Soil Mechanics, SI Version.* Wiley, New York, NY, USA.

Lee KW and Singh A (1971) Relative density and relative compaction. *Journal of the Soil Mechanics and Foundation Engineering Division, ASCE* **97(SM7)**: 1049–1052.

Moseley MP (1993) *Ground Improvement.* CRC Press, Boca Raton, FL, USA.

Olson RE (1963). Effective stress theory of soil compaction. *Journal of the Soil Mechanics and Foundation Engineering Division, ASCE* **89(SM2)**: 27–45.

Proctor RR (1933) Fundamental principles of soil compaction. *Engineering News Record* **111** (9–10, 12–13).

Seed HB and Chan CK (1959) Structure and strength characteristics of compacted clays. *Journal of the Soil Mechanics and Foundation Engineering Division, ASCE* **85(SM5)**: 87–128.

Shukla SK (1992) *Mineralogical and Strength Changes during Soil-Fly ash-Lime-Gypsum Stabilization.* Master of Technology thesis, Indian Institute of Kanpur, Kanpur, India.

Shukla SK and Gokhale KVGK (1995) Strength development during soil–fly ash–lime–gypsum stabilization. *Proceedings of the Indian Geotechnical Conference*, Bangalore, India, December 1995, vol. 1, pp. 231–233.

Shukla SK and Sivakugan N (2011) Site investigation and in situ tests. In *Geotechnical Engineering Handbook* (Das BM (ed.)). J Ross, Fort Lauderdale, FL, USA, Chapter 10, pp. 10.1–10.78.

Shukla SK, Sivakugan N, Gandhi M and Ahmed MK (2009) Improved expressions for field values of compaction test parameters. *Geotechnique, UK* **59(10)**: 851–853.

Shukla SK, Jha JN, Gill KS and Choudhary AK (2012) Estimation of field compaction parameters. *Proceedings of the Indian Geotechnical Conference*, New Delhi, India, 13–15 December 2012, vol. 2, pp. 668–671.

Standards Australia (2007) AS 1289.5.4.1-2007: Compaction control test – dry density ratio, moisture variation and moisture ratio. Standards Australia, Strathfield, NSW, Australia.

Standards Australia (2017a) AS 1289.5.1.1-2017: Determination of the dry density/moisture content relation of a soil using standard compactive effort.Standards Australia, Strathfield, NSW, Australia.

Standards Australia (2017b) AS 1289.5.2.1-2017: Determination of the dry density/moisture content relation of a soil using modified compactive effort. Standards Australia, Strathfield, NSW, Australia.

Taylor DW (1948) *Fundamentals of Soil Mechanics.* Wiley, New York, NY, USA.

Terzaghi K (1925) Erdbaumechanik auf bodenphysikalischer Grundlage. Franz Deuticke, Vienna, Austria. (In German.)

Terzaghi K, Peck RB and Mesri G (1996) *Soil Mechanics in Engineering Practice*, 3rd edn. Wiley, New York, NY, USA.

FURTHER READING

Das BM (2022) *Principles of Geotechnical Engineering*, 10th edn. Cengage, Boston, MA, USA.

Sanjay Kumar Shukla
ISBN 978-1-83608-519-5
https://doi.org/10.1108/978-1-83608-516-420252011
Emerald Publishing Limited: All rights reserved

Chapter 8
Lateral earth pressure, slope stability and bearing capacity of soil

> ## Learning aims
>
> This chapter explores the core concepts of the following topics
>
> - lateral earth pressures: earth pressure at rest, active earth pressure and passive earth pressure
> - stability analysis of infinite and finite slopes
> - bearing capacity of shallow foundations
> - bearing capacity of deep foundations.

8.1   Introduction

Chapter 1 introduced the idea that retaining walls, slopes and foundations are important geotechnical structures.

A retaining wall (see Figure 1.3(e)) is typically built to support soil, rock or other materials, such as mine tailings or coal ash, that cannot remain stable with a vertical or sloping face, especially in areas with limited space. The stability of a retaining wall or other similar structures, including abutments, basement walls and sheet pile walls, called earth-retaining structures, depend on the amount of lateral earth pressure or stress from the soil, called the backfill, supported by the wall. The lateral (or sideways) earth pressure acts due to the natural tendency of the backfill to maintain its angle of repose. Note that if water (or any other liquid) is retained, the retaining structure will be subjected to water pressure. In the practical design of retaining walls, drainage arrangements such as weep holes within the wall and/or a proper drainage system behind the wall are made to avoid any water pressure against the wall.

A soil slope can occur naturally or be created through excavation to make space, as well as through the construction of embankments and dams. As the soil has a shear strength, the soil slope may remain stable at some inclination to the horizontal. Note that the surface of a static water body does not remain sloping or vertical due to its negligible shear strength. There can be several field situations in which a soil slope may fail as slips or slides of different types, such as rotational slips, translational slips, falls and flows. In general, slope failure (also known as a slide or slip) refers to the downward and outward movement of a portion of the soil mass in a slope. When a rotational slip occurs (see Figures 1.3(b) and (c)), the failure surface in cross-section is circular in the case of a homogeneous soil, and non-circular in the case of a nonhomogeneous soil. The translation slip occurs as a slab slide or block slide with the plane failure surface, roughly parallel to the slope surface, where there is an adjacent stratum of significantly different strength at a relatively shallow depth below the slope surface. If, within a slope, a stratum of significantly different strength exists at a greater depth, the failure may take place as compound slips with a failure surface consisting of curved and plane sections. There are numerous factors, including gravitational, seepage and seismic forces, that affect the stability of a slope (Shukla, 1997). Due to these factors and also because of different failure processes, the analysis for the estimation of the stability of slopes has been a challenging task for engineers, especially under hydraulic and seismic conditions. In most civil and other ground infrastructure projects, the primary purpose of slope stability analysis is to ensure the safe and cost-effective design of earth slopes.

The term 'foundation' refers to the load-bearing structural elements of engineering systems (e.g. buildings, bridges, towers, pipelines or machinery) constructed below the ground surface, as well as the underlying soil or rock mass that ultimately supports the loads imposed by these systems. In general, the foundations are categorised into two main types: shallow foundations and deep foundations. For a shallow foundation (strip footing, spread footing or mat/raft) (see Figure 1.3(a)(i)), $D/B \leq 1$, where D and B are, respectively, the depth (or embedment depth) and width or diameter of the foundation. In practice, the ratio D/B of a foundation can be greater than unity and it can still be treated as a shallow foundation. For a deep foundation (pile or drilled pier) (see Figure 1.3(a)(ii)), $D/B > 1$. Shallow foundations are constructed when firm soil is located near the ground surface, whereas deep foundations are used when firm soil is not present close to the ground surface. The stability of a foundation depends on the load-bearing capacity or simply the bearing capacity (also known as the load-carrying capacity) of the foundation. Note that the term 'footing' is often used to refer to the structural part of the foundation. In practice, structural loads are transmitted to the soil through footings. Footings may be of the strip type (also called continuous), typically having a long rectangular shape with $L/B \geq 5$, where L is the length. Alternatively, footings may have a square or circular shape, referred to as spread footings.

This chapter introduces the fundamental mechanics of lateral earth pressure on retaining walls, the stability of soil slopes, and the bearing capacity of shallow and deep foundations.

8.2. Lateral earth pressure

A retaining wall can be either rigid or flexible and may exhibit different modes of movement, such as rotation about its base, translation, or rotation about its top or another part. The lateral earth pressure from a backfill is significantly governed by the rigidity of the wall, the type of wall movement or displacement and the type of backfill. A rigid retaining wall moves as a single unit, and the same will be considered here, with reference to the following three cases of the movement of wall while supporting the backfill (Figure 8.1).

- No movement of wall (Figure 8.1(a). In this case, the lateral pressure from the backfill is called the lateral earth pressure at rest (or lateral stress at rest), σ'_{ho}. This is possible when a wall is placed into the ground with a level surface without any disturbance or movement, or when a wall with a frictionless back surface supporting a backfill is placed with no or very light compaction. This condition is rarely observed in the field. The backfill adjacent to the wall remains in a state of elastic equilibrium.
- Movement of wall away from the backfill (Figure 8.1(b)). In this case, a small portion of backfill moves downward due to plastic failure along a surface, which is generally assumed to be a plane inclined to the horizontal to simplify the mathematical analysis. The lateral

Figure 8.1 States of stress within the homogeneous backfill with a horizontal top surface behind a retaining wall A_1A_2 having a vertical smooth back surface in contact with the backfill: (a) state of stress at rest; (b) active state of stress; (c) passive state of stress; (d) coefficient of lateral earth pressure, $K \, (= \sigma'_h/\sigma'_v)$ against rotation of the wall $\Delta x / H$

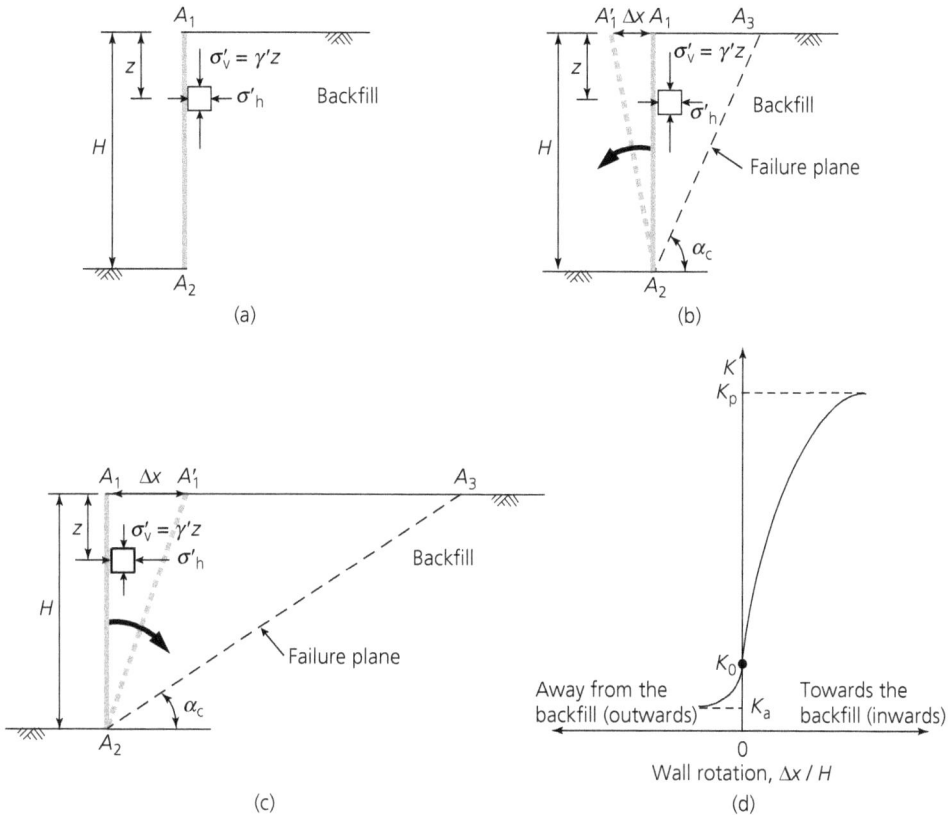

pressure from the backfill when it reaches a state of plastic equilibrium during downward movement is called the active earth pressure (or active stress), σ'_{ha}.

■ Movement of wall towards the backfill (Figure 8.1(c)). In this case, a large portion of backfill moves upward due to plastic failure along a surface, which is generally assumed to be a plane inclined to the horizontal to simplify the mathematical analysis. The lateral pressure from the backfill when it reaches a state of plastic equilibrium during upward movement is called the passive earth pressure (or passive stress), σ'_{hp}.

Note that the state of stress shown in Figure 8.1 is geostatic, and the effective horizontal normal stress, σ'_h, and the effective vertical normal stress, σ'_v, acting on the backfill element are the principal stresses, which relate each other to define the coefficient of lateral earth pressure, K (Equation 3.20, see Chapter 3) as

$$K = \frac{\sigma'_h}{\sigma'_v} \tag{8.1}$$

231

For the state of stress at rest (Figure 8.1(a)), K is called the coefficient of lateral earth pressure at rest, K_0. Thus,

$$K_0 = \frac{\sigma_h'}{\sigma_v'} = \frac{\sigma_h'}{\gamma' z} \tag{8.2a}$$

or the lateral earth pressure at rest at a depth, z, below the ground surface is

$$\sigma_h' = \sigma_{h0}' = K_0 \sigma_v' = K_0 \gamma' z \tag{8.2b}$$

where γ' ($= \gamma_{sat} - \gamma_w$) is the submerged or effective unit weight of the backfill as defined in Equation 1.13 (see Chapter 1).

For the active state of stress (Figure 8.1(b)), K is called the coefficient of active earth pressure (also referred to as the coefficient of active stress or active stress ratio), K_a. Thus,

$$K_a = \frac{\sigma_h'}{\sigma_v'} = \frac{\sigma_h'}{\gamma' z} \tag{8.3a}$$

or the active earth pressure at a depth, z, below the ground surface is

$$\sigma_h' = \sigma_{ha}' = K_a \sigma_v' = K_a \gamma' z \tag{8.3b}$$

For the passive state of stress (Figure 8.1(c)), K is called the coefficient of passive earth pressure (also referred to as the coefficient of passive stress or passive stress ratio), K_p. Thus,

$$K_p = \frac{\sigma_h'}{\sigma_v'} = \frac{\sigma_h'}{\gamma' z} \tag{8.4a}$$

or the passive earth pressure at a depth, z, below the ground surface is

$$\sigma_h' = \sigma_{hp}' = K_p \sigma_v' = K_p \gamma' z \tag{8.4b}$$

Note that a mass of soil is in a state of plastic equilibrium if every part of it is on the verge of failure (Terzaghi *et al.*, 1996). The details presented in Figures 8.1(b) and 8.1(c) consider that the backfill is in a state of plastic equilibrium with $A_2 A_3$ as the failure surface inclined at an angle, α_c, to the horizontal. Figure 8.1(d) shows a typical plot of coefficient of lateral earth pressure, K, against rotation of the wall $\Delta x / H$ about its base A_2, where Δx is the lateral displacement of the top A_1 of the wall and H is the height of the wall. Note that $K_a < K_0 < K_p$.

Figure 8.2 shows the Mohr circles for active and passive states of stress within a backfill with effective cohesion, $c' = 0$ – that is, within a cohesionless (granular or frictional) backfill or a normally consolidated clay backfill. For the active state of stress, $\sigma_v' = \sigma_1'$, and $\sigma_h' = \sigma_{ha}' = \sigma_3'$, where σ_1', and σ_3' are, respectively, the major and minor principal stresses at failure. From Equation 6.11 (see Chapter 6), or directly from the geometry of Figure 8.2

$$\frac{\sigma_{ha}'}{\sigma_v'} = \tan^2 \left(45° - \frac{\phi'}{2} \right) = \cot^2 \alpha_c$$

or

$$K_a = \tan^2 \left(45° - \frac{\phi'}{2} \right) = \frac{1 - \sin \phi'}{1 + \sin \phi'} = \cot^2 \alpha_c \tag{8.5}$$

Figure 8.2 Mohr circles for active and passive states of stress for geostatic condition

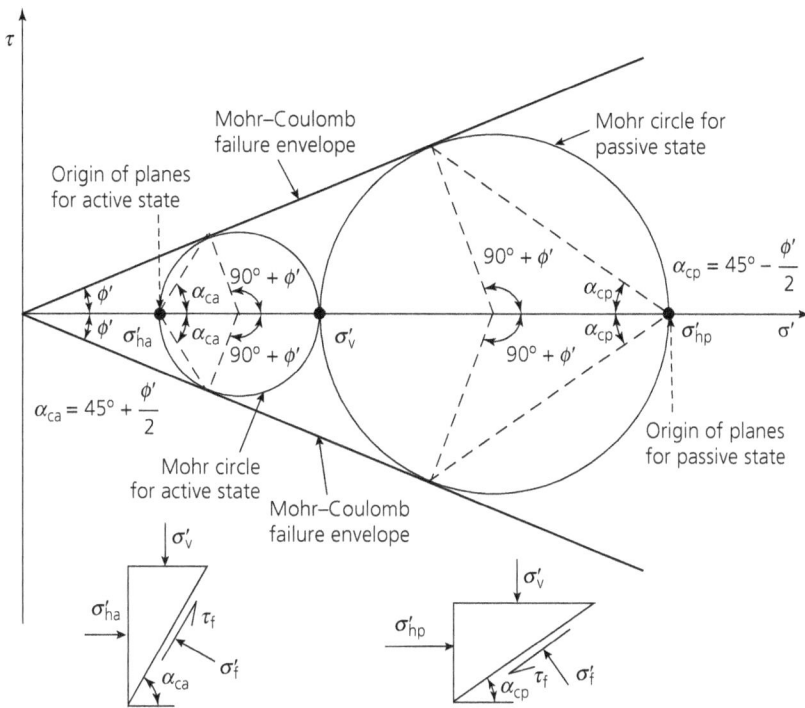

where

$$\alpha_c = \alpha_{ca} = 45° + \frac{\phi'}{2} \tag{8.6}$$

In Figure 8.2, for the passive state of stress, $\sigma'_v = \sigma'_3$ and $\sigma'_h = \sigma'_{hp} = \sigma'_1$. From Equation 6.10 (see Chapter 6), but with $\alpha_c = \alpha_{cp} = 45° - \frac{\phi'}{2}$, or directly from the geometry of Figure 8.2

$$\frac{\sigma'_{hp}}{\sigma'_v} = \tan^2\left(45° + \frac{\phi'}{2}\right) = \cot^2 \alpha_c$$

or

$$K_p = \tan^2\left(45° + \frac{\phi'}{2}\right) = \frac{1 + \sin\phi'}{1 - \sin\phi'} = \cot^2 \alpha_c \tag{8.7}$$

where

$$\alpha_c = \alpha_{cp} = 45° - \frac{\phi'}{2} \tag{8.8}$$

Similar to Figure 8.2, the Mohr circles for active and passive states of stress within the backfill with effective cohesion $c' > 0$ can also be drawn. For active state of stress within the backfill with $c' > 0$, Equation 6.11 (see Chapter 6) provides the expression for the active earth pressure as

$$\sigma'_h = \sigma'_{ha} = \sigma'_v \tan^2\left(45° - \frac{\phi'}{2}\right) - 2c' \tan\left(45° - \frac{\phi'}{2}\right) = K_a \gamma' z - 2\sqrt{K_a}\, c' \tag{8.9}$$

For passive state of stress within the backfill with $c' > 0$, Equation 6.10 (see Chapter 6) provides the expression for the passive earth pressure as

$$\sigma'_h = \sigma'_{hp} = \sigma'_v \tan^2\left(45° + \frac{\phi'}{2}\right) + 2c' \tan\left(45° + \frac{\phi'}{2}\right) = K_p \gamma' z + 2\sqrt{K_p}\, c' \tag{8.10}$$

Active and passive states of stress as explained here are two extreme situations of plastic equilibrium, known as the Rankine states of plastic equilibrium after the British engineer Rankine who presented these basic concepts in 1857 (Lambe and Whitman, 1979; Terzaghi et al., 1996). The two limiting stresses, σ'_{ha} and σ'_{hp}, are called the conjugate stresses.

Figure 8.3 illustrates a simple retaining wall with an assumed frictionless back surface under geostatic conditions, with its lower portion embedded in the foundation soil. This figure shows both the Rankine active and Rankine passive zones with slip planes or lines inclined to the horizontal at $\alpha_{ca}\, (= 45° + \phi'/2)$ and $\alpha_{cp}\, (= 45° - \phi'/2)$, respectively. The slip planes are sketched using the concept of origin of planes (poles) as shown in Figure 8.2.

Note that in the active condition, the shear strength, τ_f, opposes the effect of gravity, whereas in the passive condition, the shear strength, τ_f, acts together with gravity to oppose the large horizontal stress. Also, in the active condition, the backfill behind the retaining wall causes the failure of the wall, while the soil mass at the front of the wall resists the failure of the wall. Thus, the active pressure is always determined as the maximum value, while the passive pressure is determined as the minimum value.

Figure 8.3 Rankine active and passive failure zones

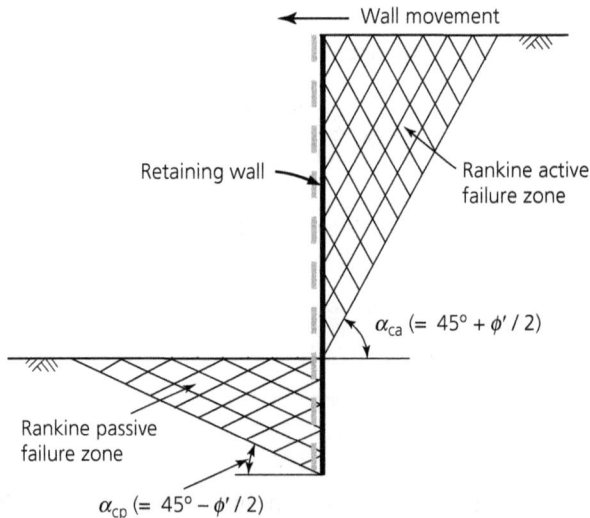

The following points regarding the lateral earth pressure are noteworthy.

- For a dry or saturated backfill with proper drainage (allowing seepage and resulting in zero pore water pressure throughout the backfill), the total unit weight, γ, is replaced by γ' in expressions of lateral earth pressure. The soil above the water table may be assumed to be completely dry for the estimation of lateral earth pressure. For a fully saturated or submerged backfill without a drainage system (not allowing seepage and resulting in existence of pore water pressure within the backfill), the lateral earth pressure must be calculated from the backfill at any point of interest with effective unit weight, γ', using the corresponding expressions, and then added to the pore water pressure at that point to determine the total pressure there.
- For loose cohesionless soils and normally consolidated clays, the coefficient of earth pressure at rest (Jaky, 1944) is

$$K_0 = 1 - \sin \phi' \tag{8.11}$$

For overconsolidated clays, the coefficient of earth pressure at rest (Mayne and Kulhawy, 1982) is

$$K_0 = (1 - \sin \phi')(OCR)^{\sin \phi'} \tag{8.12}$$

where OCR is the overconsolidation ratio, as defined in Equation 5.29 (see Chapter 5). For dense sands, typically, $K_0 \approx 0.3$ to 0.35.

- In Figure 8.1, $\Delta x / H$ ranges from 0.0005 for a sand to 0.02 for a clay to achieve the active state, while $\Delta x / H$ ranges from 0.005 for a dense sand to 0.04 for most other soils to achieve the passive state (Das, 2013).
- If a uniformly distributed surcharge per unit area (surface stress), q, is applied to the top surface of the backfill, the effective vertical stress, σ'_v increases at any depth by q, and the corresponding horizontal active and passive pressures increase by $K_a q$ and $K_p q$, respectively.
- The lateral earth pressure against a simple retaining wall with an assumed frictionless back surface from a backfill increases linearly with depth. Hence, the total active earth pressure (also referred to as the active thrust or active force) against the wall of height, H, from a backfill is

$$P_0 = \frac{1}{2}(K_0 \gamma H)H = \frac{1}{2}K_0 \gamma' H^2 \tag{8.13}$$

In a similar way, the total active earth pressure (also referred to as the active thrust or active force) against the wall of height, H, from a backfill with $c' = 0$ is

$$P_a = \frac{1}{2}(K_a \gamma' H)H = \frac{1}{2}K_a \gamma' H^2 \tag{8.14}$$

and the total passive earth pressure (also referred to as the passive resistance or the passive force) against the wall of height, H, from a backfill with $c' = 0$ is

$$P_p = \frac{1}{2}(K_p \gamma' H)H = \frac{1}{2}K_p \gamma' H^2 \tag{8.15}$$

The total lateral earth pressure acts at a height of $H / 3$ above the base of the wall.

- For the active condition within a cohesive backfill with $c' > 0$, from Equation 8.9, the active earth pressure $\sigma'_h = \sigma'_{ha}$ becomes zero at a particular depth $z = z_c$; hence, the soil is in tension up to a depth and z_c is the depth of tension crack. At a depth $z < z_c$, the pressure against the wall is negative, provided that a crack does not open between the wall and the uppermost part of the backfill. From Equation 8.), for $\sigma'_h = \sigma'_{ha} = 0$

$$z_c = \frac{2c'}{\gamma'\sqrt{K_a}} \tag{8.16}$$

As there is a tendency for tension cracks to open at the surface of the backfill behind a retaining wall, the negative pressure against the wall is generally neglected in calculating the total active earth pressure against the wall. Therefore, the total active earth pressure against the wall of height, H, from a backfill with $c' > 0$ is

$$P_a = \frac{1}{2}\left(K_a\gamma'H - 2c'\sqrt{K_a}\right)(H - z_c) = \frac{1}{2}K_a\gamma'H^2 - 2\sqrt{K_a}c'H + \frac{2c'^2}{\gamma'} \tag{8.17}$$

- In the case of active condition, the total active earth pressure is zero within a depth

$$z = H_c = 2z_c = \frac{4c'}{\gamma'\sqrt{K_a}} \tag{8.18}$$

Hence, if the height, H, of a vertical cohesive bank, say a vertical cut, is less than or equal to H_c, it should stand without a lateral support. In practice, in a cohesive soil, a cut can be excavated with vertical sides of a depth less than H_c without a support because the normal stress is zero at every point of the unsupported cut (Terzaghi *et al.*, 1996).
- For the undrained condition in a fully saturated clay, the active and passive pressures are calculated using $\phi = 0$ and $c = c_u$, and the unit weight equal to the saturated unit weight γ_{sat}. In this case, $z_c = 2c_u / \gamma_{sat}$ and $H_c = 4c_u / \gamma_{sat}$. If the crack or cut is filled with water, the water pressure acts on the wall, and therefore $z_c = 2c_u / \gamma'$ and $H_c = 4c_u / \gamma'$.
- The total passive earth pressure against a wall of height, H, from a backfill with $c' > 0$ is

$$P_p = \frac{1}{2}(K_p\gamma'H)H + (2\sqrt{K_p}c')H = \frac{1}{2}K_p\gamma'H^2 + 2\sqrt{K_p}c'H \tag{8.19}$$

- In practice, a retaining wall typically has a rough back surface. Thus, the lateral earth pressure analysis requires consideration of the roughness of the wall – that is, the wall friction – along with several other factors such as slope of the backfill top surface, slope of the back of the wall and earthquake or dynamic forces. For static lateral earth pressure analysis, Coulomb's theory, proposed by the French engineer Coulomb in 1776, is more appropriate as it considers the equilibrium of a soil mass forming a trial failure wedge behind the wall. For generalised field conditions, Shukla's generalised analytical expressions for dynamic active and passive earth pressures (Shukla 2013, 2015) can be used. Most static and dynamic analytical expressions proposed earlier for active and passive earth pressures are the special cases of Shukla's generalised expressions.
- Failure surfaces are typically curved; however, the assumption of planar failure surfaces, as considered in this discussion, tends to overestimate lateral earth pressure values. These overestimated values, particularly for passive earth pressure, should not be used directly in design without applying an appropriate reduction factor.

Example 8.1

A 10 m-high rigid retaining wall with a frictionless vertical back surface retains a granular backfill having its top surface horizontal and coinciding with the water table. The properties of the backfill are $\gamma_{sat} = 19 \, \text{kN/m}^3$, $c' = 0$ and $\phi' = 35°$. Determine the total active earth pressure and the resultant lateral force against the wall. Where is the resultant force located?

Solution

Given $H = 10$ m, $\gamma_{sat} = 19 \, \text{kN/m}^3$, $c' = 0$ and $\phi' = 35°$.

From Equation 8.5, the coefficient of active earth pressure

$$K_a = \frac{1 - \sin \phi'}{1 + \sin \phi'} = \frac{1 - \sin 35°}{1 + \sin 35°} = 0.271$$

From Equation 8.14, the total active earth pressure

$$P_a = \frac{1}{2} K_a \gamma' H^2 = \frac{1}{2}(0.271)(19 - 9.81)(10)^2 = 124.52 \, \text{kN/m}$$

The total lateral pressure from the pore water

$$P_w = \frac{1}{2}(\gamma_w H)H = \frac{1}{2}\gamma_w H^2 = \frac{1}{2}(9.81)(10)^2 = 490.5 \, \text{kN/m}$$

The resultant lateral force against the retaining wall

$$P = P_a + P_w = 124.52 + 490.5 = 615.02 \, \text{kN/m}$$

Due to the triangular distributions of both the pressures, the resultant lateral force acts at a height of $10/3 = 3.33$ m above the base of the wall.

8.3. Slope stability

The analysis of slope stability considers the core principles of soil mechanics such as the effective stress principle (Chapter 3), fluid flow through soil (Chapter 4), and shear strength and failure criteria (Chapter 6). More advanced analysis also uses the concepts of probability theory. There are several methods of slope stability analysis (Chowdhury et al., 2010). The limit equilibrium methods have been widely favoured by engineers. In these methods, a failure, slip or slide surface of a simplified geometric shape is assumed, and the soil mass above this surface is analysed as a free body by applying principles of force equilibrium, moment equilibrium or a combination of both. Based on the analysis, the available and mobilised resistances for a number of potential failure surfaces are compared to predict the critical failure surface and, consequently, the stability of the slope.

To make slope stability analysis easier to understand, soil slopes can be divided into two types: infinite and finite. Both types are briefly explained below.

8.3.1 Infinite slope

A slope is called an infinite slope if it extends over a long distance under identical conditions. Any potential failure of this slope results in the movement of a soil mass down the slope as a translational slip along a surface parallel to the slope, as shown in Figure 8.4 for a submerged slope in cohesionless soil, which may be found near the shore of a lake. If an infinite slope fails, the ratio of the maximum thickness, d, of the slide to the length, L, of the slide up the slope is very small, typically $d/L < 0.03 - 06$. Note that the curvature of the failure or slip surface remains constant along the length of the slope except at points near edges. The forces acting on the two vertical faces of any soil element or slice, say $A_1 A_2 A_3 A_4$, are equal and exactly balance each other; hence, the resultant of the forces acting on the vertical faces of the element is zero. The forces acting on the element per unit length perpendicular to the page are as follows

- submerged weight of the soil element, $W' = \gamma'bD$, where γ' is the submerged unit weight of the soil; b is the width of the soil element/slice; D is the depth or height of the soil element/slice
- normal force, $N' = W'\cos\beta = \gamma'bD\cos\beta$, on the failure plane, where β is the slope angle
- mobilised shear resistance of the soil, $T = W'\sin\beta = \gamma'bD\sin\beta$, on the failure plane.

The stability of a slope is typically expressed by defining the factor of safety (or safety factor) as

$$F = \frac{\tau_f}{\tau} \tag{8.20}$$

where τ_f is the available shear strength and τ is the mobilised shear stress on the potential failure surface.

From Equation 6.2 (see Chapter 6), the available shear strength with effective cohesion intercept, $c' = 0$ for the cohesionless soil is

$$\tau_f = \sigma_f' \tan\phi' = \frac{N'}{b\sec\beta \times 1}\tan\phi' = \frac{\gamma'bD\cos\beta}{b\sec\beta}\tan\phi' = \gamma'D\cos^2\beta\tan\phi' \tag{8.21a}$$

Figure 8.4 Submerged infinite slope in a cohesionless soil: (a) forces on a soil element; (b) force equilibrium (Note: no flow of water within the soil)

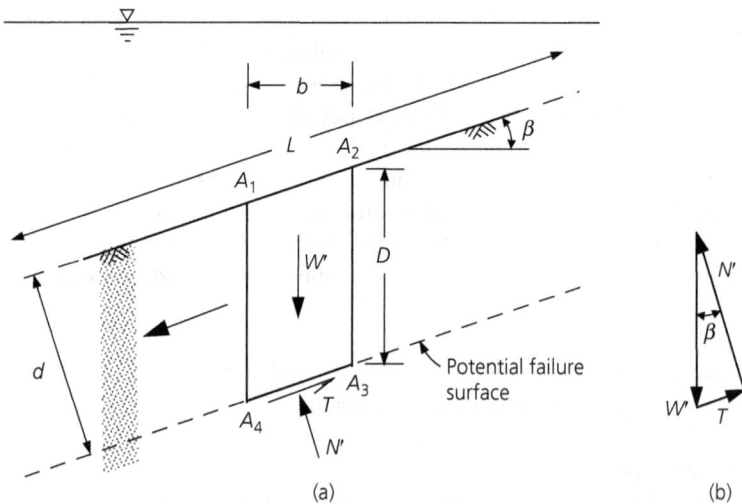

The mobilised shear stress on the failure plane is

$$\tau = \frac{T}{b\sec\beta\times 1} = \frac{\gamma'bD\sin\beta}{b\sec\beta} = \gamma'D\sin\beta\cos\beta \tag{8.21b}$$

From Equation 8.20 with Equations 8.21a and 8.21b, the safety factor of the slope is obtained as

$$F = \frac{\tau_f}{\tau} = \frac{\gamma'D\cos^2\beta\tan\phi'}{\gamma'D\sin\beta\cos\beta} = \frac{\tan\phi'}{\tan\beta} \tag{8.22}$$

The expression for F in Equation 8.22 is independent of the depth or height, D, of the soil element. Thus, the stability of a slope in cohesionless soil is identical at any depth; therefore, the slope failure or slip is equally likely to take place at any depth.

For a stable slope, $F > 1$ – that is, a slope in cohesionless soil remains stable for $\beta \leq \phi'$. A value of F less than one indicates that the slope is unstable; therefore, the slope will fail as a translational slip. A value of F equal to one refers to a critical situation in which the failure of the slope is imminent. For $F = 1$, $\beta = \phi'$, which means that the maximum slope angle in a cohesionless soil is equal to the drained friction angle of the soil.

Note that Equation 8.20 is also applicable to a slope in dry cohesionless soil. Hence, for a stable slope in cohesionless soil, $\beta \leq \phi'$, whether the slope is completely dry or completely submerged under water.

8.3.2 Finite slope

A finite slope connects two ground levels at different elevations that are relatively close to each other, so any potential failure results in the movement of a soil mass down the slope as a rotational slip, as illustrated in Figure 8.5 for a slope in cohesive soil. In the cross-section, the failure surface may appear as slope failure circle, toe failure circle or base failure circle (also referred to as the mid-point failure circle). The ratio $n_d = (H + H_1)/H$ is called the depth factor. A finite slope may exist in nature or is made in engineered structures, such as highway and railway embankments,

Figure 8.5 A finite slope with potential failure surfaces as slope failure, toe failure and base failure

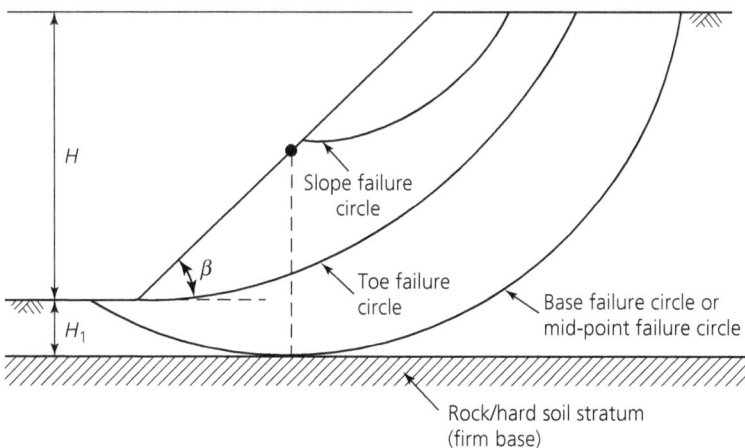

dams, canal banks and excavation for various purposes. Finite slopes are more common than infinite slopes. If a finite slope fails, the ratio of the maximum thickness, d, of the slide to the length, L, of the slide up the slope, as illustrated in Figure 8.6, is large, typically $d / L > 0.15 - 0.30$. Note that the curvature of the failure or slip surface changes from point to point along the failure surface. The forces acting on the two vertical faces of any soil element or slice are not equal and they do not balance each other.

There are several methods for analysing the stability of finite slopes; the details are presented by Chowdhury *et al.* (2010). The most commonly used methods are mainly based on the concept of the method of slices. In this method, the soil mass above the assumed failure surface is broken up into a series of vertical slices and the force equilibrium, moment equilibrium or a combination of both for each of these slices is considered (Figure 8.6). The approach is similar to the basic concept of force equilibrium of a slice, discussed in the previous section.

Figure 8.7 shows a finite slope with a potential cylindrical failure surface that appears circular in cross-section as a circular arc $A_1 A_3$ with its centre at O and radius R, in a fully saturated homogeneous clay under the undrained condition, which can occur immediately after construction. In this figure, τ is the shear stress or resistance mobilised on the failure surface; θ (in radians) is the angle, subtended by the failure arc $A_1 A_3$ at O; W is the weight of soil per unit length in the slip zone $A_1 A_2 A_3$, acting vertically downwards from the centre of gravity, G, of the failure zone; x is the horizontal distance of the line of action of W from O. Note that $W = \gamma A$, where A is the area of failure zone $A_1 A_2 A_3$ and γ is the total unit weight of soil.

Considering the moment equilibrium about the centre of rotation O

$$(\tau R \theta)R = Wx$$

or using Equation 8.20 with $\tau_f = c_u$ for the undrained condition

$$\left(\frac{c_u}{F} R \theta \right) R = Wx$$

Figure 8.6 Stability analysis of a finite slope by the method of slices

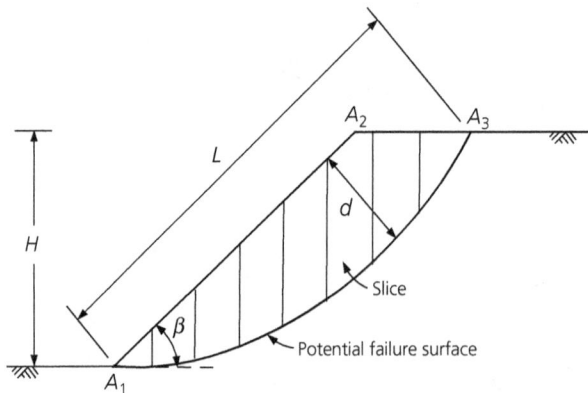

Figure 8.7 Total stress ($\phi = 0$) analysis of a finite slope in a fully saturated homogeneous clay under undrained condition

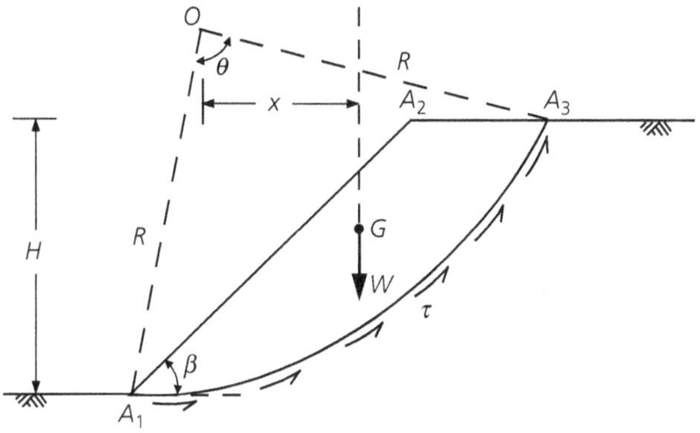

or

$$F = \frac{c_u R^2 \theta}{Wx} = \frac{c_u R^2 \theta}{\gamma Ax} \tag{8.23}$$

Example 8.2

Referring to Figure 8.7, consider the following: $W = 400$ kN/m, $R = 9$ m, $\theta = 70°$, $x = 5$ m, $c_u = 20$ kPa and $\gamma = 18$ kN/m³. Determine the factor of safety of the slope.

Solution

From Equation 8.23, the factor of safety of the slope

$$F = \frac{c_u R^2 \theta}{Wx} = \frac{(20)(9)^2 (70\pi / 180)}{(400)(5)} = 0.99$$

The following points regarding slope stability are noteworthy.

- The upstream slopes of earth dams can fail as translational slips as a result of rapid drawdown of the reservoir water.
- In Equation 8.21a, note that the effective normal stress, σ'_f, at failure on the potential failure plane is $\gamma'D\cos^2 \beta$, not simply equal to $\gamma'D$ (i.e. the depth multiplied by the unit weight), as discussed in Chapter 3.
- If the seepage of water takes place parallel to the infinite slope in a cohesionless soil as often observed in the lower portions of natural slopes, the following expression for the safety factor can be derived

$$F = \frac{\gamma'}{\gamma} \frac{\tan \phi'}{\tan \beta} \tag{8.24}$$

where γ' is the total unit weight of the soil. Note that for most soils, $\gamma' / \gamma \approx 0.5$.

■ As for most normally consolidated clays, effective cohesion intercept, $c' = 0$. Equation 8.22 also applies to infinite slopes in normally consolidated clays.

■ With reference to Figure 8.7, for the case of $\phi = 0$ analysis (or total stress analysis), Taylor (1937, 1948) presented a design chart as the stability coefficient or factor, S_n, as defined below, against the slope angle, β, for depth factor, n_d, varying from 1 to ∞. For $n_d > 10$ or $\beta = 53°$, $S_n = 0.181$.

$$S_n = \frac{c_u}{F \gamma H} \tag{8.25}$$

8.4.   Bearing capacity

8.4.1   Shallow foundations

When a footing constructed at the ground surface or at some depth below the ground surface, as is common practice, is loaded, the footing settles because the soil is pushed out from beneath the footing and the surface of the surrounding soil may heave. The failure of the foundation soil takes place as a shear failure. The bearing pressure (contact stress) – that is, the load per unit area – at the base level of the footing that causes the shear failure to occur in the foundation soil is termed the ultimate bearing capacity, q_u, of the foundation. This capacity depends on the characteristics of the soil mass and is also governed by the geometrical dimensions of the foundation, depth of foundation below the ground surface, water table location and inclination of the load applied to the foundation.

Depending on the type of soil, the foundation soil may exhibit one of the following three failure modes or mechanisms.

■ General shear failure. This is typically observed in low-compressibility soils, such as very dense/dense sands and very stiff/hard clays, with well-defined continuous slip surfaces extending to the ground level. The failure takes place suddenly with heaving or bulging around the footing at the ground surface in a tilting manner.

■ Local shear failure. This is typically observed in medium-compressibility soils, such as medium dense sands and firm/stiff clays, with well-defined continuous slip surfaces extending well below the ground surface. The failure takes place with limited heaving or bulging around the footing at the ground surface without any tilting of the footing. At large footing settlements, slip surfaces may appear at the ground surface.

■ Punching shear failure. This is typically observed in high-compressibility soils, such as loose/ very loose sands and soft/very soft clays, with well-defined continuous slip surfaces limited to the area below the footing. The failure takes place without heaving or bulging around the footing at the ground surface without any tilting of the footing.

Figure 8.8 shows typical pressure–settlement curves (also known as stress–settlement curves, load–settlement curves, or simply settlement curves) obtained from a footing load test, which is generally conducted as a plate load test (Shukla and Sivakugan, 2011). The curve C_1 is generally obtained for dense cohesionless soils or stiff cohesive soils. The curve C_2 is recorded when loose to medium dense cohesionless soils or soft to medium stiff cohesive soils are loaded. For very loose cohesionless soils or very soft clayey soils, a curve similar to C_3 is observed. For curve C_1, the failure is well defined with a distinct peak, so the ultimate bearing capacity of the soil, $(q_u)_1$, is obtained without any approximation, as shown in the figure. For curve C_2, there is no clear peak or

Figure 8.8 Typical pressure–settlement curves for three different failure modes of foundation soil (Note: curves C_1, C_2 and C_3 represent general shear failure, local shear failure and punching shear failure, respectively.)

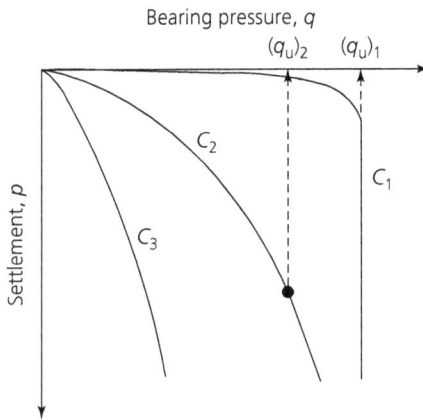

yield point and hence the ultimate bearing capacity is not always well defined. The ultimate bearing capacity of soils represented by curve C_2 is usually assumed to be equal to $(q_u)_2$, corresponding to the point at which the pressure–settlement curve becomes steep and straight (Terzaghi, 1943; Terzaghi *et al.*, 1996).

Figure 8.9 illustrates the bearing capacity failure mechanism in a soil mass under a rough rigid strip footing for the general shear failure as originally considered by Terzaghi (1943) for deriving the expression for the ultimate bearing capacity of a strip or continuous footing. The failure zone beneath the footing can be separated into the following three zones.

Figure 8.9 General shear failure mechanism considered by Terzaghi (1943) for developing the bearing capacity equation

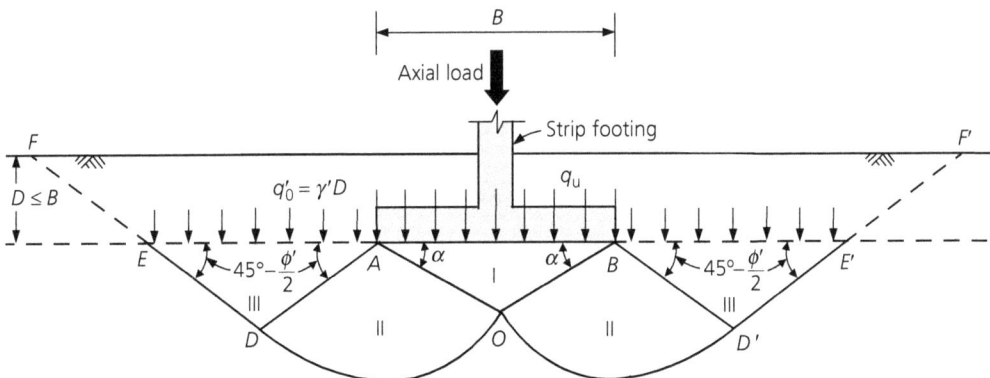

■ Zone I: soil wedge ABO immediately under the footing is an elastic zone. Both AO and BO make an angle, α, with the horizontal. Terzaghi (1943) assumed that $\alpha = \phi'$, where ϕ' is the drained angle of shearing resistance of soil under the footing. Later, it was found that $\alpha = 45° + \phi'/2$ (Vesic, 1973).

■ Zone II: soil wedges AOD and BOD' are radial shear zones, where the rupture lines OD and OD' are the arcs of a logarithmic spiral.

■ Zone III: soil wedges ADE and $BD'E'$ are the Rankine passive zones, where the rupture lines DE and $D'E'$ are the straight lines.

There are several bearing capacity equations for estimating the ultimate bearing capacity of a footing resting on soil. One of the early sets of bearing capacity equations was proposed by Terzaghi (1943) using the failure mechanism shown in Figure 8.8 with the following assumptions.

■ The foundation soil is a homogeneous, isotropic, weightless, semi-infinite and rigid-plastic material.

■ The foundation is shallow – that is, $D \leq B$ – so the shear resistance of the soil along the failure surfaces above the footing base level can be neglected without any significant error.

■ The footing has a rough base.

■ To consider the effect of the presence of soil above the footing base level, the soil may be replaced by a uniform overburden or surcharge stress, $q_0' = \gamma' D$, where γ' is the submerged unit weight of soil. This substitution simplifies the computations very considerably and the error is on the safe side.

■ The angle α is ϕ'.

■ A general shear failure occurs.

In the past, several researchers (Hansen, 1970; Meyerhof 1951, 1963; Vesic 1973, 1975) attempted to refine Terzaghi's bearing capacity equation in order to present a general bearing capacity equation incorporating certain observations of practical importance, as given below

■ shape of the footing
■ depth of the footing
■ direction of the loading
■ location of the water table with respect to the footing base level.

The general ultimate bearing capacity equation can be expressed as follows.

For general shear failure

$$q_u = cN_c s_c d_c i_c + q_0' N_q s_q d_q i_q + \frac{1}{2}\gamma B N_\gamma s_\gamma d_\gamma i_\gamma W \tag{8.26}$$

where c is the cohesion intercept; $q_0'(=\gamma'D)$ is the effective overburden or surcharge stress at the footing base level; γ is the total unit weight of the soil; B is the width or diameter of the footing; s_c, s_q and s_γ are the shape factors; d_c, d_q and d_γ are the depth factors; i_c, i_q and i_γ are the load inclination factors; W is the correction factor for water table location; N_c, N_q and N_γ are the bearing capacity factors for the general shear failure. These factors are functions of only the angle of internal friction of soil, ϕ'. Note that the values of bearing capacity factors N_c, N_q and N_γ are not the same as Terzaghi's bearing capacity factors.

Several expressions for bearing capacity factors are available in the literature; the most widely used ones are provided below.

$$N_c = (N_q - 1)\cot\phi' \tag{8.27a}$$

$$N_q = e^{\pi\tan\phi'}\tan^2\left(45^0 + \frac{\phi'}{2}\right) \tag{8.27b}$$

$$N_\gamma = 2(N_q + 1)\tan\phi' \tag{8.27c}$$

The expression for N_γ in Equation 8.27c was presented by Vesic (1973). Note that for a given value of ϕ', the values of bearing capacity factor, N_γ, obtained by different research workers vary widely due to the variation in assumption of the logarithmic spiral arc OD and exit wedge III (Figure 8.9), although values of N_c and N_q do not differ significantly. Fortunately, the term N_γ does not make a significant contribution to the computed bearing capacity, so any of the values reported in the literature or their average value can be used for practical purposes.

The values of shape, depth, load inclination and water table correction factors can be taken as provided in Table 8.1. All these factors are empirical factors based on the experimental data.

Table 8.1 Shape, depth, load inclination and water table correction factors recommended for use in general bearing capacity equations (Equations 8.26 and 8.28) (continued on next page)

Shape of the base of the footing	Shape factors		
	s_c	s_q	s_γ
Continuous strip	1.0	1.0	1.0
Rectangle (B = width, L = length)	$1 + 0.2\dfrac{B}{L}$	$1 + 0.2\dfrac{B}{L}$	$1 - 0.4\dfrac{B}{L}$
Square	1.3	1.2	0.8
Circle	1.3	1.2	0.6

Depth factors

$$d_c = 1 + 0.2\frac{D}{B}\tan\left(45° + \frac{\phi'}{2}\right)$$

$$d_q = d_\gamma = 1 \text{ for } \phi' < 10°$$

$$d_q = d_\gamma = 1 + 0.1\frac{D}{B}\tan\left(45° + \frac{\phi'}{2}\right) \text{ for } \phi' \geq 10°$$

Depth factors should be applied only when backfilling is done with proper compaction.

Table 8.1 Continued

Load inclination factors

$$i_c = i_q = \left(1 - \frac{\alpha}{90}\right)^2$$

$$i_\gamma = \left(1 - \frac{\alpha}{\phi'}\right)^2$$

where α is the inclination of the load to the vertical in degrees and ϕ' is the effective angle of shearing resistance of foundation soil in degrees.

Correction factor for location of the water table

$$W = 1 \text{ for } D_{WT} \geq (D + B)$$

$$W = 0.5 \text{ for } D_{WT} \leq D)$$

$$W = 0.5 + 0.5 \left(\frac{D_{WT} - D}{B}\right) \text{for } D < D_{WT} < (D + B)$$

where D_{WT} is the depth of the water table from the ground surface.

For local shear failure

$$q_u = \frac{2}{3}cN'_c s_c d_c i_c + q'_0 N'_q s_q d_q i_q + \frac{1}{2}\gamma B N'_\gamma s_\gamma d_\gamma i_\gamma W \tag{8.28}$$

where N'_c, N'_q and N'_γ are the bearing capacity factors for the local shear failure, which are obtained as the values of N_c, N_q and N_γ from Equations 8.27a, 8.27b and 8.27c, respectively, corresponding to $\tan^{-1}\left(\frac{2}{3}\tan\phi'\right)$ instead of ϕ'.

The ultimate bearing capacity is not directly used in foundation design. Therefore, it is necessary to understand other bearing capacity terms, as defined below.

Safe bearing capacity, q_s, is the maximum stress at the base level of the foundation that can be safely carried by the foundation soil without a shear failure of the soil mass.

The load per unit area at the foundation base level in excess of the effective overburden pressure that causes a permissible or specified settlement of the engineering system is known as the safe bearing pressure or net safe settlement pressure, q_{np}.

If the ultimate bearing capacity and the safe bearing capacity are estimated by deducting the effective overburden pressure at the base level of the foundation, they are termed net ultimate bearing capacity, q_{nu}, and net safe bearing capacity, q_{ns}, respectively. The value of q_u obtained from the bearing capacity equation is used to compute the following.

$$q_{nu} = q_u - q_0' = q_u - \gamma'D \tag{8.29}$$

$$q_{ns} = \frac{q_{nu}}{F} \tag{8.30}$$

and

$$q_s = q_{ns} + \gamma'D = \frac{q_{nu}}{F} + \gamma'D \tag{8.31}$$

where F is the factor of safety. A minimum value of 3 is commonly recommended for most bearing capacity calculations.

The safe bearing pressure, q_{np}, can be determined directly by the footing load test or indirectly using the empirical expressions involving field-determined resistance parameters, such as the standard penetration test value. The details can be found in the suggested readings. The allowable bearing pressure (or net allowable bearing pressure), q_{na}, which is the lower of net safe bearing capacity, q_{ns}, and safe bearing pressure, q_{np}, is generally recommended as the bearing capacity of the foundation for its design. Thus,

$$q_{na} = q_{ns} \text{ if } q_{np} > q_{ns} \tag{8.32}$$

$$q_{na} = q_{np} \text{ if } q_{np} < q_{ns} \tag{8.33}$$

In most foundation problems, in general, $q_{np} < q_{ns}$; thus, Equation 8.33 applies.

The following points regarding the bearing capacity of shallow foundations are noteworthy.

- For the short-term (undrained) behaviour of a saturated clay or cohesive soil, the shear strength parameters to be used for determining the bearing capacity should be the undrained shear strength parameters, $c = c_u$ and $\phi = 0$.
- For the long-term (drained) behaviour of a soil of any type, the shear strength parameters to be used for determining the bearing capacity should be the drained shear strength parameters, c' and ϕ', obtained from the drained shear tests.
- The bearing capacity equations presented here have been developed by assuming the plane strain conditions as applicable to a strip footing (strains occur only in two dimensions, namely along the depth and the width of the strip footing, no strain along its length). Thus, the equations are based on the angle of internal friction, ϕ', obtained from plane strain tests such as the direct shear tests. The value of ϕ' from the plane strain tests is about 1.1 times the value of ϕ' from the triaxial tests.
- Shallow foundations are rarely constructed on very loose/loose sands or very soft/soft clays. Such sites are typically avoided, or deep foundations are used instead. If shallow foundations are required, the weak ground must first be improved. Consequently, the bearing capacity equation for the punching shear failure mode is rarely presented.

Example 8.3

A 1.25 m-wide strip footing rests on a fully saturated clay deposit at a depth of 0.6 m. The water table is at the ground surface. Considering the undrained shear strength parameters $c = c_u = 20$ kPa and $\phi = 0$, along with $\gamma_{sat} = 19.47$ kN/m^3 for the clay, determine the ultimate bearing capacity of the footing. Assume that the failure mode is general shear failure.

Solution

Given $B = 1.25$ m, $D = 0.6$ m, $c = c_u = 20$ kPa, $= 0$ and $\gamma_{sat} = 19.47$ kN/m^3.

From Equations 8.27a–c, for $\phi = 0$

$N_c = 5.14$, $N_q = 1.0$ and $N_\gamma = 0.0$

From Table 8.1, for the strip footing

$$s_c = s_q = s_\gamma = 1.0;\ d_c = 1 + (0.2)\left(\frac{0.6}{1.25}\right) = 1.10,\ d_q = d_\gamma = 1;\ i_c = i_q = i = 1;\ \text{and}\ W = 1$$

From Equation 8.26, the ultimate bearing capacity of the footing

$$q_u = (20)(5.14)(1.0)(1.10)(1) + (19.47 - 9.81)(0.6)(1.0)(1)(1)$$
$$+ (0.5)(19.47)(1.25)(0)(1.0)(1)(1)(0.5) = 113.08 + 5.80 + 0 = 118.88 \approx 119\ \text{kPa}$$

8.4.2   Deep foundations

The most common types of deep foundations are piles and drilled piers. Piles are vertical or slightly inclined columnar foundation members made of steel, concrete and/or timber. They are constructed to transfer the load through weak, compressible soil strata or water to a firm soil or rock mass. The piles are installed into the ground by driving, boring, jetting or vibrating. A pile in sand is typically driven or pushed, as a bored hole cannot remain open without horizontal support. Placing a structure on pile foundations is significantly more expensive than on shallow foundations. A drilled pier (also referred to as a drilled shaft, drilled caisson, caisson or bored pile) is a cast-in-place pile, generally having a diameter of about 0.75 m or more. Pile foundations are the most common type of deep foundation.

If a pile rests on bedrock or a hard/dense soil stratum, its ultimate load-carrying capacity (commonly called pile capacity), Q_u, primarily depends on the ultimate bearing capacity of the bedrock or soil. Such a pile is referred to as a point-bearing or end-bearing pile (Figure 8.10(a)). Thus,

$$Q_u = Q_p = q_p A_p \tag{8.34}$$

where Q_p is the ultimate load-carrying capacity of the pile point, end, base, tip, toe or bottom, also known as the ultimate point, the end-bearing capacity, or the ultimate point/base resistance of the pile; q_p is the unit point, end or base bearing – that is, the ultimate bearing capacity of the soil below the base of the pile; and A_p is the cross-sectional area of the base.

Figure 8.10 Types of piles: (a) point-bearing pile; (b) friction pile; (c) normal pile (Note: B is the pile diameter; D is the embedment depth; L is the pile length.)

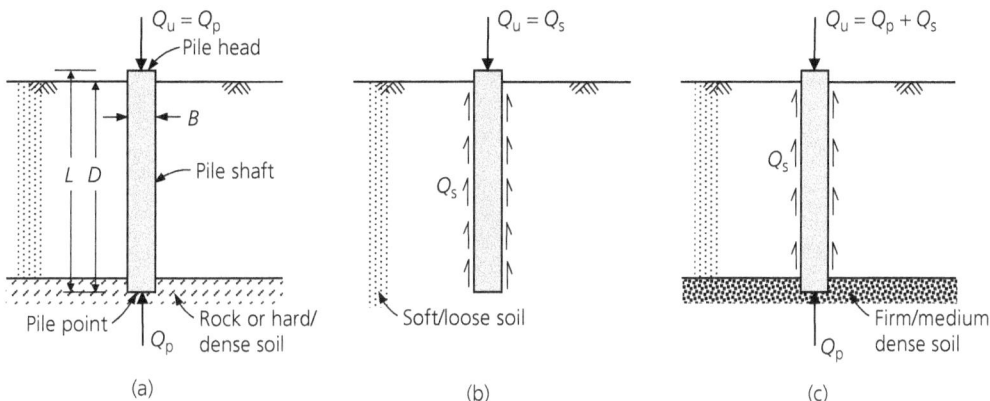

When a bedrock or a hard/dense soil stratum does not exist at a reasonable depth below the ground surface, the pile can be designed to transmit the structural load primarily through friction and/or adhesion, commonly called skin friction, to the soil adjacent to the pile stem/shaft only. Such a pile is referred to as a friction pile (Figure 8.10(b)). Thus,

$$Q_u = Q_s = f_s A_s \tag{8.35}$$

where Q_s is the ultimate frictional resistance or shaft resistance of the pile and f_s is the unit skin friction – that is, the skin friction developed at failure per unit contact area between the pile stem and the surrounding soil.

In general, a normal pile may derive its capacity from both end-bearing and frictional resistance (Figure 8.10(c)). Hence, the ultimate load-carrying capacity of the pile is given as

$$Q_u = Q_p + Q_s = q_p A_p + f_s A_s \tag{8.36}$$

Thus, there are two mechanisms of load transfer from a pile to the soil: end-bearing and skin friction. For either mechanism, some pile displacement always occurs. For a pile installed within the soil mass, typically the maximum skin friction develops at a displacement of $0.03B$ to $0.05B$, where B is the diameter of the pile stem, while the maximum end bearing develops at a displacement of about $0.15B$.

The value of Q_u can be determined theoretically or by conducting the pile load test. In the theoretical approach, Equations 8.34, 8.35 and 8.36 are generally accepted, However, there is no general agreement on how to obtain q_p and f_s. A pile load test consists of applying a static load in increments to a test pile and measuring the deflection of the pile top/head. Several recommendations from the literature should be followed and engineering judgement used to determine the appropriate value for the design. It is common practice to test a full-scale pile prior to construction of piles to validate their theoretically determined ultimate load-carrying capacity. Only a few simple theoretical expressions are provided here.

For a pile in clay or cohesive soil

$$q_p = c_u N_c = 9c_u \qquad (8.37)$$

where c_u is the undrained shear strength or undrained cohesion of the clay at the tip/base of the pile.

For a pile in sand or cohesionless soil

$$q_p = q'_0 N_q \qquad (8.38)$$

where q'_0 is the effective overburden stress at the pile tip/base and N_q is the bearing capacity factor. Note that the value of N_q for deep foundations differs from that for shallow foundations. The value of N_q typically ranges between 10 and 160 for a value of ϕ' between 25° from 45°.

As f_s depends on several factors, such as normal stress to the pile stem, friction and/or adhesion of the pile–soil interface, degree of contact between the pile stem and the soil, method of pile installation, and pile material, the determination of f_s is not an easy task.

For a pile in clay or cohesive soil

$$f_s = \bar{c}_a = \alpha \bar{c}_u \qquad (8.39)$$

where $\bar{c}_a$ is the average adhesion between the pile shaft and clay; $\bar{c}_u$ is the average undrained shear strength or undrained cohesion of clay at shaft of the pile; α is an adhesion factor that usually ranges from about 0.2 to 1, depending on the factors mentioned earlier. For soft clay, which sticks to the pile everywhere, α is taken as 1. Because a stiff clay may have incomplete contact, a lower value of α is considered.

For a pile in sand or cohesionless soil

$$f_s = \bar{\sigma}'_h \tan \delta = K \bar{\sigma}'_v \tan \delta \qquad (8.40)$$

where K is the average coefficient of earth pressure along the embedded shaft, which converts the average effective vertical stress, $\bar{\sigma}'_v$, to the average lateral stress, $\bar{\sigma}'_h$; and δ is the average pile–soil interface friction angle. The value of K for any layer depends on the factors mentioned earlier, and can exceed unity if the pile is driven into dense sand. Note that the driving of a pile displaces soil laterally and thus increases the horizontal stress on the pile shaft.

For the design of a pile foundation, the allowable load-carrying capacity, Q_a, is used, as defined below.

$$Q_a = \frac{Q_u}{F} \qquad (8.41)$$

where F is the factor of safety. Its value depends on many factors, including soil type, level of subsurface investigation, availability of pile load test results, method of pile installation, and type and importance of the structure. Typically, the value of F is taken to be between 2.5 to 6.

The following key points regarding the load-carrying capacity of deep foundations are noteworthy.

- Even a single pile is highly statically indeterminate. Therefore, the capacity of a pile cannot be precisely determined theoretically using the core principles of soil mechanics. Empirical knowledge and the results of pile load tests at the actual site are essential to accurately estimate the load-carrying capacity and perform settlement analysis.
- Usually, the weight of the pile, W_p, that acts vertically downward is small compared to Q_u; hence W_p is generally ignored. However, in marine structures in deep water where a considerable length of shaft extends above sea level, W_p should be added with a negative sign to the right side of Equations 8.34 to 8.36.
- Piles used in marine structures and slopes may be subjected to lateral or horizontal loads from water waves and lateral earth pressures. A combination of lateral and vertical loads may apply on the piles in some applications, such as piles constructed to support bridge piers and abutments.
- Piles are sometimes used to provide anchorage against uplift or horizontal pull; such piles are called anchor piles. The uplift force is resisted by the skin friction developed at the pile shaft–soil interface as well as by the weight of the pile (Das and Shukla, 2013).
- Pile-driving data with pile driving formulae are also used to estimate the load-carrying capacity of piles. This approach is called the dynamic approach, while the approach discussed here is the static approach, which uses the core principles of soil mechanics. The correlations with penetration test data are sometimes used to estimate the load-carrying capacity of piles.
- Piles are generally used in a group of at least three to support a column load, with a beam or slab, called the pile cap, connecting the head/top/butt of the piles in the group. A single pile is seldom used except when the top portion of the pile extends above the ground level to act as a column. The ultimate load-carrying capacity of a group of piles, Q_{ug}, may be less than, equal to or greater than the sum of ultimate load-carrying capacities, $\sum Q_u$, of the individual piles in the group. The ratio, $\eta = Q_{ug} / \sum Q_u$, is called the group efficiency of the pile group.

   For pile groups driven in a clay deposit, typically $\eta < 1$ for a small pile spacing, but η may be equal to one for a large spacing. In a sand deposit, the driving action of piles compacts the sand; thus, η may be greater than one for friction piles. For point-bearing piles, η is generally less than one. The value of η primarily depends on the spacing of piles in the group, type of soil and the method of pile installation. For a large spacing, the effect of interaction between piles diminishes; thus $\eta = 1$. Ideally, an attempt is made to achieve $\eta = 1$. Typically, the spacing of piles in a group is recommended between $2.5B$ to $3.5B$, where B is the diameter or width of the pile.
- There is a possibility that the soil surrounding a pile shaft may move or settle downwards relative to the pile, resulting in a downward drag, called the negative skin friction, on the pile. Negative skin friction may result due to an increase in effective stress when a fill is placed at the ground surface above a soft clay deposit or when the groundwater table is lowered. In this case, Equation 8.36 reduces to $Q_u = Q_p - Q_s$, thus resulting in a lower ultimate load-carrying capacity of the pile.

Example 8.4

A 0.25 m-diameter circular concrete pile is driven into a soft clay deposit. The embedment depth of the pile is 13 m. The clay has an average unconfined compressive strength of 30 kPa. Estimate the ultimate pile capacity. Assume that the adhesion factor is equal to 1.

Solution

Given $B = 0.25$ m, $D = 13$ m, $q_{ucs} = 30$ kPa, $\alpha = 1$.

The undrained cohesion

$$c_u = \bar{c}_u = \frac{q_{ucs}}{2} = \frac{30}{2} = 15 \text{ kPa}$$

From Equation 8.36 with Equations 8.37 and 8.39, the ultimate pile capacity

$$Q_u = Q_p + Q_s = q_p A_p + f_s A_s = (9c_u) A_p + (\alpha \bar{c}_u) A_s$$

$$= (9)(15)\left[\left(\frac{\pi}{4}\right)(0.25)^2\right] + (1)(15)\left[(\pi)(0.25)(13)\right]$$

$$= 6.63 + 153.15 = 159.78 \approx 160 \text{ kN}$$

Chapter summary

1. Depending on the movement of the retaining wall, the lateral earth pressure can be classified as earth pressure at rest, active earth pressure or passive earth pressure.
2. The stability of a slope is expressed by a factor of safety, defined as the ratio of the available shear strength to the mobilised shear stress on the potential failure surface. For a stable slope, $F > 1$.
3. Slopes can be categorised as infinite slopes and finite slopes. A slope is referred to as an infinite slope if it extends over a long distance with uniform conditions throughout.
4. Foundations can be classified as shallow ($D / B \leq 1$) or deep ($D / B > 1$), where D is the depth of the foundation and B is its width. A pile foundation is the most common type of deep foundation.
5. Equation 8.26 is the general bearing capacity equation for determining the ultimate bearing capacity of shallow foundations for the general shear failure condition.
6. The allowable bearing pressure is always the net allowable pressure, defined as the maximum stress at the foundation in excess of the effective overburden stress. It is used in the design of foundations to ensure there is no risk of shear failure or excessive settlement.
7. A pile can be point-bearing, frictional, or a combination of both. Piles are typically used in groups. The pile capacity, estimated using static or dynamic approaches, is often verified through pile load tests conducted on a limited number of piles at the project site.

Questions for practice

Select the most appropriate answer for each of the multiple choice questions from Q8.1 to Q8.10. Questions Q8.11 to Q8.25 require more detailed answers.

8.1 Select the incorrect statement.
  (a) Less strain or displacement of the wall is needed to achieve the active condition compared to reaching the passive condition.
  (b) The presence of cohesion in soil increases its strength, resulting in lower active earth pressure and higher passive resistance against the wall.
  (c) K_a and K_p are based on total stresses, not effective stresses.
  (d) K_a remains exactly the same regardless of whether the soil is dry or submerged.

8.2 The coefficient of active earth pressure for loose sand with an internal friction angle of $30°$ is
  (a) 1/3
  (b) 1/2
  (c) 2
  (d) 3.

8.3 The total passive resistance of the cohesionless backfill against a retaining wall with a vertical back face is proportional to
  (a) height of the wall only
  (b) square of the wall height
  (c) internal friction angle
  (d) none of the above.

8.4 The depth of a vertical cut in fully saturated cohesive soil with $c_u = 20\,\text{kPa}$, $\phi = 0°$ and $\gamma_{sat} = 20\,\text{kN/m}^3$ under undrained conditions, up to which no lateral support is required is
  (a) 1 m
  (b) 2 m
  (c) 3 m
  (d) 4 m.

8.5 For an infinite sandy slope inclined at an angle of $20°$ to the horizontal, with a drained angle of internal friction of $30°$, the factor of safety of this slope will be
  (a) 0.63
  (b) 1.00
  (c) 1.59
  (d) none of the above.

8.6 The base failure of a finite slope
  (a) is the most common type of failure in steep slopes
  (b) occurs when the soil above the toe or base level of the slope is relatively weak
  (c) occurs when the soil below the toe or base level of the slope is relatively strong
  (d) occurs when soil below the toe or base level of the slope is relatively weak.

8.7 The bearing capacity factors for the shallow foundations are functions of
 (a) cohesion of foundation soil
 (b) angle of internal friction of foundation soil
 (c) both cohesion and angle of internal friction
 (d) total unit weight of foundation soil.

8.8 The ultimate bearing capacity of a surface strip footing with a width of 1 m, resting on fully saturated clay under undrained condition, is 120 kPa. The ultimate bearing capacity of a surface circular footing with a diameter of 1 m, resting on the same fully saturated clay
 (a) will be less than 120 kPa
 (b) will be equal to 120 kPa
 (c) will be greater than 120 kPa
 (d) cannot be predicted.

8.9 The diameter of a drilled pier is typically
 (a) 0.25 m
 (b) 0.45 m
 (c) less than or equal to 0.75 m
 (d) equal to or greater than 0.75 m.

8.10 For the design of a pile foundation, the allowable load-carrying capacity is determined from the ultimate pile capacity by dividing it by a factor of safety, typically taken as in the range of
 (a) 1 to 1.5
 (b) 1.5 to 3
 (c) 2 to 3
 (d) 2.5 to 6.

8.11 An unsupported excavation is to be made in a clay stratum. If the properties of the clay are $\gamma = 18 \, kN/m^3$, $c' = 20$ kPa and $\phi' = 8°$, determine the depth of tension cracks and the possible maximum unsupported excavation depth in the clay.

8.12 Referring to Figure 8.7, consider the following: $W = 325$ kN/m, $R = 8.8$ m, $\theta = 67°$, $x = 5.1$ m, $c_u = 22$ kPa and $\gamma = 17.5$ kN/m³. Determine the factor of safety for the finite slope.

8.13 A 6 m-high rigid retaining wall with a frictionless vertical back face retains a granular backfill having its top surface horizontal and coinciding with the water table. The properties of the backfill are $\gamma_{sat} = 18$ kN/m³, $c' = 0$ and $\phi' = 30°$. Determine the total active earth pressure and the resultant lateral force against the wall. Where is the resultant force located?

8.14 In a technical report, the values of total active earth pressure and total passive earth pressure are reported as 50.7 kN/m and 120.8 kN/m, respectively. Round off these values to the nearest whole number and provide proper justification.

8.15 A footing load test, conducted with a footing of size 1 m × 1 m at a depth of 1.5 m below the ground surface in a saturated clay deposit, provided an ultimate load value of 150 kN.

The water table is at the ground surface. The unconfined compression test on the clay specimen provides a strength of 30 kPa. If the saturated unit weight of the clay is 16.5 kN/m³, how much does the test value of the ultimate bearing capacity differ from the value obtained using the bearing capacity equation?

8.16 A 0.3 m × 0.3 m square concrete pile is driven into a soft clay deposit. The embedment depth of the pile is 10 m. The clay has an average unconfined compressive strength of 40 kPa. Estimate the ultimate pile capacity. Assume that the adhesion factor is equal to 1.

8.17 Explain the difference between active earth pressure and passive earth pressure in terms of their definitions, conditions and applications.

8.18 Do you expect a tension crack problem in a passive stress situation? Justify your answer.

8.19 What are the typical types and mechanisms of slope failures?

8.20 List the factors affecting the factor of safety for an infinite sandy slope.

8.21 A shallow foundation is subjected to increasing loads until failure. Discuss the possible modes of failure, highlighting their typical characteristics and key differences.

8.22 In Example 8.3, using Equations 8.27a–c, for $\phi = 0$, the values obtained are $N_c = 5.14$, $N_q = 1.0$ and $N = 0.0$. Can you verify whether you can derive these values yourself? Additionally, explain all the steps that were omitted while reporting these values.

8.23 Would you define the safe bearing capacity, q_s, by directly dividing the ultimate bearing capacity, q_u, by a factor of safety, F – that is, $q_s = q_u / F$? Justify your answer.

8.24 Explain the fundamental mechanism of load transfer in pile foundations, including the roles of skin friction and end bearing.

8.25 The ultimate capacity of a single pile in clay is 10 t. Determine the ultimate capacity of a group of five piles and provide justification for your calculation.

REFERENCES

Chowdhury R, Flentje P and Bhattacharya G (2010) *Geotechnical Slope Analysis*. CRC Press, Boca Raton, FL, USA.

Das BM (2013) *Fundamentals of Geotechnical Engineering*, 4th edn. Cengage Learning, Stamford, CT, USA.

Das BM and Shukla SK (2013) *Earth Anchors*, 2nd edn. J Ross, Plantation, FL, USA.

Hansen JB (1970) *A Revised and Enlarged Formula for Bearing Capacity*. Danish Geotechnical Institute, Copenhagen, Denmark, Bulletin No. 28.

Jaky J (1944) The coefficient of earth pressure at rest. *Journal of the Society of Hungarian Architects and Engineers* 7: 355–358.

Lambe TW and Whitman RV (1979) *Soil Mechanics, SI Version*. Wiley, New York, NY, USA.

Mayne PW and Kulhawy FH (1982) K_0-OCR relations in soil. *Journal of Geotechnical Engineering, ASCE* **108(6)**: 851–872.

Meyerhof GG (1951) The ultimate bearing capacity of foundations. *Geotechnique* **2(4)**: 301–331.

Meyerhof GG (1963) Some recent research on the bearing capacity of foundations. *Canadian Geotechnical Journal* **1(1)**: 16–26.

Shukla SK (1997) A study on causes of landslides in Arunachal Pradesh. *Proceedings of Indian Geotechnical Conference*, Vadodara, India, vol. 1, pp. 613–616.

Shukla SK (2013) Generalized analytical expression for dynamic passive earth pressure from c-ϕ soil backfills. *International Journal of Geotechnical Engineering* **7(4)**: 443–446.

Shukla SK (2015) Generalized analytical expression for dynamic active thrust from c-ϕ soil backfills. *International Journal of Geotechnical Engineering* **9(4)**: 416–421.

Shukla SK and Sivakugan N (2011) Site investigation and in situ tests. In *Geotechnical Engineering Handbook* (Das BM (ed.)). J Ross, Fort Lauderdale, FL, USA, pp. 10.1–10.78.

Taylor DW (1937) Stability of earth slopes. *Journal of the Boston Society of Civil Engineers* **24(3)**: 337–386.

Taylor DW (1948) *Fundamentals of Soil Mechanics*. Wiley, New York, NY, USA.

Terzaghi K (1943) *Theoretical Soil Mechanics*. Wiley, New York, NY, USA.

Terzaghi K, Peck RB and Mesri G (1996) *Soil Mechanics in Engineering Practice*, 3rd edn. Wiley, New York, NY, USA.

Vesic AS (1973) Analysis of ultimate loads of shallow foundations. *Journal of the Soil Mechanics and Foundation Division, ASCE* **99(SM1)**: 45–73.

Vesic AS (1975) Bearing capacity of shallow foundations. In *Foundation Engineering Handbook* (Winkerton HF and Fang HY (eds.)). Van Nostrand-Reinhold, New York, NY, USA.

FURTHER READING

Atkinson J (2007) *The mechanics of Soils and Foundations*, 2nd edn. Taylor and Francis, London, UK.

Bowles JE (1996) *Foundation Analysis and Design*, 5th edn. McGraw-Hill, New York, NY, USA.

Das BM (2022) *Principles of Geotechnical Engineering*, 10th edn. Cengage, Boston, MA, USA.

Fang HY (1991) *Foundation Engineering Handbook*. Van Nostrand-Reinhold, New York, NY, USA.

Shukla SK (2025) *ICE Core Concepts: Geotechnical Engineering*, 2nd edn. Emerald/ICE Publishing, London, UK.

emerald PUBLISHING   ice

Sanjay Kumar Shukla
ISBN 978-1-83608-519-5
https://doi.org/10.1108/978-1-83608-516-420252012
Emerald Publishing Limited: All rights reserved

Chapter 9
Core principles of special topics

Learning aims

This chapter explores the core concepts of the following topics

- expansive soil
- reinforced soil
- thermal properties of soil
- electrical properties of soil
- soil liquefaction.

9.1. Introduction

Chapters 1 to 8 have covered the core principles and concepts of soil mechanics, addressing key aspects essential for students, practising engineers and construction professionals to apply in practical contexts. Topics such as expansive soil, reinforced soil, thermal and electrical properties of soil, and soil liquefaction are classified as special topics. This chapter expands on the earlier discussions by exploring the core principles of these key special topics in soil mechanics, which are equally significant for academics, industry practitioners and researchers.

9.2. Expansive soil

The clay minerals most significant to engineers are montmorillonite, illite and kaolinite, as described in Section 1.6 of Chapter 1. Montmorillonite clay minerals swell significantly when exposed to water, whereas others either do not swell or exhibit limited swelling. Soils containing a significant proportion of montmorillonite minerals tend to expand in volume when they absorb water and contract when water is removed. These soils, commonly known as expansive or swelling soils, are widespread across most continents, including Asia, Australia, Europe and North America.

Swelling pressure is the pressure exerted by expansive soil when it is not allowed to swell or when its volume change is restricted. The volumetric strains of expansive soils, characterised by swelling during wet seasons and shrinkage during dry periods, can exceed 10%, and such significant volume strains can induce differential movements, leading to surface cracking and distress in infrastructure. Lightweight structures, such as highways and single-storey residential dwellings founded on expansive soils, are particularly vulnerable. Severe structural damage may occur due to subsoil swelling, which generates considerable swelling pressure. Fixing infrastructure damage caused by expansive soil costs a lot of money worldwide every year.

As the swelling pressure is to be associated with the design of structures against possible damage, the proper estimate of possible swell volume and associated swelling pressure assumes importance. The engineering behaviour of expansive soils is primarily controlled by their physico-chemical properties, particularly clay mineralogy and clay–water interaction. The degree of swelling expected in a soil mass containing montmorillonite largely depends on its quantity as well as the type and amount of exchangeable bases (positively charged ions referred to as cations) present in the soil. Thus, the swelling pressure is dependent on several factors, including the following (IS 2720 (Part 41)-1977 (Bureau of Indian Standards, 2021a))

- type and amount of clay in the soil and nature of the clay mineral
- initial water content and dry density
- nature of pore fluid
- stress history of the soil, including the confining pressure
- drying and wetting cycles to which the soil has been subjected.

Note that the dependence of swelling pressure on volume change makes its precise measurement challenging.

The methods for estimating the swelling potential of expansive soils can generally be classified into two main categories. The first involves measuring the physical properties of soils, such as Atterberg limits, free swell and potential volume change, typically through straightforward laboratory tests. The second focuses on assessing the mineralogical and chemical properties of soils, including identification of clay minerals, clay content, cation exchange capacity and specific surface area, using advanced techniques such as microscopic examination, X-ray diffraction and differential thermal analysis. Practising engineers may not always have access to advanced facilities and often depend on simplified tests to identify expansive soils. Moreover, their primary focus is on directly observing the effects of soil expansion under loading and moisture conditions that simulate actual structural environments. Free swell and differential free swell tests are two straightforward laboratory methods used to evaluate the swelling potential of soils. While closely related, these tests are distinct, each offering unique and valuable insights into the swelling characteristics of soils.

As defined in Section 2.4, the plasticity index, I_p, and the liquidity index, I_L, commonly used to characterise fine-grained soils, are used as indicators of expansive soils. In general, soils containing a higher proportion of expansive minerals, such as montmorillonite, tend to exhibit greater plasticity. Table 9.1 presents the swelling or expansion potential based on the value of I_P (Peck et al., 1974).

Activity of clay, as defined in Section 2.4, is also an indicator of the swelling potential of soils. Active clays (activity, $A > 1.25$) have the most potential for swelling. The activity values of Ca-montmorillonite and Na-montmorillonite are 1.5 and 7.2, respectively. Certainly, montmorillonites

Table 9.1 Relationship between plasticity index and swelling potential of soils

Plasticity index, I_p : %	Swelling potential
0–15	Low
15–25	Medium
25–35	High
>35	Very high

with divalent cations (e.g. Ca-montmorillonite) usually swell less than with monovalent cations (e.g. Na-montmorillonite).

Hence, it is useful to know the type of cation as well as the cation exchange capacity of montmorillonite in order to assess the swelling potential of soils.

The swelling potential can also be indicated based on shrinkage limit and linear shrinkage (see Section 2.4 for their definitions), as presented in Table 9.2 (Gromko, 1974).

In the free swell test, a specific amount of soil is allowed to swell freely in water, and the increase in volume is measured. This test gives an indication of the total swelling capacity or potential of the soil in an unrestricted environment. The test is performed by slowly pouring 10 cc of dry soil passing through 425 micron into a 100 cc graduated cylinder filled with water. The swelled volume of the soil is recorded after it has settled at the bottom of the cylinder, a process that typically may take approximately 24 hours. The free swell index (FSI), expressed as a percentage, is calculated using the following expression (Holtz and Gibbs, 1956)

$$\text{FSI} = \left(\frac{V_\text{f} - V_\text{i}}{V_\text{i}} \right) \times 100\% \tag{9.1}$$

where V_i is the initial volume of soil and V_f is the final volume of swelled soil.

A good grade, high-swelling commercial bentonite, primarily composed of montmorillonite, may have a free swell value ranging from 1200% to 2000%. Soils having free swell values as low as 100% may exhibit considerable volume change when wetted under light loadings and should be viewed with caution, whereas soils having free swell values below 50% seldom exhibit appreciable volume changes, even under very light loadings (Holtz and Gibbs, 1956).

The differential free swell test, on the other hand, measures the relative swelling of soil in water compared to a neutral or nonpolar liquid (e.g. kerosene) which does not induce swelling. Two soil samples are prepared: one is immersed in distilled water and the other in kerosene. The difference in volume changes between the two environments is calculated as the differential free swell index. This test helps isolate the swelling caused specifically by water absorption and is particularly useful for evaluating problematic swelling soils. IS 2720 (Part 40):1977 (Bureau of Indian Standards, 2021b) outlines the procedure for determining the value of differential free swell index (DFSI), as defined below.

$$\text{DFSI} = \left(\frac{V_\text{d} - V_\text{k}}{V_\text{k}} \right) \times 100\% \tag{9.2}$$

Table 9.2 Correlation of shrinkage limit with linear shrinkage and swelling potential of soils

Shrinkage limit, w_s: %	Linear shrinkage, L_s: %	Swelling potential
<10	>8	Critical
10–12	5–8	Marginal
>12	0–5	Non-critical

where V_d is the volume of soil in distilled water and V_k is the volume of soil in kerosene.

The swelling or expansion potential (also referred to as the degree of expansiveness) of soil, as classified based on the DFSI (IS 2911 (Part 3)-2021 (Bureau of Indian Standards, 2021c)), is presented in Table 9.3.

To realistically estimate volume changes during transitions from field conditions to air-dried or saturated states, undisturbed or remoulded soil samples, with water content representative of expected conditions at the time of construction, can be subjected to a swelling test in a soil laboratory consolidometer (similar to Figure 5.4(a)) under anticipated structural loads until equilibrium is reached. Unlike index tests for estimating potential expansiveness, consolidometer tests provide quantitative uplift values under expected load conditions. These tests involve saturating undisturbed or remoulded soil specimens while simulating structural loads. Load–expansion curves are developed for various scenarios, revealing that expansions observed in similar soil specimens, when saturated under different load conditions (from zero to maximum uplift) and then brought to a common load, often vary significantly. Hence, it is essential to closely replicate anticipated field conditions of saturation and loading to ensure accurate results. ASTM D4546-14 (ASTM, 2018) presents test methods for one-dimensional (1D) swell or settlement potential of cohesive soils in their undisturbed or remoulded specimens. These methods can be used to determine (a) the magnitude of swell or settlement under known vertical (axial) pressure or (b) the magnitude of vertical pressure needed to maintain no volume change of laterally constrained, axially loaded specimens. In the 1D swell and settlement test, the per cent heave/swell or settlement is defined as

$$\text{per cent heave or settlement} = \frac{\Delta h}{h} \times 100\% \tag{9.3}$$

where Δh represents the increase (heave) or decrease (settlement) in vertical height of an in situ soil column of height, h, due to absorption of water.

Note that the percentage of heave due to water absorption at the seating pressure is referred to as the free swell. The slope of the rebound pressure against void ratio curve on a semi-log plot is known as the swell index. The pressure that prevents the specimen from swelling or the pressure required to return the specimen to its original state (void ratio and height) after swelling is the swelling pressure.

Based on a relatively simple test for practical engineering applications, the swelling/expansion potential of compacted soils can be assessed in terms of expansion index (EI) (ASTM D4829-21 (ASTM, 2021)), defined as

Table 9.3 Swelling potential of soils based on differential free swell index (DFSI)

Differential free swell index, DFSI: %	Swelling potential
<20	Low
20–35	Moderate
35–50	High
>50	Very high

$$EI = \left(\frac{h_f - h_i}{h_i}\right) \times 1000 \tag{9.4}$$

where h_i and h_f are the initial and final heights of a compacted soil specimen in the consolidometer, subjected to a vertical confining pressure of 6.9 kPa and then inundated with distilled water.

The deformation of the specimen is recorded for 24 h or until the rate of deformation decreases to less than 0.005 mm/h, whichever occurs first. A minimum recording time of 3 h is required.

Note that EI is used to measure a basic index property of soil and, therefore, the EI is comparable to other indices such as the liquid limit, plastic limit and plasticity index of soils. The EI is not used to duplicate any particular field conditions, such as soil density, water content, loading, in-place soil structure or soil water chemistry. A potentially expansive soil is classified based on its EI, as shown in Table 9.4.

The following points regarding expansive soils are noteworthy.

- Although soil plasticity may indicate the presence of expansive minerals, it is not a definitive criterion for identifying an expansive soil.
- In the literature, swelling soils are a specific category of unsaturated soils characterised by their tendency to expand on wetting.
- ASTM D5890-19 (ASTM, 2019) covers an index method that enables the evaluation of swelling properties of a clay mineral in reagent water for estimation of its usefulness in geosynthetic clay liners (GCLs).
- The increase in soil volume on wetting is a time-dependent process. In many cases, the time required for soil to reach full swelling (ultimate swell/heave) in the field can range from a few days to several years. Expansive soils can cause substantial ground uplift, posing significant challenges to the stability and performance of structures built on them. The ultimate swell can be predicted by an observational method based on the procedure proposed by Asaoka for the analysis of consolidation data (Asaoka, 1978; Prakash et al., 2021).
- Engineers should consider the design heave, which is the amount of heave that will be experienced during the design life of the foundation. Design heave is calculated based on the change in the subsurface water content profile in the design active zone. Water migration modelling can be used to predict the final water content profile at the end of the design life of the foundation. Calculations of the design free-field heave in its natural, undisturbed state

Table 9.4 Swelling or expansion potential of compacted soils (ASTM, 2021)

Expansion index (EI)	Swelling potential
0–20	Very low
21–50	Low
51–90	Medium
91–130	High
>130	Very high

should also consider the degree of wetting in the design active zone. This is the amount of heave that the foundation must be designed to tolerate within its design life (Nelson *et al.*, 2015).

■ In the literature, there are several basic and advanced models which provide reliable predictions of swell properties of expansive soils, including estimation and measurement of swelling pressure (Puppala, 2021).

■ Richards *et al.* (1983) estimated that expansive soils cover about 20% of Australia, most of which occur in the semi-arid climatic zone.

■ In India, expansive soils are known as black cotton soils because of their dark colour and their suitability for growing cotton. The residual deposits of black cotton soil, the most commonly found type in India, typically have a thickness of less than 5 m. In contrast, transported deposits, which have also been reported, can be much thicker, reaching up to 9 m or more. These soils have a liquid limit ranging from 50% to 100%, a plasticity index varying between 20% and 65%, and a shrinkage limit of 9% to 15%. The clay-sized fraction typically ranges between 40% and 75%. In the dry state, the soil is very hard and exhibits high shear strength. However, it becomes soft with low shear strength on water ingress. During dry periods, shrinkage cracks, which can be up to 10 cm wide and extend to depths of 3 m or more, often form a distinctive hexagonal columnar structure. Figure 9.1 shows shrinkage cracks in an expansive soil at a project site in Gujarat, India.

Figure 9.1 Site view of expansive soil in Gujarat, India

Example 9.1

A soil was tested to determine its swelling potential and the following data were obtained

volume of soil in distilled water, $V_d = 63\,cc$
volume of soil in kerosene, $V_k = 50\,cc$

Calculate the differential free swell index and classify the swelling potential.

Solution

From Equation 9.2, the differential free swell index of soil

$$\text{DFSI} = \left(\frac{V_d - V_k}{V_k}\right) \times 100\% = \left(\frac{63 - 50}{50}\right) \times 100\% = 26.0\%$$

As DFSI = 26%, from Table 9.3, the soil falls in the range of moderate swelling potential.

9.3.   Reinforced soil

In modern construction practices across civil and environmental engineering, as well as in certain mining, agricultural and aquacultural engineering projects, various types and forms of reinforcement materials are introduced to enhance the engineering properties of soil and similar materials, such as strength, stiffness, permeability and compressibility. This process results in the creation of a relatively new material known as reinforced soil. Reinforced soil is a composite material in which tension-resisting elements are embedded within a soil mass that is inherently weak in tension.

The concept of soil reinforcement is not new as soil has been reinforced since prehistoric times. In nature, even some insects, birds and other animals use soil reinforcement in suitable forms to create structures to meet their requirements. However, the modern form of soil reinforcement was first developed by Vidal (1966, 1969) in the form of steel or metal strips, and the reinforced soil was patented under the name of reinforced earth (Schlosser and Long, 1974). The steel strips are available with smooth or ribbed surfaces. To avoid the problem of corrosion, the steel strips are often galvanised suitably to maintain their design life within the soil mass.

In addition to metal strips, geosynthetics are also widely used to strengthen and enhance soil properties due to their versatile functionality. Their primary functions are outlined below (Shukla, 2016, 2022; Shukla and Yin, 2006)

▨ reinforcement: enhancing soil strength and stiffness while minimising settlement or deformation
▨ separation: preventing the mixing of dissimilar soil materials, such as separating subgrade soil from the granular subbase course at the pavement subgrade level
▨ drainage: facilitating the removal of water from areas where water accumulation could cause problems, such as in pavements, backfills behind retaining walls, and basements
▨ filtration: allowing water to pass from one area to another while preventing the migration of base soil particles – for example, in coastal protection, a geotextile filter ensures the stability of revetments by retaining the base soil
▨ fluid barrier: acting as a liner or barrier to prevent the migration of liquids and gases from one area to another

■ protection: preventing damage to surfaces caused by overstressing or punctures from sharp edges of soil particles or other materials.

Note that geosynthetic is a generic name representing a broad range of mostly planar products, but also three-dimensional (3D) products, such as geotextiles, geogrids, geonets, geomembranes, geocells, geofoams and geocomposites, which are manufactured mainly from synthetic polymeric materials (e.g. polypropylene, polyester, polyethylene and polyamide) and used in contact with soil, rock and/or any civil engineering-related material as an integral part of a human-made project, structure or system.

Similar to geosynthetics, products made from natural fibres, such as jute, coir, cotton and wool, are also used in contact with soil, rock or other civil engineering materials, particularly in temporary applications such as erosion control and pavements. These products, known as geonaturals, have a shorter lifespan when used with earth materials due to their biodegradable nature. Consequently, they do not have as many field applications as geosynthetics have. Although geonaturals are significantly different from geosynthetics in material characteristics, they can be considered a complementary companion of geosynthetics, rather than a replacement, mainly because of some common field application areas. In fact, geonaturals are also polymeric materials since they contain a large proportion of naturally occurring polymers, such as lignin and cellulose.

In most applications, soil reinforcement is often used nowadays in the form of continuous geosynthetic or geonatural reinforcement inclusions (e.g. sheets, nets, meshes, grids, strips or bars) within the soil mass in a definite pattern, resulting in systematically reinforced soil. For more details on such systematically reinforced soils, known as geosynthetic-reinforced soils, which are studied under the subject of geosynthetic engineering, see Shukla (2016, 2022, 2025) and Koerner (2012). Note that geosynthetic or geonatural reinforcements as well as metal strips are normally oriented in a preferred direction (thus the reinforcement orientation is generally 1D) and are installed sequentially in alternating layers as per the design requirements of the specific application. Figure 9.2 illustrates an example of systematically reinforced soil. The geosynthetic-reinforced backfill helps reduce the lateral earth pressure significantly, thereby decreasing the thickness of the wall. With a geosynthetic-reinforced soil backfill, there is a possibility of completely avoiding the construction of a thick wall by providing a thin skin face to mainly prevent surface soil erosion. Woven geotextiles and geogrids are commonly used as reinforcement materials to improve ground infrastructure in a cost-effective, environmentally friendly, energy-efficient and sustainable manner. Applications

Figure 9.2 A retaining wall with a systematically reinforced soil backfill (after Shukla, 2017)

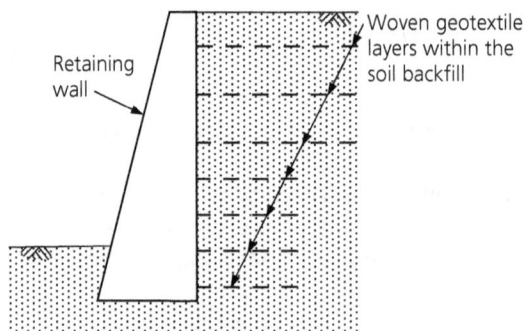

include constructing reinforced walls, stabilising backfills behind retaining walls, enhancing foundation soils and stabilising slopes.

In the past few decades, experimental and mathematical studies have been conducted to investigate the behaviour of soil reinforced randomly with different types of discrete, flexible, synthetic and natural fibres, called randomly distributed or oriented fibre-reinforced soils (RDFRS), or simply fibre-reinforced soils (FRS), with the intention of improving soil strength and other engineering characteristics (Shukla, 2017; Shukla *et al.*, 2009). Within fibre-reinforced soil, the fibre arrangement is basically 2D or 3D, depending on their placement or mixing. In field applications, the 1D arrangement of fibres in any specific direction within the soil mass is a difficult task, and so no effort is made to consider this 1D arrangement. In general, the reinforced soil is obtained by mixing the soil and fibres in such a way that the fibres are introduced into the soil mass randomly and homogeneously. The quality of mixing is very important to avoid any planes of weakness, such as those parallel to the oriented reinforcement, or even very small portions with insufficient fibre content. Figure 9.3 illustrates an example of randomly distributed fibre-reinforced soil. Fibre-reinforced backfill significantly reduces lateral earth pressure, allowing for a decrease in wall thickness, as the systematically reinforced soil backfill provides enhanced stability. The subject of fibres and their applications is known as fibre-reinforced soil engineering (Shukla, 2017).

Several laboratory tests (e.g. permeability test, 1D consolidation test, direct shear test, triaxial compression test and unconfined compression test) and some field tests (e.g. plate load tests) have been used to investigate the behaviour of geosynthetic-reinforced and fibre-reinforced soils. All these tests, which were originally developed to assess the characteristics of unreinforced or plain soils, are also conducted for reinforced soils in accordance with relevant standards (e.g. ASTM standards) with required modifications, as applicable, for the specific study or project. Based on the observations in the experimental studies, attempts have also been made by researchers to present analytical and numerical models to predict the behaviour of reinforced soil.

If the reinforced soil is considered a homogeneous material, but with anisotropic characteristics, the Mohr–Coulomb failure criterion can be applied to explain the basic mechanism of reinforced soil. Consider a simplified situation, shown in Figures 9.4(a) and 9.4(b), where two cylindrical specimens of a cohesionless soil are subjected to the same triaxial compression loading with the minor principal stress

Figure 9.3 A retaining wall with randomly distributed fibre-reinforced soil backfill (after Shukla, 2017)

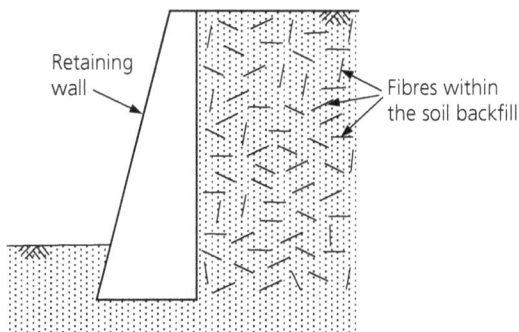

Figure 9.4 Basic soil reinforcing mechanism: (a) unreinforced cylindrical soil specimen; (b) fibre-reinforced cylindrical soil specimen; (c) magnified view of the fibre-reinforced soil element shown in (b) (after Shukla, 2017; Shukla et al., 2009)

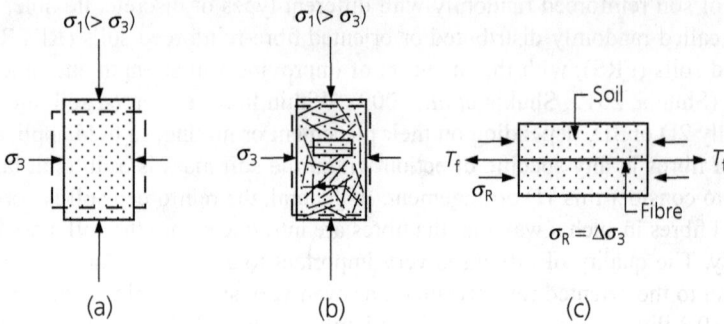

(confining stress), σ_3, and the major principal stress, σ_1. The first soil specimen is not reinforced, but the second one is reinforced with fibres. Figure 9.4(c) shows a magnified view of the reinforced soil element, as indicated in Figure 9.4(b), with a horizontal fibre. Because of skin friction and/or adhesion between the fibre and the soil, the fibre applies a confining stress, σ_R ($=\Delta\sigma_3$, increase in the minor principal stress), on the soil and, in this process, the fibre gets stretched with mobilisation of a tensile force, T_f, as shown in Figure 9.4(c). Note that the tensile force, T_f, and hence the confining stress, σ_R, will vary with orientation of the fibre within the soil mass.

Figure 9.5 shows the Mohr–Coulomb failure envelopes for cohesionless soil where l_U denotes the unreinforced case and l_{RC} denotes the reinforced case with inextensible reinforcement, such as steel strips.

The strength increase caused by reinforcement can be characterised by a constant cohesion intercept, c_R, as an apparent cohesion, introduced due to reinforcement (Schlosser and Vidal, 1969), as given below

$$c_R = \frac{\sigma_{RCmax} \tan(45° + \phi/2)}{2} = \frac{\sigma_{RCmax}\sqrt{K_p}}{2} = \frac{\sigma_{RCmax}}{2\sqrt{K_a}} \qquad (9.5)$$

where

$$K_a = \tan^2(45° - \phi/2) \qquad (9.6a)$$

and

$$K_p = \tan^2(45° + \phi/2) \qquad (9.6b)$$

where ϕ is the angle of internal friction of unreinforced cohesionless soil; σ_{RCmax} is the maximum value of lateral restraint σ_R, depending on the strength of the reinforcement, when failure takes place by rupture of the reinforcement; K_a and K_p are the Rankine's coefficients of active and passive lateral earth pressures, respectively, as defined in Section 8.2 (see Chapter 8).

Figure 9.5 Typical Mohr–Coulomb failure envelopes for cohesionless soil (adapted from Shukla, 2017; Shukla *et al.*, 2009)

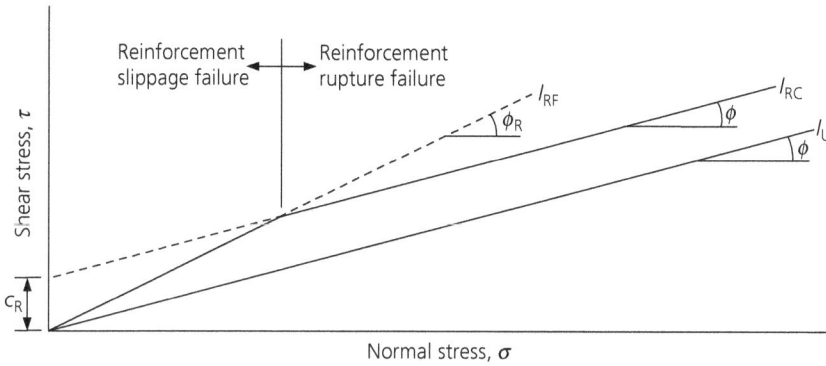

Note: l_U is the unreinforced case, and l_{RC} and l_{RF} are the reinforced cases.

Note that anisotropic cohesion is produced in the direction of reinforcement orientation; this concept is based on laboratory shear strength studies on reinforced soil specimens.

The results obtained from both triaxial compression tests and direct shear tests on sand specimens reinforced with tensile inclusions have shown that the apparent cohesion of reinforced soil is a function of the orientation of the inclusions with respect to the direction of the maximum extension in the soil. Thus, the strength envelope for a reinforced cohesionless soil for the reinforcement rupture condition can be interpreted in terms of the Mohr–Coulomb failure envelope, l_{RC}, for homogeneous cohesive soil, as shown in Figure 9.5.

The strength envelope for a reinforced cohesionless soil for the reinforcement slippage condition can be interpreted in terms of the Mohr–Coulomb failure envelope, l_{RF}, for homogeneous cohesionless soil, as shown in Figure 9.5. In this case, the strength increase can be characterised by an increased angle of internal friction, ϕ_R, given as

$$\sin \phi_R = \frac{1 + F - K_a}{1 - F + K_a} \tag{9.7}$$

where $F \left(= \sigma_{RF} / \sigma_{10} \right)$ is a friction factor that depends on the cohesionless soil–reinforcement interface characteristics; σ_{RF} is the maximum value of lateral restraint, σ_R, under a constant vertical stress, σ_{10}, at the soil–reinforcement interface, occurring when failure is caused by slippage between the reinforcement and the soil.

Note that the behaviour of soil reinforced with extensible reinforcements, such as geosynthetic reinforcements, does not fall entirely within the concepts described above. For details of reinforcing mechanisms and modes of failure of geosynthetic-reinforced soils, see Shukla (2016, 2022) and Shukla and Yin (2006).

Example 9.2

For sand strengthened with reinforcement, consider the following

angle of shearing resistance of unreinforced sand, $\phi = 30°$
friction factor, $F = 0.15$

Determine the angle of internal friction of the reinforced sand.

Solution

From Equation 9.6a, Rankine's active earth pressure coefficient

$$K_a = \tan^2(45° - \phi/2) = \tan^2(45° - 30°/2) = 0.333$$

From Equation 9.7

$$\sin \phi_R = \frac{1 + F - K_a}{1 - F + K_a} = \frac{1 + 0.15 - 0.333}{1 - 0.15 + 0.333} = 0.691$$

Therefore, angle of internal friction of the reinforced sand

$$\phi_R = \sin^{-1}(0.691) = 43.71°$$

The improvement in the engineering properties of soil due to the systematic inclusion of reinforcement depends on several parameters related to both soil and reinforcement characteristics. These include the type of reinforcement, the length or width of reinforcement, the spacing of reinforcement layers (if multiple layers are used) and the depth of the reinforced zone. Similarly, the enhancement of soil properties through the random inclusion of discrete fibres depends on numerous factors, such as soil and fibre characteristics, fibre concentration and orientation, the presence of admixtures (if any), the method of mixing, the type and degree of compaction, and the test and field conditions.

The improvement in specific soil properties or parameters due to the inclusion of reinforcement is often expressed as an improvement factor. Examples of improvement factors with their definitions are provided below.

Based on direct shear tests, the shear strength improvement factor, I_s, is defined as

$$I_s = \frac{\Delta\tau}{\tau_{fU}} = \frac{\tau_R - \tau_{fU}}{\tau_{fU}} = \frac{\tau_R}{\tau_{fU}} - 1 \tag{9.8}$$

where τ_{fU} is the shear strength of the unreinforced soil and τ_R is the shear stress on the reinforced soil specimen corresponding to the shear displacement $\delta(= \delta_{fU})$ of the unreinforced soil specimen at failure.

This factor, I_s, represents the relative gain in shear strength of soil due to the inclusion of reinforcement and is typically expressed as a percentage.

Note that in Equation 9.8, τ_R can be the shear stress on the reinforced soil specimen at any selected or permissible shear displacement, as required to ensure the safety of the specific structure under consideration. For $\tau_R = \tau_{fR}$ (shear strength of reinforced soil) at $\delta = \delta_{fR}$ (shear displacement of reinforced soil specimen at failure), Equation 9.8 reduces to the shear strength improvement factor at failure, I_{sf}, given as

$$I_{sf} = \frac{\Delta \tau_f}{\tau_{fU}} = \frac{\tau_{fR} - \tau_{fU}}{\tau_{fU}} = \frac{\tau_{fR}}{\tau_{fU}} - 1 \tag{9.9}$$

Based on triaxial compression tests, the deviator stress improvement factor, I_d, is defined as

$$I_d = \frac{\Delta(\sigma_1 - \sigma_3)}{(\sigma_1 - \sigma_3)_{fU}} = \frac{(\sigma_1 - \sigma_3)_R - (\sigma_1 - \sigma_3)_{fU}}{(\sigma_1 - \sigma_3)_{fU}} = \frac{(\sigma_1 - \sigma_3)_R}{(\sigma_1 - \sigma_3)_{fU}} - 1 \tag{9.10}$$

where $(\sigma_1 - \sigma_3)_{fU}$ is the failure deviator stress of the unreinforced soil specimen and $(\sigma_1 - \sigma_3)_R$ is the deviator stress of the reinforced soil specimen corresponding to the axial strain $\varepsilon(= \varepsilon_{fU})$ of the unreinforced soil specimen at failure.

This factor, I_d, represents the relative gain in deviator stress of the soil specimen due to inclusion of reinforcement and is normally expressed as a percentage.

Note that in Equation 9.10, $(\sigma_1 - \sigma_3)_R$ can be the deviator stress of the reinforced soil specimen at any selected or permissible axial strain as required to ensure the safety of the specific structure under consideration. For $(\sigma_1 - \sigma_3)_R = (\sigma_1 - \sigma_3)_{fR}$ at $\varepsilon = \varepsilon_{fR}$ (axial strain of reinforced soil specimen at failure), Equation 9.10 reduces to the deviator stress improvement factor at failure I_{df}, given as

$$I_{df} = \frac{\Delta(\sigma_1 - \sigma_3)_f}{(\sigma_1 - \sigma_3)_{fU}} = \frac{(\sigma_1 - \sigma_3)_{fR} - (\sigma_1 - \sigma_3)_{fU}}{(\sigma_1 - \sigma_3)_{fU}} = \frac{(\sigma_1 - \sigma_3)_{fR}}{(\sigma_1 - \sigma_3)_{fU}} - 1 \tag{9.11}$$

The improvement in shear strength of soil resulting from the inclusion of reinforcement can also be expressed in terms of deviator stress ratio (DSR) or deviator stress ratio at failure (DSR$_f$), as defined below.

$$DSR = \frac{(\sigma_1 - \sigma_3)_R}{(\sigma_1 - \sigma_3)_{fU}} \tag{9.12}$$

$$DSR_f = \frac{(\sigma_1 - \sigma_3)_{fR}}{(\sigma_1 - \sigma_3)_{fU}} \tag{9.13}$$

Based on unconfined compressive strength, the unconfined compressive strength improvement factor, I_{UCS}, is defined as

$$I_{UCS} = \frac{q_{UR} - q_{UU}}{q_{UU}} = \frac{q_{UR}}{q_{UU}} - 1 \tag{9.14}$$

where q_{UR} is the unconfined compressive strength (UCS) of reinforced soil and q_{UU} is the UCS of unreinforced soil.

Basically, the factor I_{UCS} represents the relative gain in UCS of soil due to the inclusion of reinforcement and is normally expressed as a percentage. Note that $I_{UCS} = I_{df}$ for $\sigma_3 = 0$ in Equation 9.11.

Note that the unconfined compressive strength improvement factor can be defined for the same selected or permissible axial strains for reinforced and unreinforced soils as the shear strength improvement factors defined earlier. Hence, the unconfined compressive strength improvement factor can also be expressed as

$$I_{UCS} = \frac{q_R - q_{UU}}{q_{UU}} = \frac{q_R}{q_{UU}} - 1 \tag{9.15}$$

where q_R is the unconfined compressive stress on the reinforced soil specimen corresponding to the axial strain $\varepsilon(= \varepsilon_{fU})$ of the unreinforced soil specimen at failure.

The factor I_{UCS} basically represents the relative gain in shear strength of soil due to the inclusion of reinforcement and is normally expressed as a percentage.

The effect of reinforcement on swelling behaviour of soil can be expressed as the swelling potential ratio (SPR), which is defined as a ratio of the swelling potential of the reinforced soil specimen $(\Delta H / H)_R$ to the swelling potential of the unreinforced soil specimen $(\Delta H / H)_U$, where ΔH is the change in the thickness of the soil specimen. Thus,

$$SPR = \frac{\left(\dfrac{\Delta H}{H}\right)_R}{\left(\dfrac{\Delta H}{H}\right)_U} \tag{9.16}$$

The load-bearing capacity and settlement characteristics of reinforced soil have been extensively studied by several researchers, mainly through model plate load tests. For the analysis and design of shallow foundations on reinforced soil beds, the effect of reinforcement on the load-bearing capacity of a footing resting on reinforced soil is typically represented by the bearing capacity ratio (BCR), a nondimensional parameter, defined as

$$BCR = \frac{q_R}{q_{uU}} \tag{9.17}$$

where q_{uU} is the ultimate load-bearing capacity of the unreinforced soil and q_R is the load-bearing pressure on the reinforced soil corresponding to the settlement $\rho(= \rho_{uU})$ of the footing resting on unreinforced soil at the ultimate load-bearing capacity, q_{uU}.

Figure 9.6 shows typical load–settlement curves for a soil with and without reinforcement and illustrates how q_{uU} and q_R are determined. The procedure for determining the ultimate load-bearing capacity, q_{uR}, of reinforced soil and the corresponding settlement, ρ_{uR}, is also explained in Figure 9.6.

Figure 9.6 Typical load–settlement curves for unreinforced and reinforced soils

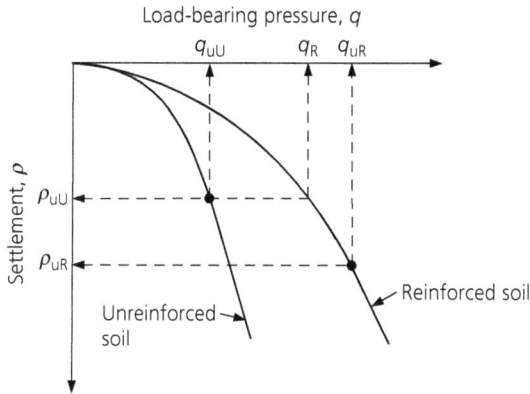

Note that in Equation 9.17, q_R can be the load-bearing pressure on the reinforced soil at any selected or permissible settlement of the footing as required to ensure the safety of the specific structure.

For $q_R = q_{uR}$ at $\rho = \rho_{uR}$, Equation 9.17 reduces to ultimate bearing capacity ratio as

$$\mathrm{BCR}_u = \frac{q_{uR}}{q_{uU}} \tag{9.18}$$

Hence, it is important to clarify the definition of BCR being used in discussion as the values obtained from Equations 9.17 and 9.18 can be quite different. In the author's opinion, it is better to use Equation 9.17, considering the same level of permissible settlement levels in both reinforced and unreinforced cases, to describe the benefits of reinforcement in terms of an increase in load-bearing capacity compared to the load-bearing capacity of unreinforced soil.

The improvement in the load-bearing capacity of soil resulting from the inclusion of reinforcement can also be expressed in terms of the ultimate bearing capacity improvement factor, I_{UBC}, defined as

$$I_{UBC} = \frac{\Delta q_u}{q_{uU}} = \frac{q_{uR} - q_{uU}}{q_{uU}} = \frac{q_{uR}}{q_{uU}} - 1 = \mathrm{BCR}_u - 1 \tag{9.19}$$

This factor, I_{UBC}, essentially represents a relative gain in load-bearing capacity of soil due to the inclusion of reinforcement and is normally expressed as a percentage.

Example 9.3

In a triaxial compression test conducted on unreinforced and reinforced soils, consider the following values for the deviator stress, $(\sigma_1 - \sigma_3)$, at failure.

For unreinforced soil, $(\sigma_1 - \sigma_3)_{fU} = 171 \, \mathrm{kPa}$

For fibre-reinforced, $(\sigma_1 - \sigma_3)_{fR} = 325 \, \mathrm{kPa}$

Determine the deviator stress improvement factor at failure. What does this indicate?

Solution

From Equation 9.11, the deviator stress improvement factor at failure

$$I_{df} = \frac{(\sigma_1 - \sigma_3)_{fR}}{(\sigma_1 - \sigma_3)_{fU}} - 1 = \frac{325}{171} - 1 \approx 0.91 \text{ or } 91\%$$

This value of I_{df} indicates that inclusion of reinforcement in the soil results in a 91% increase in the deviator stress at failure, significantly improving the strength of the soil.

The following points regarding the reinforced soils are noteworthy.

- Geosynthetics are generally described commercially by their polymer type, element type, manufacturing process, geosynthetic type, mass per unit area and/or thickness, as applicable.
- The design of a structure incorporating geosynthetics aims to ensure strength, stability and serviceability throughout its intended lifespan.
- Previously, soil was described as a three-phase system consisting of solid (mineral particles), liquid (water), and gas (air) (see Chapter 1). In reinforced soils, especially soils reinforced with fibres, however, solids exist in two different forms: soil solids and reinforcement solids. Therefore, reinforced soil may be considered a four-phase system, comprising mineral solid (soil particles), reinforcement solid (fibres), liquid (water) and gas (air). The phase relationships and inter-relationships can be addressed in a suitable manner, as applicable (Shukla, 2017).
- Soil–geosynthetic interaction through interface friction and/or interlocking characteristics is the key element in the performance of geosynthetic-reinforced soil structures. The direct shear test and the pullout/anchorage test are commonly used to evaluate soil–geosynthetic interaction.
- Incorporation of fibres within the soil tends to decrease its maximum dry unit weight and increase its optimum water content, mainly due to increased voids caused by fibre separation of the soil particles. The increased compactive effort, from standard to modified compaction, results in increased strength of fibre-reinforced soil and somewhat improved interfacial fibre– soil bond (Hoover *et al.*, 1982).
- Shear strength increases resulting from the inclusion of fibre reinforcement are approximately the same for loose and dense sands. However, larger strains are required to reach the peak shear resistance in the loose case. Relatively low modulus fibre reinforcements (natural and synthetic fibres) behave as 'ideally extensible' inclusions; they do not rupture during shear. Their main role is to increase the peak shear strength and to limit the magnitude of post-peak reduction in shear resistance in a dense sand (Gray and Ohashi, 1983).
- Failure surfaces in a triaxial compression test on randomly distributed fibre-reinforced sand are planar and oriented in the same manner as predicted by the Mohr–Coulomb failure criterion – that is, at the angle of obliquity, or $(45° + \phi/2)$, where ϕ is the angle of internal friction of sand. This finding suggests an isotropic reinforcing action with no development of preferred planes of weakness or strength (Maher and Gray, 1990).
- Inextensible reinforcements generally have rupture strains smaller than the maximum tensile strains in unreinforced soil, thereby resulting in catastrophic failures, while extensible reinforcements may have rupture strains larger than the maximum tensile strains in unreinforced soil; thus they seldom rupture.

9.4.   Thermal properties of soil

The thermal properties of soil and rock are important for designing and maintaining structures and systems in cold areas and places with thermal loads, such as energy piles, geothermal systems, underground transmission lines, oil and gas pipelines, radioactive waste disposal, geothermal components, nuclear facilities, hydro projects and solar thermal storage facilities. Key thermal properties (conductivity, specific heat capacity and diffusivity) govern heat transfer, soil–structure interactions and long-term performance under temperature variations. In nuclear projects, accurate thermal analysis ensures efficient heat dissipation, preventing overheating in reactors and waste repositories, while in hydro projects, it addresses seepage, frost action and structural integrity. Artificial ground freezing is a ground improvement technique that requires the determination of the amount of heat to be removed and the rate at which the frozen barrier is established. Thus, knowledge of thermal properties of soils enables sustainable designs, prevents thermally induced failures and enhances the efficiency and safety of geotechnical and geoenvironmental applications in addition to their applications in ground improvement, geology, geophysics and agriculture.

The fundamentals of thermodynamics are covered in most basic physics textbooks. Some essential terms and their definitions are provided here to facilitate a quick understanding, particularly in the context of learning about the thermal properties of soils discussed in this section. For comparison with other common materials, their specific property values are also provided.

Temperature, θ, is defined as a degree of hotness or coldness of a material. It determines the thermal state of a material whether it can give or receive heat. It is commonly measured in degrees Celsius (°C) but can also be expressed in kelvins (K), the SI unit of temperature.

Heat, Q, is a form of energy that transfers through systems or objects solely due to a temperature difference $(\theta_1 - \theta_2)$, as illustrated in Figure 9.7. It is measured in joules (J). Note that the term 'heat' is meaningful only when it is being transferred. Once this energy is transferred, it becomes the internal energy of the receiving body. Thus, expressions like 'heat in a body' or 'heat of a body' are meaningless. Two bodies are said to be in thermal equilibrium if, and only if, they have the same temperature, resulting in no heat transfer between them.

The problem of heat transfer in soils is very complicated. Heat flow in soils involves multiple mechanisms, including conduction, convection and radiation. Conduction transfers heat through direct molecular interaction, moving thermal energy from higher to lower temperatures without bulk material movement. Convection involves heat transfer by way of the bulk movement of fluids, driven naturally by density differences or externally by forces like fans. Radiation transfers energy as electromagnetic waves emitted by all bodies due to their temperature and can occur without a medium, even in a vacuum. Heat conduction through soil is analogous to water flow through soil and serves as the primary mechanism of heat transfer in soils.

Figure 9.7 Steady-state heat flow in a soil mass or any other material

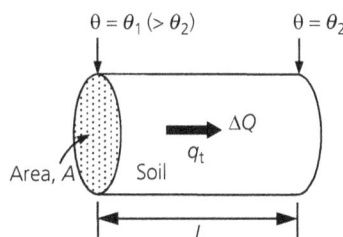

Because of the analogy between heat flow and water flow, concepts from fluid mechanics are often used to describe heat transfer. For instance, heat is equivalent to the volume of water in flow problems, while the temperature gradient, $\Delta\theta/L$, drives heat flow within a material just as the hydraulic gradient, $\Delta h/L$, drives water flow within a soil mass (see Chapter 4), where L is the flow path length.

Experiments show that if ΔQ amount of heat passes through any cross-section of a cylindrical soil mass or any other material in a time interval, Δt, as illustrated in Figure 9.7, the heat transfer rate (also referred to as the heat current or conduction rate) q_t $(= \Delta Q/\Delta t)$ is directly proportional to the area of cross-section, A, directly proportional to the temperature difference $(\theta_1 - \theta_2)$ with θ_1 and θ_2 as constant end temperatures, and inversely proportional to the heat flow path length, L. These proportions may be converted to an equation as

$$\frac{\Delta Q}{\Delta t} = k_t \left(\frac{\theta_1 - \theta_2}{L} \right) A = -k_t \left(\frac{\Delta\theta}{L} \right) A \tag{9.20}$$

or

$$q_t = k_t i_t A \tag{9.21}$$

or

$$v_t = k_t i_t \tag{9.22}$$

where $v_t \left(= q_t / A\right)$ is the rate of flow of heat; $i_t = \left((\theta_1 - \theta_2)/L = -\Delta\theta/L\right)$ is the thermal gradient over flow path length, L; k_t is a constant of proportionality, which represents the property of the material, called its thermal conductivity. The SI units for q_t, v_t, i_t and k_t are joules per second (Js^{-1}) or watts (W), joules per second per square metre ($Js^{-1}m^{-2}$ or Wm^{-2}), per kelvin per metre ($K^{-1}m^{-1}$) and joules per second per metre per kelvin ($Js^{-1}m^{-1}K^{-1}$ or $Wm^{-1}K^{-1}$), respectively. As a temperature change of 1 K and a temperature change of 1°C are the same, the unit of k_t may also be written as $Wm^{-1}°C^{-1}$.

From Equation 9.22, for $i_t = 1K^{-1}m^{-1}$, $v_t = k_t$. Hence, the thermal conductivity of a soil or any other material is defined as the heat energy passing in unit time through a unit cross-sectional area of the soil under a unit temperature gradient applied in the direction of this heat flow. Both Equation 9.21 and Equation 9.22 are known as Fourier's law of heat flow.

The thermal conductivity, k_t, is analogous to permeability (hydraulic conductivity), k, in fluid flow systems, and is an indication of the rate at which heat flows through soil or any other material under a given temperature gradient. Hence, Equation 9.22, known as Fourier's law, proposed in 1822 for heat flow, is analogous to Darcy's law (Equation 4.8, see Chapter 4) for fluid flow through soil. This equation may be used to compute the rate of flow of heat through soil mass having a uniform cross-section perpendicular to the direction of flow, provided the flow has attained steady-state conditions and the ends are kept at constant values of θ_1 and θ_2. Note that analogy helps simplify the understanding and modelling of heat transfer processes in soils and other materials.

The thermal conductivity of soil depends on many parameters, the most important of which include mineral composition, soil texture, density, moisture content, degree of saturation, organic content and temperature. Table 9.5 provides the values of thermal conductivity of several materials, including building and geomaterials, at about room temperature and pressure levels. For most materials, thermal conductivity increases slightly with increasing temperature, but the variation is small and can often be neglected.

Table 9.5 Typical values of thermal conductivity for materials

Materials	Thermal conductivity, k_t: Wm^{-1}K^{-1}
Common solids	
Stainless steel	14
Iron	67
Brass	109
Aluminium	235
Copper	401
Silver	428
Ice with variable density from 917 to 920 kg/m³	2.0–2.6
Snow (100 kg/m³)	0.06
Snow (500 kg/m³)	0.59
Common fluids	
Oxygen	0.023
Air (dry)	0.026
Hydrogen	0.18
Water with variable density from 960 to 1000 kg/m³	0.5–0.8
Building and geomaterials	
Polyurethane foam	0.024
Polystyrene	0.03–0.06
Cork	0.04
Rock wool	0.043
Fibre glass	0.048
Rocks with variable porosity, moisture and composition	0.1–6.0
Wood (pine)	0.116
Wood (oak)	0.146
Insulating brick	0.15
Soils with variable porosity, moisture and composition	0.2–4.7
Celotex (sugarcane fibre or bagasse)	0.50
Red brick	0.6
Concrete	0.8
Window glass	1.0
Shale	1.56
Permafrost with variable ice content	1.5–2.5
Granite	2.76

Note: based mainly on Briaud (2013), Mitchell and Soga (2005), Sears *et al.* (1982), Verma (2000) and Walker (2008).

Equation 9.21 shows that the larger the value of k_t, the larger the heat current, other factors being equal. A material for which k_t is large is, therefore, a good heat conductor, while if k_t is small, the material is a poor conductor or a good insulator. Note that $k_t \to \infty$ for a perfect heat conductor, while $k_t \to 0$ for a perfect insulator. Metals have much greater thermal conductivity than non-metals, while the thermal conductivity of gases is extremely small. In ordinary circumstances, a perfect heat conductor or a perfect insulator does not exist.

Example 9.4

Referring to Figure 9.7, consider the following

$$A = 2.5\,\text{m}^2, L = 1.5\,\text{m}, \theta_1 = 20\,°\text{C}, \theta_2 = -10\,°\text{C} \text{ and } k_t = 0.5\,\text{Wm}^{-1}\text{C}^{-1}$$

Determine conduction rate, q_t, and the rate of flow of heat, v_t, through the soil mass, assuming no heat transfer occurs through the cylindrical surfaces.

Solution

From Equation 9.21, the conduction rate through the soil mass

$$q_t = k_t i_t A = k_t \left(\frac{\theta_1 - \theta_2}{L} \right) A = (0.5) \left[\frac{20 - (-10)}{1.5} \right] (2.5) = 25\,\text{W}$$

From Equation 9.22, the rate of flow of heat through the soil mass

$$v_t = k_t i_t = k_t \left(\frac{\theta_1 - \theta_2}{L} \right) = (0.5) \left[\frac{20 - (-10)}{1.5} \right] = 10\,\text{Wm}^{-2}$$

For insulation applications, the concept of thermal resistance, R, defined as $R \left(= L / (k_t A) \right)$, has been introduced into engineering practice. With this definition, Equation 9.21 becomes $\theta_1 - \theta_2 = q_t R$, which is analogous to Ohm's law, as discussed in the following section of this chapter. The SI unit of R is KW^{-1}.

Heat can flow through stratified soil deposits or any other layered system, such as an asphalt concrete pavement over soil in the heat of the summer or a layer of snow covering the soil surface in the winter. Similar to water flow parallel to the bedding planes in a stratified or layered soil deposit, as shown in Figure 4.7 (see Chapter 4), if heat flows parallel to bedding planes of a stratified or layered soil deposit that consists of n strata of thicknesses H_1, H_2, ..., H_n, with respective thermal conductivities k_{t1}, k_{t2}, ..., k_{tn}, then the thermal gradient (i_t) remains the same in all the layers; hence, the equivalent thermal conductivity (k_{tII}) can be expressed as

$$k_{tII} = \frac{\sum_{j=1}^{n} k_j H_j}{H} \tag{9.23}$$

For heat flow perpendicular to the bedding planes, the heat transfer rate (q_t) remains the same in all the layers; hence, the equivalent thermal conductivity $(k_{t\perp})$ can be expressed as

$$k_{t\perp} = \frac{H}{\sum_{j=1}^{n} \frac{H_j}{k_{tj}}} \tag{9.24}$$

In Equations 9.23 and 9.24

$$H = \sum_{j=1}^{n} H_j \tag{9.25}$$

In addition to thermal conductivity, specific heat capacity and thermal diffusivity are two other important thermal properties of interest to geotechnical engineers.

If ΔQ represents the heat supplied to a soil or any other material having a mass, m, causing a temperature increase of Δ, then

$$\Delta Q = mc\Delta\theta \tag{9.26}$$

where c is a constant, called the specific heat capacity of the material under the given surrounding conditions. Its SI unit is $Jkg^{-1}K^{-1}$. The quantity, mc, is called the heat capacity of the material and its SI unit is JK^{-1}. Typical values of c for several materials are given in Table 9.6.

Fourier's law is the constitutive law for heat flow as Darcy's law is for fluid flow through soil. Similar to Terzaghi's 1D consolidation equation (Equation 5.28, see Chapter 5), a differential equation is applicable to the transient heat flow problem, as given below.

$$\alpha \frac{\partial^2 \theta}{\partial z^2} = \frac{\partial \theta}{\partial t} \tag{9.27}$$

with

$$\alpha = \frac{k_t}{\rho c} \tag{9.28}$$

where ρ is the density of material and α is a material property, called thermal diffusivity, that describes how quickly heat is conducted through a material relative to how much heat is stored in it.

Thermal diffusivity is basically a measure of how efficiently a material can transfer thermal energy. The SI unit of thermal diffusivity is m^2/s. It is used in analysing heat transfer in building materials, soils and rocks, and insulation. A solution to partial differential Equation 9.27 uses the Fourier series expansion method for a given set of initial and boundary conditions as applicable for the heat flow problem concerned.

Typical ranges of thermal diffusivity for air, water, ice, soil and rock are 13–30, 0.13–0.17, 1.24–1.52, 0.1–4.2 and 1.1–3.0 mm^2/s, respectively (Briaud, 2013).

Materials with high thermal diffusivity (e.g. metals) can quickly adjust their temperature to match changes in their surroundings because they conduct heat efficiently and store less heat, while materials with low thermal diffusivity (e.g. soils, wood) conduct heat more slowly and are better insulators because they store more heat relative to their ability to conduct it.

Table 9.6 Typical values of specific heat capacity for some materials

Material	Specific heat capacity, c: $Jkg^{-1}K^{-1}$
Common solids	
Silver	236
Brass	380
Copper	386
Iron	470
Aluminium	900
Snow (100 kg/m³)	210
Snow (500 kg/m³)	1050
Ice (−10°C)	2220
Common fluids	
Steam	1926
Seawater	3900
Water	4180
Building and geomaterials	
Clay (unfrozen)	750–920
Clay (frozen)	650–800
Sand (unfrozen)	630–1460
Sand (frozen)	500–1200
Granite	790
Glass	840
Rock	800–2200

Note: based mainly on Briaud (2013), Mitchell and Soga (2005), Verma (2000) and Walker (2008).

Note that high thermal conductivity indicates that heat travels easily through the material; a high specific heat capacity means a large amount of heat is required to raise the temperature of the material; a high diffusivity implies the temperature will rise quickly within the material. These thermal properties depend on several factors, including temperature, pressure, moisture content and density.

It is also important to note that the temperature on the Earth's surface varies from about −80°C to 55°C, depending on the location and conditions. The temperature inside the Earth typically ranges from the surface temperatures to approximately 5500°C at its centre. Rocks and soil particles melt at temperatures ranging between 600°C and 1200°C, depending on their composition. Below the Earth's surface, the temperature increases downwards at an average rate of 30°C/km. This temperature gradient is higher near a source of heat, such as an active volcanic centre, and is also affected by the thermal conductivity of the rocks at a particular locality.

The measurement of soil thermal properties in the laboratory and field can be conducted using suitable methods, such as steady-state or transient tests, as described by Farouki (1981). In general, to measure thermal conductivity, it is necessary to set up a temperature gradient across the soil specimen being tested, and this may induce appreciable moisture migration in unsaturated soils. While some moisture migration may also occur in situ, it is likely to be less than in the laboratory because of the generally smaller temperature gradients existing in situ. There may be a critical temperature gradient below which excessive migration does not occur.

The most important steady state method for measuring the thermal conductivity of soils is the guarded hot plate (GHP) test. With transient methods the temperature of the soil varies with time. Such methods are more versatile than steady-state methods and can be more easily performed. Compared to the GHP test, the probe method, in particular, requires much less time. ASTM D5334-22 (ASTM, 2022a) presents a procedure for determining the thermal conductivity (λ) of soil and rock using the transient heat method. This test method is applicable for both intact specimens of soil, rock and reconstituted soil specimens, and is effective in both the laboratory and field. The test is applicable to dry, unsaturated or saturated materials that can sustain a hole for the sensor. It is valid over temperatures ranging from <0 to $>100°C$, depending on the suitability of the thermal needle probe to temperature extremes. However, care must be taken to prevent significant errors from (a) redistribution of water due to thermal gradients resulting from heating of the needle probe; (b) redistribution of water due to hydraulic gradients (gravity drainage for high degrees of saturation or surface evaporation); and (c) phase change of water in specimens with temperatures near 0°C or 100°C.

ASTM E2585-09 (ASTM, 2022b) presents the flash method for measuring the values of thermal diffusivity of a wide range of primarily homogeneous isotropic solid materials. It is particularly advantageous because of simple specimen geometry, small specimen size requirements, rapidity of measurement and ease of handling. Thermal diffusivity values ranging from 10^{-7} to $10^{-3}\,m^2/s$ are readily measurable by this test from about 75 K to 2800 K.

Note that various methods have also been developed to calculate the thermal conductivity of soils. These methods are either purely empirical or partially based on theoretical principles. Complete details can be found in Farouki (1981).

The following points regarding the fundamentals of heat and temperature, and thermal properties of soils are noteworthy.

- The Kelvin scale begins at absolute zero, equivalent to $-273°C$, which represents the theoretical point where all molecular motion ceases. This is the lower limit of the temperature scale, as no temperature can exist below absolute zero. In contrast, there is no known upper limit to the temperature scale.
- For water, the phase transition from solid to liquid occurs at 0°C, and the transition from liquid to gas occurs at 100°C, both under standard atmospheric pressure (1 atmosphere). However, these temperatures vary with changes in pressure. The heat required for a phase transition is known as the latent heat, typically expressed as $Q = mL$, where Q is the total heat required, m is the mass of the substance and L is the latent heat of the material. The SI unit of L is J/kg. The freezing and thawing processes of soils are slowed down by the effects of latent heat.

- In general, solids are better thermal conductors than liquids, while liquids conduct heat more effectively than gases. Metals, in particular, are far superior conductors compared to non-metals due to the abundance of free electrons in metals. These free electrons can move freely, efficiently transferring thermal energy.

- A dry soil will have a thermal conductivity lower than the same soil once saturated, because air has a lower thermal conductivity than water. Also, sand in a very dense state will have a higher thermal conductivity than the same sand in a very loose state, because soil particles have a higher thermal conductivity than air or water.

- Geotechnical engineers often work on deep offshore platforms for oil retrieval, with water depths reaching several kilometres. At the ocean floor, temperatures are only a few degrees Celsius, while oil reservoirs at depths of up to 15 km can reach temperatures of 100°C when extracted. Despite these extremes, soil particles and rocks remain solid. However, at Earth's surface temperatures, water exists as either liquid or solid (frozen), and the thermal properties of soil vary depending on whether it is frozen or unfrozen, influencing geotechnical considerations in design and analysis (Briaud, 2013).

- Previously, soil was described as a three-phase system consisting of solid (mineral particles), liquid (water) and gas (air) (see Chapter 1). In frozen soil, however, water can exist in both frozen and unfrozen states. Therefore, frozen soil may be considered a four-phase system, comprising mineral solid (particles), ice solid (frozen water acting as a cementing material), liquid (unfrozen water) and gas (air). When discussing water content, it is important to distinguish between unfrozen water content and frozen water (or ice) content. All other phase relationships and inter-relationships should be addressed in a similar manner, as applicable. For instance, parameters such as the degree of saturation, void ratio and porosity can be defined either by including or excluding the ice, depending on the context.

- The unit weight of ice is about 10% less than the unit weight of water. The temperature at which the water in the voids will freeze depends on many factors, including the salt content. The higher the salt content, the lower the temperature required for the water to freeze. Generally, freezing starts at around −1°C, and at −20°C, most soils are completely frozen. Almost all frozen soils retain some amount of liquid water. The thermal properties of frozen soil result from the combined properties of water, ice, air and the soil skeleton (Briaud, 2013).

- Frozen soil exhibits higher thermal conductivity, lower specific heat and higher thermal diffusivity compared to unfrozen soil. In other words, heat flows more quickly through frozen soil, making it easier to extract heat from it.

- Frost susceptibility is the tendency of soil to retain water and undergo frost heave, requiring both capillary action and hydraulic conductivity. It is lowest in clean gravels, sands and high-plasticity clays but highest in silts, which balance capillary potential and water flow. Gravels and sands allow easy water movement but lack capillary action, while clays have strong capillary action but low hydraulic conductivity, making water movement difficult. Silts achieve a balance of these properties, making them highly frost susceptible and often causing structural damage and instability in freezing conditions. Based on the core principles of thermal properties of soils as presented earlier, the depth of frost penetration within a soil mass subjected to freezing can be estimated.

- A soil may have a much greater thermal diffusivity when frozen than when unfrozen because of two factors: the higher thermal conductivity of the frozen soil and the lower specific heat of the ice as compared with liquid water. In frozen soils, temperatures can, therefore, change much more rapidly and to a greater extent than in unfrozen soils. The depth of heat penetration into the soil and the amplitude of daily and seasonal temperature variations are influenced by the thermal properties of soil. Diurnal temperature penetration typically ranges from 0.3 to 0.8 m, while the annual temperature wave can penetrate to a depth of approximately 10 m (Farouki, 1981).

▨ The thermal conductivity of a soil may depend on the direction of heat flow. This may be related to structural effects, such as those in mica, giving rise to a thermal conductivity along the planes of cleavage about ten times as great as that across these planes (Goldsmid and Bowley, 1960). Other clay minerals may show similar differences.

▨ When the ground temperature rises above the prevailing freezing point, the soil begins to thaw. During this process, the strength of the soil decreases.

9.5.   Electrical properties of soil

Understanding electrical resistivity is important for ground infrastructure development and assessment because it helps assess soil properties like water content, salinity and compaction. It is widely used in geotechnical investigations to identify suitable ground conditions for construction, to determine depth to bedrock and anomalies in the stratigraphic profile, to detect underground water and to locate weak zones. Resistivity measurements are required to design grounding systems and corrosion protection for buried structures like pipelines and cables. The subsurface distribution of electrical resistivity of earth materials as determined from field measurements by the electrical resistivity method on the ground surface ensures safe and cost-effective planning and maintenance of infrastructure projects, especially in challenging ground conditions.

Electric charge, Q, is an intrinsic property of the fundamental particles, such as electrons and protons, of matter. It can be positive or negative; an electron has a negative charge while a proton has a positive charge. Its SI unit is coulomb (C). Note that electric charge is neither generated nor created; it can only be transferred from one body to another.

The electric potential (or simply potential), V, at a point in an electric field is the potential energy per unit charge at that point. The electric potential difference between any two points i and f ($\Delta V = V_f - V_i$), often called the voltage, in an electric field (E) is equal to the difference in the potential energy per unit charge between these two points. The SI unit of potential is volt (V). Note that 1 volt (V) = 1 joule per coulomb (JC^{-1}).

Experiments show that if a net electric charge, ΔQ, passes through the cross-section of a cylindrical soil mass or any material in a time interval, Δt, as illustrated in Figure 9.8, the rate of flow of electric charge, called the electric current I ($= \Delta Q / \Delta t$), is directly proportional to the area of cross-section, A, directly proportional to the voltage ($V_1 - V_2$) with V_1 and V_2 as constant end electric potentials, and inversely proportional to the charge flow path length, L. These proportions may be converted to an equation as

$$\frac{\Delta Q}{\Delta t} = \sigma_e \left(\frac{V_1 - V_2}{L} \right) A = -\sigma_e \left(\frac{\Delta V}{L} \right) A \tag{9.29}$$

or

$$I = \sigma_e i_e A \tag{9.30}$$

or

$$J = \sigma_e i_e \tag{9.31}$$

Figure 9.8 Steady-state electric charge flow in a soil mass or any other material

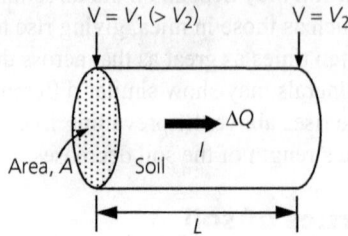

Figure 9.8 Steady-state electric charge flow in a soil mass or any other material

where $J (= I/A)$ is the current density; $i_e = ((V_1 - V_2)/L = -\Delta V/L)$ is the electric potential gradient over flow path length, L, referred to as the electric field (E); σ_e is a constant of proportionality, which represents the property of the material, called its electrical conductivity.

Both Equation 9.30 and Equation 9.31 are known as Ohm's law of electrical flow.

From Equation 9.29

$$\Delta V = \left(\frac{1}{\sigma_e}\right)\left(\frac{L}{A}\right)I$$
$$= \rho\left(\frac{L}{A}\right)I$$
(9.32)

or

$$\Delta V = RI$$
(9.33)

where

$$\rho = \frac{1}{\sigma_e}$$
(9.34)

and

$$R = \rho\frac{L}{A}$$
(9.35)

The parameters R and ρ are referred to as the electrical resistance and resistivity of the material, respectively. The SI units of R, ρ, i_e and σ_e are ohms (Ω), ohm-metres (Ωm), volts per metre (Vm^{-1}) and (Ω^{-1}m^{-1}) or siemens per metre (Sm^{-1}), respectively. Equation 9.33 is another form of Ohm's law, which is widely used in electrical circuit analysis.

From Equation 9.31, for $i_e = 1\,\text{Vm}^{-1}$, $J = k_t$. Hence, the electrical conductivity of a soil or any other material is defined as the electric charge passing per unit time through a unit cross-sectional area of the soil under a unit potential gradient applied in the direction of the charge flow.

Each soil or rock has its own resistivity depending on water content, compaction and composition. Certain minerals, such as native metals and graphite, conduct electricity through the passage of electrons. Most rock-forming minerals are, however, insulators, and electric current is carried through soil or rock mass mainly by the passage of ions in the pore water. Thus, most soils and rocks conduct electricity primarily through electrolytic rather than electronic processes. Porosity is a major factor influencing the resistivity of soils and rocks, with resistivity generally increasing as porosity decreases. However, even crystalline rocks with negligible intergranular porosity are conductive along cracks and fissures. The range of resistivity among earth materials is enormous, extending from 10^{-5} to $10^{15}\,\Omega$m – for example, resistivity is low for saturated clays and high for loose, dry gravel or solid rock, as seen in Table 9.7. Since there is considerable overlap in resistivity values between different earth materials, identification of a soil or rock is not possible solely based on resistivity data. Strictly, Equation 9.35 refers to electronic conduction, but it may still be used to describe the effective resistivity of soils and rocks – that is, their resistivities with resistivity of pore water if present. Archie (1942) proposed an empirical expression for effective resistivity as

$$\rho = an^{-b}S^{-c}\rho_w \tag{9.36}$$

Table 9.7 Typical values of electrical resistivity of materials

Material	Electrical resistivity, ρ: Ωm
Common solids	
Iron	9.71×10^{-8}
Aluminium	2.63×10^{-8}
Gold	2.35×10^{-8}
Copper	1.72×10^{-8}
Silver	1.47×10^{-8}
Common fluids	
Water	20–60
Landfill leachate	0.5–10
Sea water	0.18–0.24
Building and geomaterials	
Marble	10^{12}
Quartz	10^{10}
Glass	10^{10}–10^{14}
Wood	10^{8}–10^{11}
Rock salt	10^{6}–10^{7}
Granite	5000–10^{6}
Sandstone	35–4000
Limestone	120–400
Saturated sandy soils	15–300
Saturated clays and slits	1–120

Note: based mainly on Sears et al. (1982), Shukla and Sivakugan (2011) and Verma (2000).

where n is the porosity; S is the degree of saturation; ρ_w is the resistivity of water or ionic solution in the pores; a, b and c are empirical constants, known as tortuosity factor, cementation exponent and saturation exponent, respectively.

The tortuosity factor a depends on pore geometry, while the cementation exponent b depends on the degree of consolidation of the rock. In the case of sands, the empirical parameters a and b range from 0.35 to 4.78 and from 1.14 to 2.52, respectively (Bassiouni, 1994). The value ρ_w can vary considerably according to the quantity and conductivity of dissolved materials.

Equation 9.36, commonly known as Archie's equation, can be presented as

$$\rho_0 = an^{-b}\rho_w = F\rho_w \tag{9.37}$$

where ρ_0 is resistivity of saturated porous material and

$$F = an^{-b} \tag{9.38}$$

is the formation resistivity factor, or simply formation factor of the porous material.

Normally, one would expect a fairly uniform increase of resistivity with geologic age because of the greater compaction associated with increasing thickness of overburden. There is no consistent difference between the range of resistivity of igneous and sedimentary rocks, although metamorphic rocks appear to have higher resistivity, statistically, than either of the other rocks (Dobrin, 1976).

Example 9.5

Consider the following parameters for a sandy formation

   tortuosity factor, $a = 0.45$
   cementation parameter, $b = 1.1$
   formation factor, $F = 2$.

What is the porosity of the sandy formation?

Solution

From Equation 9.38, the porosity of the rock

$$n = \left(\frac{a}{F}\right)^{1/b} = \left(\frac{0.45}{2}\right)^{1/1.1} = 0.258 \text{ or } 25.8\%$$

In the electrical resistivity method for determining resistivity at a project site, the test involves injecting direct currents or low-frequency alternating currents into the ground and measuring the resulting potential differences at the surface. For this purpose, four metal spikes are driven into the ground at the surface along a straight line, generally at equal distances; one pair serves as current electrodes and the other pair as potential electrodes, as illustrated in Figure 9.9. The resistivity can be estimated using the following (Kearey et al., 2002).

$$\rho = \frac{2\pi V}{I\left[\left(\dfrac{1}{r_1} - \dfrac{1}{r_2}\right) - \left(\dfrac{1}{R_1} - \dfrac{1}{R_2}\right)\right]} \tag{9.39}$$

where V is the potential difference between electrodes P_1 and P_2; r_1 and r_2 are the distances from potential electrode P_1 to current electrodes C_1 and C_2, respectively; R_1 and R_2 are the distances from potential electrode P_2 to current electrodes C_1 and C_2, respectively.

When the ground is uniform, the resistivity calculated using Equation 9.39 should remain constant, regardless of electrode spacing or surface location. However, if subsurface inhomogeneity exists, resistivity will vary depending on the relative positions of the electrodes. Any computed value is then known as the apparent resistivity, ρ_a, and will be a function of the form of inhomogeneity. Equation 9.39 is thus the basic equation for calculating the apparent resistivity for any electrode configuration. The current electrode separation must be selected to ensure the ground is energised to the required depth and should be at least equal to that depth. This places practical limits on the depths of penetration attainable by normal resistivity methods due to the difficulty in laying long lengths of cable and the generation of sufficient power. Depths of penetration of about 1 km are the limit for normal equipment.

There are several electrode configurations, but the Wenner configuration is the simplest, as the current and potential electrodes are maintained at an equal spacing, l (Figure 9.10). Substituting this condition (i.e. $r_1 = l$, $r_2 = 2l$, $R_1 = 2l$ and $R_2 = l$) with $\rho = \rho_a$ into Equation 9.39 yields

$$\rho_a = 2\pi\, l\frac{V}{I} \tag{9.40}$$

In the study of horizontal or near-horizontal overburden soil–bedrock interfaces, the spacing, a, is gradually increased about a fixed central point. Consequently, readings are taken as the current reaches progressively greater depths. This technique, known as vertical electrical sounding, also referred to as electrical drilling or expanding probe, is extensively used to determine overburden thickness and to define horizontal zones of porous media. To study the lateral variation of resistivity, the current and potential electrodes are maintained at a fixed separation and progressively moved along a profile. This technique, known as constant separation traversing, also referred to

Figure 9.9 Generalised form of the electrode configuration used in the electrical resistivity method

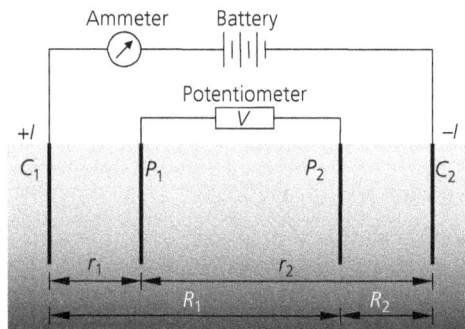

Note: C_1 and C_2 are current electrodes; P_1 and P_2 are potential electrodes.

Figure 9.10 The Wenner electrode configuration used in the electrical resistivity method

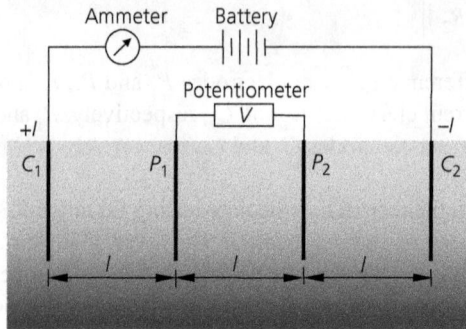

Note: C_1 and C_2 are current electrodes; P_1 and P_2 are potential electrodes.

as electrical profiling, is used to determine variations in bedrock depth and the presence of steep discontinuities.

ASTM G57-06 (ASTM, 2020) provides detailed guidelines on the equipment and procedures for measuring soil resistivity in the field using the Wenner four-electrode method, applicable both in situ and for samples removed from the ground, to control the corrosion of buried structures. ASTM G187-23 (ASTM, 2023) outlines the two-electrode soil box method, which is more accurate and faster than the four-pin soil box method and often complements the in situ four-pin soil resistivity method.

In the laboratory, electrical resistivity tests on soils can be conducted in accordance with AS 1289.4.4.1-1997 (Standards Australia, 1997). Pandey and Shukla (2018) fabricated a soil box from a 10 mm-thick Perspex sheet, with internal dimensions of 200 mm length, 40 mm width and 30 mm depth, as shown in Figure 9.11 along with the test arrangement. The box was fitted with two 10 mm-thick brass plate electrodes, C1 and C2, of the same cross-sectional area as the box, and two brass potential measuring pins, P1 and P2. The diameter of the pins was 3 mm and the distance between their axes was 120 mm. As the connections are shown, the four-terminal ground resistance testing machine was used to measure the electrical resistivity of the soil mixture. This test set-up was used to study the effect of the state of compaction on the electrical resistivity of sand, bentonite and their mixtures. Figure 9.12 compares the electrical resistivities of sand, bentonite and their mixtures at different water contents. It can be observed that the effect of bentonite addition on the electrical resistivity of the sand–bentonite mixture is most pronounced at 20%. Negligible impact on the resistivity is observed at bentonite contents greater than 20%.

The following points regarding the electrical properties of soil are noteworthy.

- The typical ranges of flow parameters for saturated fine-grained soils with porosity varying from 10% to 70% are (Mitchell and Soga, 2005)

  hydraulic conductivity, k: 1×10^{-11} to 1×10^{-6} ms^{-1}
  thermal conductivity, σ_t: 0.25 to 2.5 Wm^{-1}K^{-1}
  electrical conductivity, σ_e: 0.01 to 1.0 Sm^{-1}

  These values may be much less in partly saturated soil.

Figure 9.11 Experimental set-up for the measurement of electrical resistivity (Pandey and Shukla, 2018)

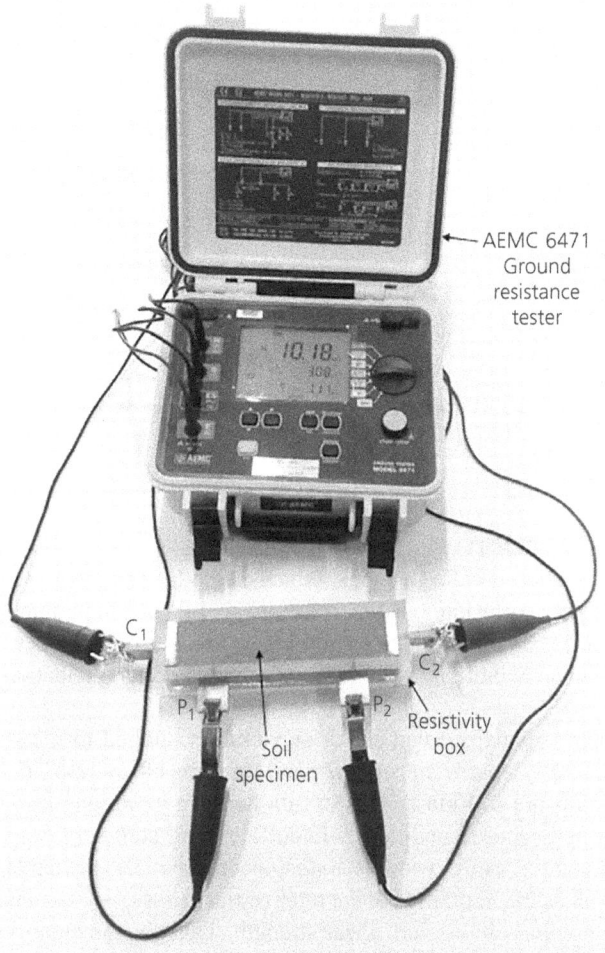

- For sandy soils, relative density and water content can be used to predict electrical resistivity and vice versa (Pandey *et al.*, 2015).
- In dry soil, electrical resistivity is very high because there is little interaction between the electric charge and the ions in the soil. However, in moist soil, a thin layer of water around the soil particles helps connect the electric charge to the ions, increasing conductivity and lowering resistivity.
- By applying a direct current through soft, waterlogged soil, the water content can be reduced, leading to improved strength and bearing capacity. This phenomenon is known as electro-osmosis, and the technique that utilises this process to enhance the ground is called electrokinetic dewatering. More details can be found in Shukla (2025).

Figure 9.12 Comparison of electrical resistivity values of sand, bentonite and their mixtures at different water contents

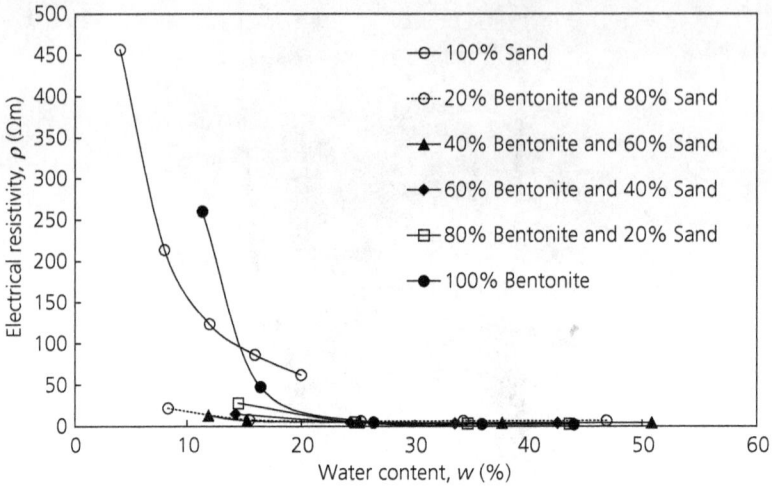

9.6. Soil liquefaction

A load of very short duration is called a shock, which can induce shaking or vibration. Shocks are caused by sudden impact, explosion, blasting, earthquake, pile driving, rapid load application, sea dashing, traffic movement or abrupt changes in force or velocity. The sudden drawdown of reservoir water may also induce vibrations in both the dam body and its foundation.

Section 4.7 of Chapter 4 explained that shock or vibration applied to a certain loose, saturated cohesionless soil, generally coarse silt or fine sand, may result in zero effective vertical stress ($\sigma'_v = \sigma_v - u = 0$) within the soil mass. This occurs because vibrations tend to increase the soil density due to its initially loose condition, without allowing sufficient time for water to drain, leading to undrained conditions. As water is almost incompressible, the tendency of densification raises the pore water pressure, u, relative to the total vertical stress, σ_v. As a result, the soil behaves like a dense fluid, which has almost zero shear strength. This phenomenon is called liquefaction, which may result in significant ground instability, leading to settlement, sinking, tilting or collapse of structures, flotation of buried structures and slope failures, as well as lateral spreading and sand boils where soil and water are ejected to the surface. Once shaking or vibration ceases, the excess pore water pressure gradually dissipates and the soil regains its strength, with settlement into a more compact state.

Both laboratory investigations and observations of field performance have shown that the liquefaction potential of a soil deposit to earthquake motions depends on several factors, including soil type, relative density or void ratio, initial confining pressure, intensity of ground shaking and duration of ground shaking (Seed and Idriss, 1971).

Liquefaction can be classified into the following two categories: flow liquefaction and cyclic mobility. Flow liquefaction occurs when a soil mass suddenly loses its shear strength due to a rapid increase in pore water pressure, causing it to behave like a fluid and flow under gravitational forces.

This happens when the static shear stresses within the soil mass exceed the shear strength of liquefied soil. Cyclic mobility occurs when the static shear stresses within a soil mass are slightly less than the shear strength of liquefied soil but the static plus cyclic shear stresses are greater than the shear strength of liquefied soil. Unlike flow liquefaction, which results in a sudden, catastrophic failure, cyclic mobility leads to gradual deformation, lateral spreading and settlement due to cyclic stress-induced pore water pressure build-up.

To assess whether a soil will undergo liquefaction during seismic loading during earthquakes, engineers use several key criteria or approaches, including the following: historical, geological and compositional approaches based on some rules of thumb, and rational approaches based on testing and analysis, such as the cyclic stress approach, cyclic strain approach, energy dissipation approach, effective stress-based response analysis approach and probabilistic approach. If liquefaction has occurred at a site in the past and the soil and water table conditions remain unchanged, it can be assumed that liquefaction is likely to occur again (historical criterion). Younger geologic deposits are more susceptible to liquefaction than older deposits (geologic criterion). Well-graded coarse-grained soils are less susceptible to liquefaction than poorly graded soils, and soils with rounded particles are more likely to liquefy than those with flaky particles (compositional criteria).

One of the first attempts to explain the liquefaction phenomenon in sandy or cohesionless soils was made by Casagrande (1936). This is based on the concept of critical void ratio. As discussed in Chapter 6, dense sand, when subjected to shear, tends to dilate, while loose sand, under similar conditions, tends to decrease in volume. The void ratio at which the sand does not change in volume when subjected to shear is referred to as the critical void ratio. According to Casagrande, sand deposits with a void ratio greater than critical void ratio tend to decrease in volume when subjected to vibration caused by seismic loading. If drainage is unable to occur, the pore water pressure increases, and it may become equal to total stress that does not change, resulting in zero effective stress; hence liquefaction occurs. Note that the critical void ratio concept may not be sufficient for a quantitative evaluation of soil liquefaction potential of sand deposits for the following reasons (Das and Luo, 2017).

▨ Critical void ratio is not a constant value, but changes with confining pressure.
▨ Volume changes due to dynamic loading conditions are different from the 1D static load conditions realised in the laboratory by direct shear and triaxial compressions tests, as presented in Chapter 6.

For this reason, intensive investigations have been conducted worldwide since the mid-1960s to determine the soil parameters that influence liquefaction.

Earthquake loading is defined by cyclic shear stress amplitude, which correlates with earthquake ground motion amplitude. Cyclic stresses can be estimated through detailed site-specific ground response analyses or a simplified approach. Ground response analyses can predict shear stress histories at different depths within a soil deposit. These analyses require precise site characterisation, representative input motions and an appropriate computer program, all of which demand time and cost.

The cyclic stress approach, developed by Seed and Idriss (1971), is the simplified approach used to evaluate the liquefaction potential of soil by comparing the induced cyclic stress ratio (CSR) due to seismic loading with the cyclic resistance ratio of the soil. For level sites, as illustrated in Figure 9.13, the maximum cyclic shear stress amplitude (peak cyclic shear stress amplitude), $(\tau_{max})_r$, at a

Figure 9.13 Maximum cyclic shear stress amplitude below a soil column with a unit cross-sectional area

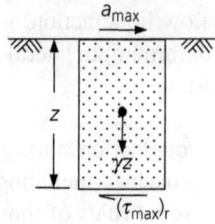

depth, z, due to a maximum ground surface acceleration (peak ground acceleration), a_{max}, induced by an earthquake can be given by

$$\left(\tau_{max}\right)_r = \left(\frac{a_{max}}{g}\right)\gamma z \tag{9.41}$$

where γ is the total unit weight of soil and g is the acceleration due to gravity.

Note that γz in Equation 9.41 represents the total weight of a soil column as a rigid body with height, z, and a unit cross-sectional area. This corresponds to the total vertical stress, σ_{v0}, at depth, z.

As the soil column is not rigid, its flexibility must be considered by treating it as a deformable body. Therefore, Equation 9.41 may be modified to

$$\left(\tau_{max}\right)_d = r_d\left(\frac{a_{max}}{g}\right)\gamma z \tag{9.42}$$

where r_d is the stress reduction coefficient with a value less than 1.

In the literature, it has been reported that the value of τ_{max} determined from the shear stress–time history during an earthquake can be converted into an equivalent number of significant stress cycles. Considering this, for practical purposes, Seed and Idriss (1971) suggest that the average equivalent uniform cyclic shear stress amplitude can be taken as

$$\tau_{av} = 0.65 r_d\left(\frac{a_{max}}{g}\right)\gamma z \tag{9.43}$$

The appropriate number of significant stress cycles, N_c, depends on the duration of ground shaking, and thus on the magnitude, M, of the earthquake. Representative numbers of stress cycles are given in Table 9.8.

As stated earlier, the loading applied to an element of soil in the field is typically expressed in terms of a dimensionless CSR, defined as the ratio of horizontal cyclic shear stress amplitude, τ_{av}, to initial effective vertical stress, (σ'_{v0}). Thus,

$$CSR = \frac{\tau_{av}}{\sigma'_{v0}} = 0.65 r_d\left(\frac{a_{max}}{g}\right)\frac{\gamma z}{\sigma'_{v0}} \tag{9.44}$$

Table 9.8 Significant number of stress cycles, N_c, corresponding to τ_{av}

Earthquake magnitude, M	N_c
7	10
7.5	20
8	30

The CSR value at which liquefaction initiates is called the cyclic resistance ratio (CRR). It is typically determined through empirical correlations based on past earthquakes and in situ test measurements, such as the SPT, cone penetration test or shear wave velocity. Such correlations are available in the literature and relevant design standards and should be referred to as applicable in the particular workplace or country.

For different soil profiles, r_d typically varies as follows: 0.92–0.97 at depth 25 feet (7.62 m); 0.61–0.88 at depth 50 feet (15.24 m); 0.35–0.76 at depth 75 feet (22.86 m); and 0.30–0.69 at depth 100 feet (30.48 m) (Seed and Idriss, 1971). The r_d parameter can be adequately expressed as a function of depth and earthquake magnitude, M, as given below (Idriss and Boulanger, 2008).

$$r_d = e^{\alpha(z)+\beta(z)M} \tag{9.45}$$

where

$$\alpha(z) = -1.012 - 1.126\sin\left(\frac{z}{11.73} + 5.133\right) \tag{9.46}$$

and

$$\beta(z) = 0.106 + 0.118\sin\left(\frac{z}{11.28} + 5.142\right) \tag{9.47}$$

in which z is depth in metres; M is earthquake magnitude; and the arguments inside the sine terms are in radians.

Equations 9.45–9.47 are mathematically applicable to a depth of $z \leq 34$ m. However, the uncertainty in r_d increases with increasing depth, so these equations should be applied only for depths that are less than about 20 m. Liquefaction evaluations at greater depths often involve special conditions for which more detailed analyses can be justified. For these reasons, it is recommended that the CSR (or equivalent r_d values) at depths greater than about 20 m should be considered based on detailed site response studies.

In geotechnical design, factors of safety are commonly used, as defined in Chapter 8 for slopes and foundations. The factor of safety against liquefaction triggering, F_L, is defined as the ratio of CRR to CSR. Thus,

$$F_L = \frac{CRR}{CSR} \tag{9.48}$$

Liquefaction is not expected to occur for $F_L > 1$. The values of CSR and CRR vary across sites and within a site as they depend on several site-specific factors, including earthquake magnitude, initial effective vertical stress, and initial, static shear stress. For a reliable value of F_L, these variations must be accounted for to ensure consistency in calculations of CSR and CRR. For specific details, see Das and Luo (2017) and Kramer and Steward (2025).

Example 9.6

Referring to Figure 9.13, consider the following

$z = 8\,\text{m}, \gamma = 19.5\,\text{kN}/\text{m}^3$ and $a_{\text{max}} = 0.6g$

If the soil deposit consists of fine sand with water table coinciding with ground surface, and the earthquake magnitude, $M = 6.5$, determine the cyclic stress ratio (CSR). If the cyclic resistance ratio (CRR) based on the standard penetration test value is reported as 0.25, do you expect liquefaction to occur?

Solution

Given $z = 8\,\text{m}, \gamma = 19.5\,\text{kN}/\text{m}^3, a_{\text{max}} = 0.6g, M = 6.5$ and $CRR = 2.5$.

From Equation 9.46

$$\alpha(z) = -1.012 - 1.126\sin\left(\frac{z}{11.73} + 5.133\right) = -1.012 - 1.126\sin\left(\frac{8}{11.73} + 5.133\right) = -0.504$$

From Equation 9.47

$$\beta(z) = 0.106 + 0.118\sin\left(\frac{z}{11.28} + 5.142\right) = 0.106 + 0.118\sin\left(\frac{8}{11.28} + 5.142\right) = 0.056$$

From Equation 9.45

$$r_d = e^{\alpha(z) + \beta(z)M} = e^{-0.504 + (0.056)(6.5)} = 0.869$$

From Equation 9.44, the cyclic stress ratio

$$CSR = \frac{\tau_{av}}{\sigma'_{v0}} = 0.65 r_d\left(\frac{a_{\text{max}}}{g}\right)\frac{\gamma z}{\sigma'_{v0}} = (0.65)(0.869)(0.6)\left[\frac{(19.5)(8)}{(19.5 - 9.81)(8)}\right] = 0.682$$

From Equation 9.48, the factor of safety against liquefaction triggering

$$F_L = \frac{CRR}{CSR} = \frac{0.25}{0.682} = 0.37$$

As $F_L < 1$, the soil will liquefy.

The following points regarding soil liquefaction are noteworthy.

- Clays do not liquefy during earthquakes because vibrations are not effective in bringing clay particles together into a closer packing. Additionally, the shear strength of clay does not become zero when effective stress is zero.
- Liquefaction is more likely when the water table is close to the ground surface.
- Quicksand, which is a hydraulic condition, as discussed in Section 4.7 of Chapter 4, occurs during steady state seepage, whereas liquefaction is induced by transient loading and lasts only for a few seconds. Hence, quicksand and liquefaction are not the same phenomenon.
- As discussed in Section 6.7 of Chapter 6, quick clays have a very high sensitivity (>16). Vibrations during earthquakes can cause a collapse of the soil structure.
- Sometimes, the soil may liquefy partially and possess some residual strength. In this case, the soil behaviour is governed by a complex constitutive (stress–strain–time) relationship.
- Cyclic mobility is an earthquake-related phenomenon; however, flow liquefaction can be initiated in a variety of ways, such as by monotonic loading (static liquefaction) and non-seismic sources of vibrations (e.g. pile driving, traffic movement, blasting and geophysical exploration).
- Mitigation measures such as soil densification, drainage improvement, stone columns, cement or chemical grout injection, and deep foundations are commonly used to reduce the risk of liquefaction-induced damage. See Shukla (2025) for detailed information on various ground improvement techniques.

Chapter summary

1. Expansive or swelling soils are one of the most problematic soils. The swelling potential of soil depends on the amount of montmorillonite present, as well as the type and quantity of exchangeable cations. Microscopy, X-ray diffraction and differential thermal analysis help identify expansive soils by analysing the types and amounts of minerals present.
2. Common laboratory tests, such as plasticity index and shrinkage limit, are effective in assessing the expansiveness of soils. Additionally, simple free swell and differential free swell tests offer a quick and qualitative measure of soil expansiveness.
3. A good grade, high-swelling commercial bentonite, primarily composed of montmorillonite, may have a free swell value ranging from 1200–2000%.
4. Reinforced soil is a composite material that consists of soil mass (or other similar material) and some form of reinforcement material (inextensible such as steel strips, or extensible such as geosynthetics or fibres) which provides an improvement in the engineering characteristics (e.g. strength, stiffness, permeability, compressibility and other properties) of soil.
5. Reinforcement can be included into a soil mass systematically or randomly, resulting in systematically reinforced soil (e.g. geosynthetic-reinforced soil) and randomly distributed fibre-reinforced soil, respectively.
6. The behaviour of reinforced soils is generally studied by conducting tests developed for unreinforced soils with suitable changes as applicable in specific cases.
7. Conduction analogies in porous media for fluid flow, heat flow and electrical flow are governed by Darcy's law ($q = kiA$, Equation 4.7, see Chapter 4); Fourier's law ($q_t = k_t i_t A$, Equation 9.21) and Ohm's law ($I = \sigma_e i_e A$, Equation 9.30), respectively.

8. The amount and condition of water in the soil play a crucial role in its thermal behaviour. Latent heat effects caused by changes in the state of water, as well as moisture migration, can significantly contribute to heat transfer in certain situations.

9. Thermal conductivity, specific heat capacity and diffusivity are three important thermal properties of soils used by geotechnical engineers. Thermal conductivity measures how well a material conducts heat. Specific heat capacity indicates the amount of heat required to raise the temperature of a unit mass of the material by 1°C. Thermal diffusivity indicates how quickly temperature changes propagate within a material.

10. The resistivity of each soil or rock varies depending on its water content, level of compaction and composition. The range of resistivity among earth materials is enormous, extending from 10^{-5} to 10^{15} Ωm – for example, resistivity is low for saturated clays and high for loose, dry gravel or solid rock.

11. Archie's equation is used to estimate the porosity of a rock or soil and its water saturation level by relating electrical resistivity to these properties in porous, water-bearing formations.

12. The electrical resistivity method with the Wenner electrode configuration is often used at project sites to determine subsurface conditions, such as soil or rock resistivity, layer thickness, groundwater presence and variations in material properties.

13. Liquefaction in a loose, saturated, non-plastic soil causes loss of strength and stiffness, typically triggered by earthquakes but also by monotonic loading. It occurs due to an increase in pore water pressure under undrained conditions.

14. The cyclic stress approach, the most common method for evaluating liquefaction potential, represents earthquake loading and soil resistance in terms of cyclic stresses. Transient motion is converted into equivalent shear stress cycles, linked to earthquake magnitude.

Questions for practice

Select the most appropriate answer for each of the multiple choice questions from Q9.1 to Q9.10. Questions Q9.11 to Q9.32 require more detailed answers.

9.1 Soil may exhibit significant swelling and shrinkage characteristics due to the presence of
   (a) kaolinite
   (b) illite
   (c) montmorillonite
   (d) halloysite.

9.2 If the differential free swell index of a soil is 28%, its swelling potential is considered
   (a) low
   (b) moderate
   (c) high
   (d) very high.

9.3 Select the incorrect statement.
   (a) Black cotton soils are organic in nature.
   (b) The lower the initial water content of the soil, the greater the swelling on saturation.
   (c) Swelling becomes less pronounced when a heavy load is imposed on soil.
   (d) A soil with a high plasticity index and low shrinkage index is likely to exhibit a high degree of swelling

9.4　Which of the following is an extensible reinforcement?
(a) steel
(b) geotextile
(c) polypropylene fibre
(d) both (b) and (c).

9.5　A geosynthetic can improve the soil characteristics by performing as a
(a) separator
(b) reinforcement
(c) filter
(d) all of the above.

9.6　Select the incorrect statement.
(a) Within fibre-reinforced soil, the fibre arrangement is basically one-dimensional.
(b) Incorporation of fibres within soil tends to decrease its maximum dry unit weight and increase its optimum water content.
(c) The shear strength increases resulting from the inclusion of fibre reinforcement are approximately the same for loose and dense sands.
(d) Bearing capacity ratio is a nondimensional parameter used in the design of reinforced soil foundations.

9.7　The SI unit of thermal diffusivity is
(a) J/s
(b) $K^{-1}m^{-1}$
(c) m^2/s
(d) $Js^{-1}m^{-2}$.

9.8　Among the following options, which material has the highest specific heat capacity?
(a) soil
(b) water
(c) ice
(d) steam.

9.9　Which of the following materials has the highest electrical resistivity?
(a) granite
(b) iron
(c) saturated clay
(d) water.

9.10　Select the incorrect statement.
(a) Electric charge is neither generated nor created; it can only be transferred from one body to another.
(b) Ohm's law for electrical flow is analogous to Darcy's law for fluid flow.
(c) The SI unit of electrical conductivity is Ωm.
(d) Liquefaction is more likely when the water table is close to the ground surface.

9.11 A soil was tested for its plasticity index by determining its liquid limit and plastic limit, and the result was found to be 47%. Describe its swelling potential.

9.12 A soil was tested to determine its swelling potential and the following data were obtained

volume of soil in distilled water, $V_d = 42\,cc$
volume of soil in kerosene, $V_k = 30\,cc$

Calculate the differential free swell index and classify the swelling potential.

9.13 A plate load test was conducted on unreinforced and reinforced soils. For a 25 mm settlement of the footing, the applied pressures measured were 100 kPa and 180 kPa, respectively. Calculate the bearing capacity ratio for the reinforced soil.

9.14 Given that the Earth's crust temperature increases by about 30°C for every kilometre of depth, and assuming the surface temperature is 25°C, what would be the estimated temperature at a depth of 30 km? Do you believe the rock mass at this depth would melt? Justify your answer with supporting reasoning.

9.15 Referring to Figure 9.7, consider the following

$A = 2.0\,m^2$, $L = 1.2\,m$, $\theta_1 = 25°C$, $\theta_2 = 2°C$ and $k_t = 0.8\,Wm^{-1}°C^{-1}$

Determine conduction rate, q_t, and the rate of flow of heat, v_t, through the soil mass, assuming no heat transfer occurs through the cylindrical surfaces.

9.16 Referring to Figure 9.10, consider the following, as the test was conducted at a project site

$l = 10\,m$, $I = 2\,A$ and $V = 30\,mV$

Determine the apparent resistivity of the subsurface material.

9.17 If the cyclic resistance ratio (CRR) based on the standard penetration test value for a saturated sand deposit is reported as 0.1, what should be the cyclic stress ratio (CSR) value to prevent liquefaction?

9.18 Do you think the soil in the area where you live is expansive? Provide justification for your answer based on observations, local soil characteristics or relevant data.

9.19 How do the free swell test and the differential free swell test differ?

9.20 What field conditions generally favour the swelling of expansive soils?

9.21 Differentiate between systematically and randomly distributed fibre-reinforced soils.

9.22 How can the concepts of reinforced soil engineering benefit the community? Provide an explanation with justification.

9.23 What are the primary functions of geosynthetics? Can direct shear tests and triaxial compression tests be conducted on geosynthetic-reinforced soils using the available testing facilities in the soil mechanics laboratory? Is there any modification required?

9.24 An unreinforced soil has an internal friction angle of 30°. Can this value be increased by reinforcing the soil? If yes, explain how.

9.25 What are the key factors that influence the thermal conductivity of soil? Do they also affect the electrical properties of soil? Provide an explanation.

9.26 The unit weight of ice is about 10% less than the unit weight of water. How does this behaviour affect the properties and performance of the soil when it freezes? Discuss the implications for design and performance.

9.27 Describe how electrical resistivity measurements can be used to assess subsurface soil conditions.

9.28 Is it possible to determine the electrical resistivity of soil in the laboratory? What type of apparatus can be used, if applicable?

9.29 What type of soil exists in the area where you live? Can you find it listed in Table 9.7? If yes, compare its tabulated value with the measured value, as found in local technical documents.

9.30 How do soil composition and saturation influence liquefaction susceptibility?

9.31 A sand deposit at a site is known to be susceptible to liquefaction. If a heavy structure is built on it, will its resistance to liquefaction increase or decrease? Justify your answer.

9.32 In which types of location is liquefaction most likely to occur? Are they always near water bodies?

REFERENCES

Archie GE (1942) The electrical resistivity log as an aid in determining some reservoir characteristics. *Transactions of the American Institute of Mining and Metallurgical Engineers* **146**: 54–62.

Asaoka A (1978) Observational procedure of settlement prediction. *Soils and Foundations* **18(4)**: 87–101.

ASTM (2018) ASTM D4546-14: Standard test methods for one-dimensional swell or collapse of soils. ASTM International, West Conshohocken, PA, USA.

ASTM (2019) ASTM D5890-19: Standard test method for swell index of clay mineral component of geosynthetic clay liners. ASTM International, West Conshohocken, PA, USA.

ASTM (2020) ASTM G57-06: Standard test method for field measurement of soil resistivity using the Wenner four-electrode method. ASTM International, West Conshohocken, PA, USA.

ASTM (2021) ASTM D4829-21: Standard test method for expansion index of soils. ASTM International, West Conshohocken, PA, USA.

ASTM (2022a) ASTM D5334-22: Standard test method for determination of thermal conductivity of soil and rock by thermal needle probe procedure. ASTM International, West Conshohocken, PA, USA.

ASTM (2022b) ASTM E2585-09(2022): Standard Practice for thermal diffusivity by the flash method. ASTM International, West Conshohocken, PA, USA.

ASTM (2023) ASTM G187-23: Standard test Method for measurement of soil resistivity using the two-electrode soil box method. ASTM International, West Conshohocken, PA, USA.

Bassiouni Z (1994) *Theory, Measurement, and Interpretation of Well Logs*. Society of Petroleum Engineers, Richardson, TX, USA.

Briaud J-L (2013) *Geotechnical Engineering: Unsaturated and Saturated Soils*. Wiley, Hoboken, NJ, USA.

Bureau of Indian Standards (2021a) IS 2720 (Part 41)-1977: Measurement of swelling pressure of soils. Bureau of Indian Standards, New Delhi, India.

Bureau of Indian Standards (2021b) IS 2720 (Part 40)-1977: Determination of free swell index of soils. Bureau of Indian Standards, New Delhi, India.

Bureau of Indian Standards (2021c) IS 2911 (Part 3)-2021: Design and construction of pile foundations – Code of practice: Part 3 Under-reamed piles (second revision). Bureau of Indian Standards, New Delhi, India.

Casagrande A (1936) Characteristics of cohesionless soils affecting the stability of slopes and earth fills. *Journal of the Boston Society of Civil Engineers* **23**: 257–276.

Das BM and Luo Z (2017) *Principles of Soil Dynamics*, 3rd edn. Cengage Learning, Boston, MA, USA.

Dobrin MB (1976) *Introduction to Geophysical Prospecting*. McGraw-Hill, New York, NY, USA.

Farouki OT (1981) *Thermal Properties of Soils*. United States Army Corps of Engineers, Cold Regions Research and Engineering Laboratory, Hanover, NH, USA.

Goldsmid HJ and Bowley AE (1960) Thermal conduction in mica along the planes of cleavage. *Nature* **187**: 864–865.

Gray DH and Ohashi H (1983) Mechanics of fibre reinforcement in sand. *Journal of Geotechnical Engineering, ASCE* **109(3)**: 335–353.

Gromko GJ (1974) Review of expansive soils. *Journal of Geotechnical Engineering Division, ASCE* **100(GT6)**: 667–687.

Holtz WG and Gibbs HJ (1956) Engineering properties of expansive clays. *Transactions of the ASCE*, Paper No. 2814, **121(1)**: 641–663.

Hoover JM, Moeller DT, Pitt JM, Smith SG and Wainaina NW (1982) *Performance of Randomly Oriented Fiber-Reinforced Roadway Soils – a Laboratory and Field Investigation*. Iowa DOT Project Report HR-211, Department of Civil Engineering, Engineering Research Institute, Iowa State University, Ames, IO, USA.

Idriss IM and Boulanger RW (2008) *Soil Liquefaction during Earthquakes*. Earthquake Engineering Research Institute, Oakland, CA, USA.

Kearey P, Brooks M and Hill I (2002) *An Introduction to Geophysical Exploration*. Blackwell, London, UK.

Koerner RM (2012) *Designing with Geosynthetics*, 6th edn. Xlibris, Bloomington, IN, USA, volumes 1 and 2.

Kramer SL and Steward JP (2025) *Geotechnical Earthquake Engineering*, 2nd edn. CRC Press, Taylor and Francis, NY, USA.

Maher MH and Gray DH (1990) Static response of sands reinforced with randomly distributed fibres. *Journal of Geotechnical Engineering, ASCE* **116(11)**: 1661–1677.

Mitchell JK and Soga K (2005) *Fundamentals of Soil Behaviour*, 3rd edn. Wiley, Hoboken, NJ, USA.

Nelson JD, Chao KC, Overton DD and Nelson EJ (2015) *Foundation Engineering for Expansive Soils*. Wiley, NJ, USA.

Pandey LMS and Shukla SK (2018) Effect of state of compaction on the electrical resistivity of sand-bentonite materials. *Journal of Applied Geophysics* **155(1)**: 208–216.

Pandey LMS, Shukla SK and Habibi D (2015) Electrical resistivity of sandy soil. *Geotechnique Letters* **5(3)**: 178–185.

Peck RB, Hanson WE and Thornburn TH (1974) *Foundation Engineering*, 2nd edn. Wiley, New York, NY, USA.

Prakash K, Prabhu NS and Sridharan A (2021) Observational procedure for the prediction of ultimate swell. *Geotechnical and Geological Engineering* 39: 4669–4676.

Puppala AJ (2021) Performance evaluation of infrastructure on problematic expansive soils: characterization challenges, innovative stabilization designs, and monitoring methods. *Journal of Geotechnical and Geoenvironmental Engineering, ASCE*, Paper No. 04021053, **147(8)**: 1–15.

Richards BG, Peter P, and Emerson WW (1983) The effects of vegetation on the swelling and shrinking of soils in Australia. *Geotechnique* 33(2): 127–139.

Schlosser F and Long N-T (1974) Recent results in French research on reinforced earth. *Journal of the Construction Division, ASCE* **100(3)**: 223–237.

Schlosser F and Vidal H (1969) Reinforced earth. *Bulletin de Liaison des Laboratoires des Ponts et Chaussees* 41, Paris, France.

Sears FW, Zemansky MW and Young HD (1982) *University Physics*, 6th edn. Addison-Wesley, Reading, MA, USA.

Seed HB and Idriss IM (1971) Simplified procedure for evaluating soil liquefaction potential. *Journal of the Soil Mechanics and Foundations Division, ASCE* **97(SM9)**: 1249–1273.

Shukla SK (2016) *An Introduction to Geosynthetic Engineering*. CRC Press, Taylor and Francis, London, UK.

Shukla SK (2017) *Fundamentals of Fibre-Reinforced Soil Engineering*. Springer Nature, Singapore.

Shukla SK (2022) *ICE Handbook of Geosynthetic Engineering*, 3rd edn. ICE Publishing, London, UK.

Shukla SK (2025) *ICE Core Concepts: Geotechnical Engineering*, 2nd edn. Emerald/ICE Publishing, London, UK.

Shukla SK and Sivakugan N (2011) Site investigation and in situ tests. In *Geotechnical Engineering Handbook* (Das BM (ed.)). J Ross, Fort Lauderdale, FL, USA, pp. 10.1–10.78.

Shukla SK and Yin J-H (2006) *Fundamentals of Geosynthetic Engineering*. Taylor and Francis, London, UK.

Shukla SK, Sivakugan N and Das BM (2009) Fundamental concepts of soil reinforcement – an overview. *International Journal of Geotechnical Engineering* 3(3): 329–343.

Standards Australia (1997) AS 1289.4.4.1:1997 Determination of the electrical resistivity of a soil – Method for fine granular materials. Standards Australia, Strathfield, NSW, Australia.

Verma HC (2000) *Concepts of Physics*. Bharati Bhawan, Patna, India.

Vidal H (1966) La terre Armée. Annales de l'Institut Technique de Batiment et de Travaux Publics, Paris, France.

Vidal H (1969) The principle of reinforced earth. *Highway Research Record* 282: 1–16.

Walker J (2008) *Fundamentals of Physics*, 8th edn. Wiley, Hoboken, NJ, USA.

FURTHER READING

Das BM and Luo Z (2017) *Principles of Soil Dynamics*, 3rd edn. Cengage Learning, Boston, MA, USA.

Farouki OT (1981) *Thermal Properties of Soils*. United States Army Corps of Engineers, Cold Regions Research and Engineering Laboratory, Hanover, NH, USA.

Nelson JD, Chao KC, Overton DD and Nelson EJ (2015) *Foundation Engineering for Expansive Soils*. Wiley, NJ, USA.

Shukla SK (2016) *An Introduction to Geosynthetic Engineering*. CRC Press, Taylor and Francis, London, UK.

Shukla SK (2017) *Fundamentals of Fibre-Reinforced Soil Engineering*, Springer Nature, Singapore.

Sanjay Kumar Shukla
ISBN 978-1-83608-519-5
https://doi.org/10.1108/978-1-83608-516-420252003
Emerald Publishing Limited: All rights reserved

Answers to selected questions

Chapter 1

1.1 (b)

1.3 (a)

1.5 (d)

1.7 (d)

1.9 (b)

1.11 1.39

1.13 7.2%

1.15 (a) 0.79, (b) 99%

Chapter 2

2.1 (d)

2.3 (b)

2.5 (a)

2.7 (c)

2.9 (c)

2.13 0.74, No

2.15 Slope I

2.17 (a) CL, (b) 0.26, 0.74, (c) 0.70

Chapter 3

3.1 (b)

3.3 (d)

3.5 (b)

3.7 (c)

3.9 (a)

3.11 149.95 kPa

3.13 6 kPa, 60 kPa, 600 kPa

3.15 72.3 kPa

3.19 97.43 kPa

3.21 (a) 60 kPa, 30.6 kPa (b) 80 kPa (c) 140 kPa, 110.6 kPa; total and effective vertical stresses before and after construction of the embankment increase equally by 80 kPa.

Chapter 4

4.1 (c)

4.3 (b)

4.5 (a)

4.7 (a)

4.9 (b)

4.11 0 m, 1.0 m, −1.0 m, 0 m

4.13 0.00178 m^3/s, 0.00178 m^3/s, 0.7 m, 1.47

4.15 10.18 m

4.17 1.11 × 10^{-7} m^3/s/m

4.19 66.59°

4.21 $h_2 = \sqrt{h_1 h_3}$; Yes

Chapter 5

5.1 (d)

5.3 (c)

5.5 (b)

5.7 (a)

5.9 (c)

5.11 (a) 3.61 × 10^{-3} m^2/kN, (b) 1.08, (c) 1.37 × 10^{-3} m^2/kN, (d) 1.34 × 10^{-7} m^2/s

5.13 38.6 days

5.15 (a) 2.10 × 10^{-8} m^2/s, (b) 0.170 m^2/kN

5.17 109 mm

5.19 5 cm

Chapter 6

6.1   (b)

6.3   (a)

6.5   (a)

6.7   (c)

6.9   (c)

6.11 71.24 kPa

6.13 12.5 kPa, 36.87°, 387.5 kPa

6.15 34.7 kPa, 17.35 kPa

6.17 32.34 kPa

Chapter 7

7.1   (b)

7.3   (a)

7.5   (a)

7.7   (d)

7.9   (c)

7.11 596 kJ/m³, 590 kJ/m³, 594 kJ/m³, 580 kJ/m³

7.13 (a) 19.10 kN/m³ (b) 0.36 (c) 86.9% (d) 3.6% (e) 13.1%

7.15 17.73 kN/m³, 13.5%, 21.14 kN/m³, 13.5%

Chapter 8

8.1   (c)

8.3   (b)

8.5   (c)

8.7   (b)

8.9   (d)

8.11 2.56 m, 5.12 m

8.13 49.14 kN/m, 225.72 kN/m, 2 m

8.15 7.7 kPa

Chapter 9

9.1 (c)

9.3 (a)

9.5 (d)

9.7 (c)

9.9 (a)

9.11 Very high swelling potential

9.13 1.8

9.15 30.67 W, 14.33 Wm^{-2}

9.17 CSR < 0.1

Sanjay Kumar Shukla
ISBN 978-1-83608-519-5
https://doi.org/10.1108/978-1-83608-516-420252004

Index

www.ingramcontent.com/pod-product-compliance
Lightning Source LLC
Chambersburg PA
CBHW081052220326
41598CB00038B/7067

* 9 7 8 1 8 3 6 0 8 5 1 9 5 *